Optics of Charged Particles

Optics of Charged Particles

Hermann Wollnik

Physikalisches Institut
Justus Liebig-Universität
Giessen, Federal Republic of Germany

1987

ACADEMIC PRESS, INC.
Harcourt Brace Jovanovich, Publishers
Orlando   San Diego   New York   Austin
Boston   London   Sydney   Tokyo   Toronto

PHYSICS

5388 7025

ACADEMIC PRESS, INC.
Orlando, Florida 32887

United Kingdom Edition published by
ACADEMIC PRESS INC. (LONDON) LTD.
24–28 Oval Road, London NW1 7DX

Library of Congress Cataloging in Publication Data

Wollnik, Hermann.
  Optics of charged particles.

  Includes bibliographies and index.
  1. Electron optics. 2. Electromagnetic lenses.
3. Particle beams. I. Title.
QC793.5.E622W64   1987   539.7'2112   86-14033
ISBN 0–12–762130–X (alk. paper)

PRINTED IN THE UNITED STATES OF AMERICA

87 88 89 90   9 8 7 6 5 4 3 2 1

Contents

v

8   Image Aberrations

9   Design of Particle Spectrometers and Beam Guide Lines

Preface

Some introductory books as well as a multitude of articles in scientific journals and research reports have been published on the subject of charged-particle optics. However, except for basic aspects, it is very difficult to obtain a general overview from these sources, since most concentrate on isolated problems and normally use different mathematical formalisms. Furthermore, since the results are applied to quite different problems, the solutions obtained are often incompatible. This book unifies such approaches, resulting in a description of how charged particles move in the main and fringing fields of magnetic or electrostatic dipoles, quadrupoles, hexapoles, etc., using the same type of formulation and consistent nomenclature throughout. Besides the description of particle trajectories and beam shapes, guide lines are given for designing particle optical instruments.

This book does not require the reader to have any knowledge of charged-particle optics; however, the equivalent of an undergraduate education in physics and mathematics is needed. It is written neither to carry everyone to the mountain tops of scientific findings, nor to stop at the foothills of the mountain range. Rather, it should lead the interested reader to a high plateau of understanding from which he can reach the mountain tops by his own strength.

I have tried to supply a comprehensive set of references with each chapter, normally quoting an early and a recent publication for any given problem. However, not all possible references are given since this would have exceeded the available space.

At this point I should like to acknowledge the many fruitful discussions I have had with M. Berz, H. Matsuda, T. Matsuo, and H. Nestle. I am also greatly indebted to R. Kosempel for the skilled and careful drawing of the many figures in this book and to M. Gowans for the experienced and patient typing and retyping of the manuscript.

1

Gaussian Optics and Transfer Matrices

Charged particle optics is very similar to light optics. Therefore, in Chapter 1 we first discuss the more familiar light optical systems. The results obtained can later be applied to the discussion of ion and electron optical systems.

1.1 METHOD OF TRANSFER MATRICES

For geometric light optics it has been customary since the time of Newton to use an algebraic formulation for all equations involved; however, this method has been replaced in many cases by the use of *transfer matrices*. For simple optical systems the use of transfer matrices has no particular advantage. For complex systems, on the other hand, it offers an unexcelled simplicity and clarity, as has been shown by Cotte (1938), Penner (1961), Brown *et al.* (1964), and Wollnik (1967) for particle optics and by Herzberger (1958) or Halbach (1964) for light optics.

1.1.1 Description of Straight Light Rays

Assume the z axis of a Cartesian coordinate system to represent the optic axis of a bundle of rays. The deviation of any ray of this bundle from the optic axis (Fig. 1.1) can then be described by

$$x(z_2) = x_1 + (z_2 - z_1) \tan \alpha_1, \tag{1.1a}$$

$$y(z_2) = y_1 + (z_2 - z_1) \tan \beta_1, \tag{1.1b}$$

as long as the refractive index does not vary between z_1 and z_2. Though not important here, it is generally useful to describe not only the deviation $x(z)$ and $y(z)$ but also the angle of inclination $\alpha(z)$ and $\beta(z)$ of this ray relative to the optic axis. This inclination stays constant in a region of constant refractive index:

$$\tan \alpha(z) = \tan \alpha_1, \tag{1.2a}$$

$$\tan \beta(z) = \tan \beta_1. \tag{1.2b}$$

Using matrix notation, we can rewrite Eqs. (1.1) and (1.2) describing the motion of rays over the drift length $l = z_2 - z_1$ (see Fig. 1.1) as

$$\begin{pmatrix} x_2 \\ \tan \alpha_2 \end{pmatrix} = \begin{pmatrix} 1 & l \\ 0 & 1 \end{pmatrix} \begin{pmatrix} x_1 \\ \tan \alpha_1 \end{pmatrix}, \tag{1.3a}$$

$$\begin{pmatrix} y_2 \\ \tan \beta_2 \end{pmatrix} = \begin{pmatrix} 1 & l \\ 0 & 1 \end{pmatrix} \begin{pmatrix} y_1 \\ \tan \beta_1 \end{pmatrix}. \tag{1.3b}$$

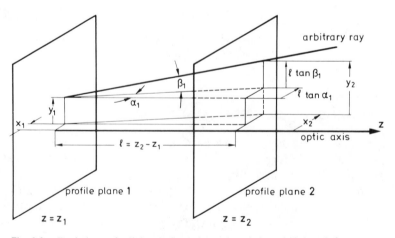

Fig. 1.1. Deviations of a light ray from the optic axis in a drift length $l = z_2 - z_1$.

Equations (1.3a) and (1.3b) express the fact that *position vectors* $X_1 = (x_1, \tan \alpha_1)$ and $Y_1 = (y_1, \tan \beta_1)$ are transformed by transfer matrices to new position vectors $X_2 = (x_2, \tan \alpha_2)$ and $Y_2 = (y_2, \tan \beta_2)$. Along with the term *position vector*, we introduce the term *profile plane*. A profile plane is defined as a plane that is perpendicular to the optic axis and positioned at a certain path coordinate z, as shown in Fig. 1.1. Thus, the transfer matrices introduced in Eqs. (1.3a) and (1.3b) transfer a ray from one profile plane to another.

In case of rotationally symmetric systems, the x and y transfer matrices are identical. In nonrotationally symmetric systems, these transfer matrices must be determined independently. In order to simplify the description, in this chapter we discuss only rotationally symmetric systems so that the x transfer matrix describes rays in the xz plane as well as in the yz plane.

1.1.2 Properties of an Ideally Thin Lens

A thin lens is defined as an infinitely thin system that causes a bundle of parallel rays to be focused to a point, as shown in Fig. 1.2. In other words, on passing through this thin lens, each ray experiences a bend $\Delta \alpha$ toward the optic axis, whereas its x coordinate stays unchanged. This bend is proportional to the distance x_2, that the ray under consideration deviates from the optic axis shortly before the thin lens. From Fig. 1.2 we can see that

$$\tan(\Delta \alpha) = -x_2/f. \qquad (1.4)$$

This relation can also be written as

$$x_3 = x_2, \qquad \tan \alpha_3 = -x_2/f,$$

with α_3 characterizing the angle of inclination of a ray relative to the optic axis at profile plane 3 immediately behind the thin lens.

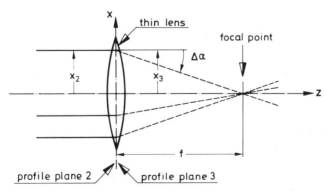

Fig. 1.2. Principle of a thin lens that focuses a bundle of incoming parallel rays to a point.

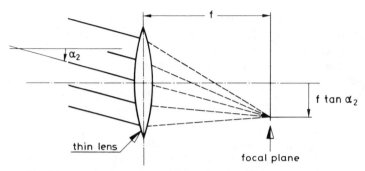

Fig. 1.3.   The focusing of an oblique bundle of initially parallel rays.

Assuming that the rays in Fig. 1.2 do not move from left to right but from right to left, it is evident that the bend $\Delta\alpha$ is the same for a trajectory that is originally parallel to the z axis or for a trajectory that is originally inclined. Thus, we may state that the angle of bend $\Delta\alpha$ always has the same value, independent of the initial angle of inclination α_2. This statement can be expressed as

$$x_3 = x_2, \tag{1.5a}$$

$$\tan\alpha_3 = (\tan\alpha_2) - (x_2/f). \tag{1.5b}$$

In the form of transfer matrices, Eqs. (1.5a) and (1.5b) can be rewritten to express the optical properties of a thin lens:

$$\begin{pmatrix} x_3 \\ \tan\alpha_3 \end{pmatrix} = \begin{pmatrix} 1 & 0 \\ -1/f & 1 \end{pmatrix} \begin{pmatrix} x_2 \\ \tan\alpha_2 \end{pmatrix}; \tag{1.6}$$

that is, the transformation of the position vector $\mathbf{X}_2 = (x_2, \tan\alpha_2)$ at profile plane 2 to the position vector $\mathbf{X}_3 = (x_3, \tan\alpha_3)$ at profile plane 3 of Fig. 1.2.

The transfer matrix of Eq. (1.6) expresses all first-order properties of a thin lens. Note that the *focal length f* or the *refractive power* $1/f$ is the only quantity characterizing the properties of a thin lens. Note also that in Eq. (1.5b) or (1.6), $\tan\alpha_3$ equals $\tan\alpha_2$ for $x_2 = 0$; i.e., an oblique bundle of parallel rays will stay oblique. Consequently, the point at which all these rays are focused is not the focal point, but a point approximately in the focal plane, as shown in Fig. 1.3.

1.2   TRANSPORT THROUGH AN OPTICAL SYSTEM OF ONE THIN LENS

Let us now determine the relation between the position vector $\mathbf{X}_1$ of a ray at profile plane 1, at distance l_1 upstream from a thin lens, and the position vector $\mathbf{X}_4$ characterizing the same ray at profile plane 4, at distance l_2 downstream from the thin lens in Fig. 1.4. Knowing the position vector $\mathbf{X}_1 = (x_1, \tan \alpha_1)$ at profile plane 1 in Fig. 1.4, we find the position vector $\mathbf{X}_2 = (x_2, \tan \alpha_2)$ at profile plane 2 just before the thin lens by use of a drift-length transfer matrix

$$T_{21} = \begin{pmatrix} 1 & l_1 \\ 0 & 1 \end{pmatrix}$$

[Eq. (1.3a)]. This position vector can be transformed to a third one $\mathbf{X}_3 = (x_3, \tan \alpha_3)$ at profile plane 3 just behind the thin lens with a thin-lens transfer matrix

$$T_{32} = \begin{pmatrix} 1 & 0 \\ -1/f & 1 \end{pmatrix},$$

as given by Eq. (1.6). Applying this position vector to another drift-length transfer matrix

$$T_{43} = \begin{pmatrix} 1 & l_2 \\ 0 & 1 \end{pmatrix},$$

we obtain the desired position vector $\mathbf{X}_4 = (x_4, \tan \alpha_4)$ at profile plane 4 in Fig. 1.4. As we know from matrix algebra, it would have been possible to obtain the same result by applying the vector $\mathbf{X}_1$ to the product of the previous three transfer matrices

$$\mathbf{X}_4 = T_{43} \cdot T_{32} \cdot T_{21} \cdot \mathbf{X}_1. \tag{1.7}$$

Writing Eq. (1.7) explicitly yields

$$\begin{pmatrix} x_4 \\ \tan \alpha_4 \end{pmatrix} = \begin{pmatrix} 1 & l_2 \\ 0 & 1 \end{pmatrix} \begin{pmatrix} 1 & 0 \\ -1/f & 1 \end{pmatrix} \begin{pmatrix} 1 & l_1 \\ 0 & 1 \end{pmatrix} \begin{pmatrix} x_1 \\ \tan \alpha_1 \end{pmatrix},$$

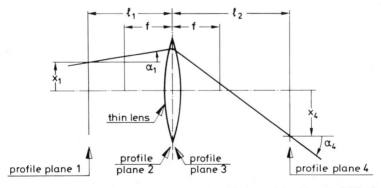

Fig. 1.4. A light ray is shown passing through a thin lens of focal length f. Of primary importance is the final relation between the position and inclination of this light ray in profile planes 1 and 4.

or after multiplying these three transfer matrices, the relation between the profile planes 1 and 4 in Fig. 1.4,

$$\begin{pmatrix} x_4 \\ \tan \alpha_4 \end{pmatrix} = \begin{pmatrix} 1 - (l_2/f) & l_1 + l_2 - (l_1 l_2/f) \\ -1/f & 1 - (l_1/f) \end{pmatrix} \begin{pmatrix} x_1 \\ \tan \alpha_1 \end{pmatrix}. \tag{1.8}$$

For $l_1 = l_2 = f$, Eq. (1.8) describes the important relation between profile planes at distance f in front and in back of the thin lens:

$$\begin{pmatrix} x_4 \\ \tan \alpha_4 \end{pmatrix} = \begin{pmatrix} 0 & f \\ -1/f & 0 \end{pmatrix} \begin{pmatrix} x_1 \\ \tan \alpha_1 \end{pmatrix}. \tag{1.9}$$

1.2.1 Formation of an Image

We assume now that l_1 and l_2 are chosen such that the second element in the first row of Eq. (1.8) vanishes; i.e., $l_1 + l_2 - (l_1 l_2/f) = 0$. The first line of Eq. (1.8) then reads

$$x_4 = [1 - (l_2/f)]x_1. \tag{1.10}$$

In this case, x_4 depends only on the position x_1 of the rays in Fig. 1.5 but not on their inclinations α_1 in profile plane 1. Different rays leaving a specific point x_1 of an object under different angles α_1 will thus be focused to a corresponding point x_4 of an image.

The condition $l_1 + l_2 - (l_1 l_2/f) = 0$ can also be written in the form of the familiar *lens equation*:

$$(1/l_1) + (1/l_2) = 1/f. \tag{1.11}$$

Equation (1.11) relates the object distance l_1 and the image distance l_2 with the focal length f of the lens under consideration (see Fig. 1.5). If Eq. (1.11) is fulfilled, we say that *between the profile planes* 1 *and* 4 *there exists an object–image relation.*

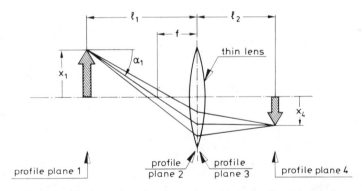

Fig. 1.5. The formation of an image by a thin lens of focal length f.

1.2.2 Object and Image Distances

An object–image relation between the profile planes 1 and 4 in Fig. 1.5 requires Eq. (1.11) to be fulfilled. Consequently, the magnification $M = x_4/x_1$ is obtained from Eqs. (1.10) and (1.11) as

$$\frac{x_4}{x_1} = M = 1 - \frac{l_2}{f} = \left(1 - \frac{l_1}{f}\right)^{-1}, \tag{1.12a}$$

which can be rewritten as

$$(x_4/x_1) = M = (-l_2/l_1). \tag{1.12b}$$

For negative magnifications M the ratios l_1/f and l_2/f are both positive and larger than 1. For positive magnifications M, one of these ratios must be negative and the other positive and less than 1.

For a focusing lens ($f > 0$) and positive values of l_1 and l_2, Eqs. (1.12) yield a negative magnification $M = x_4/x_1$. The range of possible combinations of l_1 and l_2 is thus found from Eq. (1.12a) as

$$\infty > l_1 > 2f \quad \text{for} \quad f < l_2 < 2f \quad \text{and} \quad |M| < 1,$$

which is illustrated in Fig. 1.6a. The limiting case, $x_4 = -x_1$, is reached for $l_1 = l_2 = 2f$. Exchanging object and image, we find analogously,

$$2f > l_1 > f \quad \text{for} \quad 2f < l_2 < \infty \quad \text{with} \quad |M| > 1.$$

According to Eq. (1.12b), for a positive magnification M, either the object distance l_1 or the image distance l_2 must be negative. Thus, either the image or the object is virtual and not accessible. For a real object and a virtual image, Eq. (1.12a) yields the ranges of l_1 and l_2 as

$$f > l_1 > 0 \quad \text{for} \quad -\infty < -l_2 < 0 \quad \text{with} \quad |M| > 1,$$

which is illustrated in Fig. 1.6b. In the limiting case $l_1 = f$ and $l_2 = \infty$, we find $M = -\infty$, which causes rays that originated from a point of the object to leave the lens as a parallel bundle. Exchanging object and image so that the rays in front of the lens seem to go to a virtual object behind the lens, Eq. (1.12a) yields

$$-\infty < -l_1 < 0 \quad \text{for } f > l_2 > 0 \quad \text{with} \quad |M| < 1.$$

For a defocusing lens ($f < 0$), Eq. (1.12a) shows for a positive magnification M that either the object distance l_1 or the image distance l_2 must be negative. Thus, either the object is real and the image is virtual or vice versa. For a real object, we find

$$\infty > l_1 > 0 \quad \text{for} \quad -f < -l_2 < 0 \quad \text{with} \quad |M| < 1$$

as illustrated in Fig. 1.7a. Exchanging object and image results in

$$0 > -l_1 > -f \quad \text{for} \quad 0 < l_2 < \infty \quad \text{with} \quad |M| > 1.$$

8

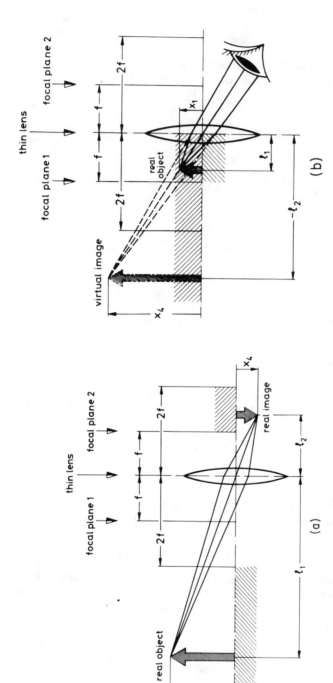

Fig. 1.6. (a) A real object placed in front of a focusing thin lens between $l_1 = f$ and $l_1 = 2f$ behind the thin lens. This optical arrangement is used in a photographic camera. Exchanging the positions of object and image we find that a real object placed in front of a focusing thin lens between $l_1 = 2f$ and $l_1 = f$ causes an inverted and magnified real image located between $l_2 = 2f$ and $l_2 = \infty$ behind the thin lens. (b) A real object placed in front of a focusing thin lens at distance $l_1 < f$ causes the rays to be deflected such that they appear to come from an upright and magnified virtual image located somewhere in front of the thin lens. This optical arrangement is used as an eyepiece for a Kepler telescope or a microscope. Exchanging the positions of object and image, we find that a virtual object somewhere behind a thin lens (i.e., a real image formed by some other focusing lens system in the absence of the focusing lens under consideration) causes an upright and demagnified real image located at $l_2 < f$ behind the thin lens.

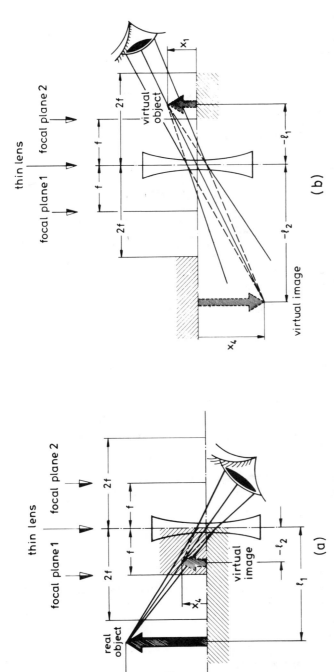

Fig. 1.7. (a) A real object placed somewhere in front of a defocusing thin lens causes an upright and demagnified virtual image located at $l_2 < f$ in front of the thin lens. Exchanging the positions of object and image, we find that a virtual object placed at $l_1 < f$ behind a defocusing thin lens (i.e., a real image formed by some other lens system in the absence of the defocusing lens under consideration) causes an upright and magnified real image located behind the thin lens. (b) A virtual object between $l_1 = f$ and $l_1 = 2f$ behind the defocusing thin lens (i.e., a real image formed by some other lens system in the absence of the defocusing lens under consideration) causes an inverted and magnified virtual image between $l_2 = -\infty$ and $l_2 = -2f$ in front of the thin lens. This optical arrangement is used as an eye piece for a Galilean telescope. Exchanging the position of object and image, we find that a demagnified virtual image occurs between $l_2 = -2f$ and $l_2 = -f$ in front of the thin lens if a virtual object was placed between $l_1 = -2f$ and $l_1 = -\infty$ behind a defocusing thin lens (i.e., a real image formed by some other focusing lens system in the absence of the defocusing lens under consideration).

According to Eq. (1.12a), for negative magnifications M, both object and image are virtual (as indicated in Fig. 1.7b) so that l_1 and l_2 must both be negative,

$$-f > -l_1 > -2f \quad \text{and} \quad -\infty < -l_2 < -2f \text{ with } |M| > 1,$$

or exchanging object and image,

$$-2f > -l_1 > -\infty \quad \text{and} \quad -2f > -l_2 > -f \text{ with } |M| < 1.$$

A limiting case also exists here: $l_1 = l_2 = -2f$ and $x_4 = -x_1$.

1.2.3 Newton's Formulation of the Lens Equation

Instead of defining object and distances (l_1, l_2) relative to the principal plane, we can also define object and image distances (w_1, w_2) relative to the focal planes of a thin lens. With $w_1 = l_1 - f$ and $w_2 = l_2 - f$, Eq. (1.12a) transforms to

$$M = -w_2/f = -f/w_1, \tag{1.13}$$

or $w_1 w_2 = f^2$, as first stated by Isaac Newton.

1.3 TRANSPORT THROUGH A GENERAL OPTICAL SYSTEM

The relation between position vectors $\mathbf{X}_1$ and $\mathbf{X}$ in profile planes at z_1 and z of an arbitrary optical system can always be written as

$$\begin{pmatrix} x(z) \\ \tan \alpha(z) \end{pmatrix} = \begin{pmatrix} (x_2|x_1) & (x_2|\tan \alpha_1) \\ (\tan \alpha_2|x_1) & (\tan \alpha_2|\tan \alpha_1) \end{pmatrix} \begin{pmatrix} x_1 \\ \tan \alpha_1 \end{pmatrix}. \tag{1.14a}$$

These matrix elements completely describe the optical system between the two profile planes under consideration. In what follows we shall abbreviate the transfer matrix of Eq. (1.14a) as

$$\begin{pmatrix} x(z) \\ a(z) \end{pmatrix} = \begin{pmatrix} (x|x) & (x|a) \\ (a|x) & (a|a) \end{pmatrix} \begin{pmatrix} x_1 \\ a_1 \end{pmatrix}. \tag{1.14b}$$

In these matrix elements all indices have been omitted. Furthermore, $\tan \alpha$ has been replaced by a quantity a, which for the moment can be interpreted as a shorthand notation for $\tan \alpha$.

To be more specific about the quantity a, we define

$$a = v_x/c.$$

Analogously, we also define a quantity b:

$$b = v_y/c.$$

Here, c is the velocity of light, and v_x and v_y describe the x and y components of the velocity of the ray.

For an oblique light ray, we find $\tan \alpha = v_x / v_z$ and $\tan \beta = v_y / v_z$ (Fig. 1.1), and thus,

$$a = \frac{\tan \alpha}{\sqrt{1 + \tan^2 \alpha + \tan^2 \beta}} = \frac{\sin \alpha}{\sqrt{1 + \tan^2 \beta \cos^2 \alpha}}$$

$$= \sin \alpha \left(1 - \frac{\beta^2}{2} + \frac{\beta^2(4\alpha^2 + 3\beta^2)}{8} + \cdots \right),$$

$$\tag{1.15a}$$

$$b = \frac{\tan \beta}{\sqrt{1 + \tan^2 \alpha + \tan^2 \beta}} = \frac{\sin \beta}{\sqrt{1 + \tan^2 \alpha \cos^2 \beta}}$$

$$= \sin \beta \left(1 - \frac{\alpha^2}{2} + \frac{\alpha^2(4\beta^2 + 3\alpha^2)}{8} + \cdots \right),$$

$$\tag{1.15b}$$

so that we find to second order $a \approx \sin \alpha \approx \tan \alpha$ and $b \approx \sin \beta \approx \tan \beta$. We may note that the Eqs. (1.15a) and (1.15b) yield for a ray in the xz plane ($\beta = 0$),

$$a = \sin \alpha, \qquad b = 0;$$

and for a ray in the yz plane ($\alpha = 0$),

$$a = 0, \qquad b = \sin \beta.$$

1.3.1 Characteristic Trajectories $C(z)$ and $S(z)$

A lens bends rays (Section 1.1.2) toward the optic axis with the bending angle proportional to the distance x between the ray under consideration and the optic axis. For a homogeneous thick lens, each ring section within the lens, i.e., each region $\pm(x + dx)$ in Fig. 1.2, must act equally. Thus, we may state that $d(\tan \alpha)/dz$, i.e., the change of $\tan \alpha$ caused by a ring of radius x and extension dx of the lens, must be proportional to x, the distance between the ray under consideration, and the optic axis. Defining the proportionality constant to be $-k^2$, we find

$$d(\tan \alpha)/dz = -k^2 x$$

or with $\tan \alpha = dx/dz$ and $x'' = d^2x/dz^2$,

$$x'' = -k^2 x. \tag{1.16}$$

In the most general case this proportionality constant varies with z, and the differential [Eq. (1.16)] becomes Hill's equation (named after the astronomer George Hill). Assume now that the refractive index of the glass within a lens does not vary with z, although it may have different values in different lenses and be equal to 1 between lenses. In this case, the k in Eq. (1.16) is piecewise constant, and the differential equation [Eq. (1.16)] is solved for each section with constant $k^2 > 0$ by

$$x(z) = c \cos(kz) + d \sin(kz). \tag{1.17a}$$

From this solution we find $\tan \alpha(z) = dx(z)/dz$, or approximating $\tan \alpha(z)$ by $a(z)$,

$$a(z) = -kc \sin(kz) + kd \cos(kz) \tag{1.17b}$$

Introducing the initial conditions $x(z_1) = x_1$ and $a(z_1) = a_1$ into Eqs. (1.17a) and (1.17b) we can determine the coefficients c and d and find $X(z) = TX(z_0)$. Explicitly, this reads

$$\begin{pmatrix} x(z) \\ a(z) \end{pmatrix} = \begin{pmatrix} c_x & s_x \\ -k^2 s_x & c_x \end{pmatrix} \begin{pmatrix} x_1 \\ a_1 \end{pmatrix}, \tag{1.18}$$

where for $k^2 > 0$,

$$c_x = \cos kx, \qquad s_x = k^{-1} \sin kx,$$

and for $k^2 < 0$,

$$c_x = \cosh|k|x, \qquad s_x = |k|^{-1} \sinh|k|x.$$

Equation (1.18) is a solution of Hill's equation and thus the most general first-order description of rays. Inspecting the transfer matrix of Eq. (1.18), we find, remarkably, that the determinants of these transfer matrices always equal 1. This may be expressed as

$$|T| = \begin{vmatrix} (x|x) & (x|a) \\ (a|x) & (a|a) \end{vmatrix} = 1, \tag{1.19a}$$

or

$$(x|x)(a|a) - (x|a)(a|x) = 1. \tag{1.19b}$$

Since the determinant of the product of several matrices is the product of the determinants of the individual matrices, the statement of Eqs. (1.19a) and (1.19b) holds true for the transfer matrix of any complex optical system (Banford, 1966).

Special solutions for Eqs. (1.17a) and (1.17b) are found for an arbitrary z with $x_1 = 1$ and $a_1 = 0$ as in Fig. 1.8 as

$$x(z) = \cos(kz) = (x|x) = C(z), \tag{1.20a}$$

$$a(z) = -k \sin(kz) = (a|x) = C'(z), \tag{1.20b}$$

with $C'(z) = dC(z)/dz$ and $(x|x)$ and $(a|x)$ being elements of the general transfer matrix connecting profile planes at z_1 and z. Analogously, we find for $x_1 = 0$ and $a_1 = 1$ as in Fig. 1.8,

$$x(z) = k^{-1} \sin(kz) = (x|a) = S(z), \tag{1.21a}$$

$$a(z) = \cos(kz) = (a|a) = S'(z), \tag{1.21b}$$

with $S'(z) = dS(z)/dz$ and $(x|a)$ and $(a|a)$ being elements of the general transfer matrix connecting profile planes at z_1 and z.

Because of Eqs. (1.20a) and (1.20b), the function $C(z)$ is a solution of Eq. (1.16) with the initial conditions

$$C(z_1) = 1, \qquad C'(z_1) = 0. \tag{1.22a}$$

Similarly, the function $S(z)$ is a solution of Eq. (1.16) with the initial conditions

$$S(z_1) = 0, \qquad S'(z_1) = 1. \tag{1.22b}$$

The corresponding two rays $C(z)$ and $S(z)$, with the initial conditions of Eqs. (1.22a) and (1.22b), are called the characteristic rays of a bundle. These characteristic rays are shown in Fig. 1.8 for the thin-lens example of Fig. 1.5 and in Fig. 1.9 for two general optical systems.

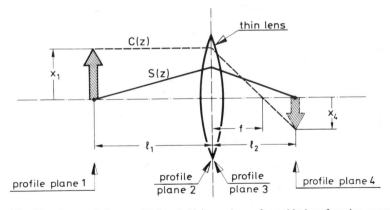

Fig. 1.8. The characteristic rays $C(z)$ and $S(z)$ are shown for a thin-lens focusing system.

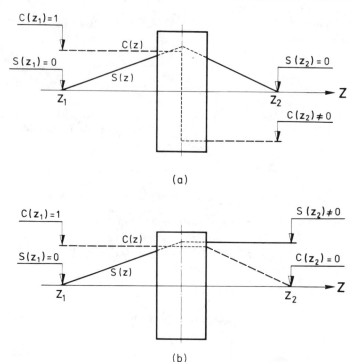

(a)

(b)

Fig. 1.9. (a) The characteristic rays $S(z)$ and $C(z)$ are shown for a general optical system extending from z_1 to z_2 for which the matrix elements $(x|a)$ and $(a|x)$ in the corresponding transfer matrix vanish simultaneously. (b) The characteristic rays $S(z)$ and $C(z)$ are shown for a general optical system extending from z_1 to z_2 for which the elements $(x|x)$ and $(a|a)$ in the corresponding transfer matrix vanish simultaneously.

1.3.2 Significance of the Disappearance of Elements of a Transfer Matrix

In order to understand the meaning of the different matrix elements in more detail, let us discuss what their individual disappearance implies (Halbach, 1964). Note here that the determinant of any transfer matrix equals unity according to Eqs. (1.19a) and (1.19b) so that (at most) two matrix elements can vanish simultaneously [i.e., either $(x|x) = (a|a) = 0$ or $(x|a) = (a|x) = 0$].

1.3.2.1 The Condition (x|a) = 0

For $(x|a) = 0$, the first row of Eq. (1.14b) reads

$$x(z_2) = (x|x)x_1,$$

expressing the fact that $x(z_2)$ does not depend on a_1; in other words, all rays starting from a point at $z = z_1$ will be concentrated to a point at $z = z_2$ (Fig. 1.10a). This relation was discussed previously in Section 1.21, where it was found that $(x|a) = 0$ characterizes an object–image relation between the two profile planes in question with a lateral magnification $x_2/x_1 = (x|x)$. Note that the condition $(x|a) = 0$ is identical to postulating that the characteristic function $S(z)$, which was zero at z_1, crosses the z axis again at z_2 in Fig. 1.9a.

1.3.2.2 The Condition (a|x) = 0

For $(a|x) = 0$, the second row of Eq. (1.14b) reads

$$a(z_2) = (a|a)a_1,$$

showing that $a(z_2)$ does not depend on x_1. For a bundle of parallel rays at z_1, as in Fig. 1.10b we also find a bundle of parallel rays at z_2. However, the angle $\alpha_1 \approx a_1$ under which these rays were inclined relative to the z axis at z_1 is changed so that $a_2 = (a|a)a_1$ at z_2 or $\alpha_2 \approx (a|a)\alpha_1$. Consequently, the angular magnification is to first order $\alpha_2/\alpha_1 \approx (a|a)$.

An optical system with arbitrary $(x|a)$, for which the coefficient $(a|x)$ vanishes, is called a telescope focused at infinity. However, an optical system for which the coefficients $(x|a)$ and $(a|x)$ vanish simultaneously is called a telescopic system. Note that in a telescopic system, indicated in Fig. 1.9a, the characteristic function $C(z)$ is parallel to the z axis at z_1 as well as at z_2, whereas the characteristic function $S(z)$ crosses the optic axis at z_1 as well as at z_2.

1.3.2.3 The Condition (x|x) = 0

For $(x|x) = 0$, the first row of Eq. (1.14b) reads

$$x(z_2) = (x|a)a_1,$$

showing that $x(z_2)$ does not depend on x_1. All rays of a parallel bundle inclined under the angle $\alpha_1 \approx a_1$ relative to the z axis at z_1 are focused to the same point at $z = z_2$ (Figs. 1.3 and 1.10c). Consequently, we may state that $(x|x) = 0$ *indicates that the second profile plane of the corresponding optical system is a focal plane* [Eq. (1.9)]. Note that the condition $(x|x) = 0$ is identical to stating that the characteristic function $C(z)$, which is parallel to the z axis at z_1, crosses the z axis again at z_2 in Fig. 1.9b.

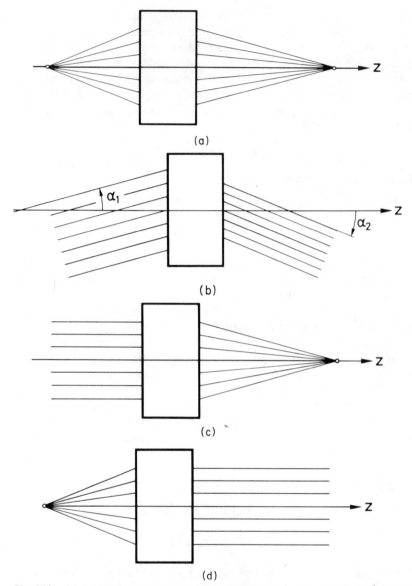

Fig. 1.10. (a) Schematic representation of an optical system with vanishing $(x|a)$. Such a system is also referred to as point-to-point focusing. (b) Schematic representation of an optical system with $(a|x) = 0$. Such a system is also referred to as parallel-to-parallel focusing (it is also called a telescope). (c) Schematic representation of an optical system with $(x|x) = 0$. Such a system is also referred to as parallel-to-point focusing. (d) Schematic representation of an optical system with $(a|a) = 0$. Such a system is also referred to as point-to-parallel focusing.

1.3.2.4 The Condition (a|a) = 0

For $(a|a) = 0$, the second row of Eq. (1.14b) reads

$$a(z_2) = (a|x)x_1,$$

showing that $a(z_2)$ does not depend on a_1. In other words, all rays starting from a point at $z = z_1$ will be parallel at $z = z_2$ (Fig. 1.10d). Thus $(a|a) = 0$ *indicates that the first profile plane of the corresponding optical system is a focal plane* [Eq. (1.9)]. Note that the condition $(a|a) = 0$ is identical to stating that the derivative of the characteristic function $S(z)$ vanishes at z_2, in other words, that the characteristic function $S(z)$, which crossed the z axis at z_1, is parallel to the z axis at $z = z_2$ Fig. 1.9b.

1.3.3 Transfer Matrices of General Focusing Systems

The focal length of a general focusing system is determined somewhat differently from that of a thin lens. Assume that two parallel rays enter a lens system as in Fig. 1.11. The focusing action of the lens causes these rays to meet at a point in the focal plane 2 of the lens. The relation between x_1 and $\tan \alpha_2$ then defines the focal length f_2 as

$$f_2 = -x_1/\tan \alpha_2 \approx -x_1/a_2. \qquad (1.23a)$$

For a reversed system, the corresponding focal length is obtained by

$$f_1 = -x_2/\tan \alpha_1 \approx -x_2/a_1. \qquad (1.23b)$$

A transfer matrix that relates the two focal planes of a general focusing system is therefore characterized by $(x|x) = (a|a) = 0$ according to the

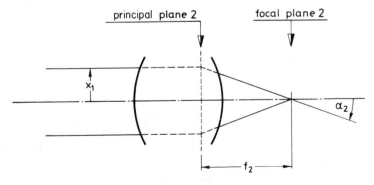

Fig. 1.11. Definition of the focal length of a general focusing system.

arguments of Sections 1.3.2.3 and 1.3.2.4. Because of Eqs. (1.23a) and (1.23b) this matrix reads:

$$T_{21} = \begin{pmatrix} 0 & f_1 \\ -1/f_2 & 0 \end{pmatrix}. \tag{1.24}$$

For the case $f_1 = f_2$, the determinant of the tansfer matrix of Eq. (1.24) equals 1. In this case, the transfer matrices presented in Eqs. (1.9) and (1.24) are identical, and the lens equation [Eq. (1.13)] applies not only to a thin lens, but to any focusing system.

1.3.4 Transfer Matrices between Object and Image Profile Planes

At this point we should determine the relation between profile planes a distance w_1 upstream from focal plane 1 and a distance w_2 downstream from focal plane 2 of a general focusing system. This relation is found by multiplying the transfer matrix of Eq. (1.24) by a drift-length matrix from both left and right, yielding for $f_1 = f_2 = f$,

$$\begin{pmatrix} (x|x) & (x|a) \\ (a|x) & (a|a) \end{pmatrix} = \begin{pmatrix} -w_2/f & f - (w_1 w_2/f) \\ -1/f & -w_1/f \end{pmatrix}$$

$$= \begin{pmatrix} 1 & w_2 \\ 0 & 1 \end{pmatrix} \begin{pmatrix} 0 & f \\ -1/f & 0 \end{pmatrix} \begin{pmatrix} 1 & w_1 \\ 0 & 1 \end{pmatrix}. \tag{1.25}$$

Note that the elements $(x|x)$, $(x|a)$, and $(a|a)$ depend on w_1 or w_2, whereas the element $(a|x)$ is simply $-1/f$. By postulating $(x|a) = 0$ in Eq. (1.25), that is, by postulating the existence of an object–image relation between the profile planes in question, we obtain Newton's lens equation [Eq. (1.13)]. Writing M_x for $-w_2/f$, Eq. (1.25) transforms with $w_1 w_2 = f^2$ to

$$\begin{pmatrix} (x|x) & (x|a) \\ (a|x) & (a|a) \end{pmatrix} = \begin{pmatrix} M_x & 0 \\ -1/f & 1/M_x \end{pmatrix}. \tag{1.26}$$

This states that if an object–image relation exists between two profile planes, the lateral magnifiation $(x|x)$ and the angular magnification $(a|a)$ are reciprocal.

1.3.5 Principal Planes of General Focusing Systems

Normally, the sum of the two focal lengths $f_1 + f_2$ is smaller than the distance between the two focal planes as is the case in Fig. 1.12, so that the simple formation of an image as shown in Fig. 1.5 does not apply. Instead of one profile plane characterizing the position of the thin lens,

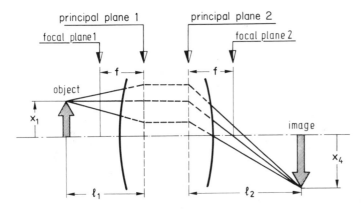

Fig. 1.12. Principal planes of a general focusing system.

here we must use two separated *principal planes* of the system, with both principal planes being distances f_1 and f_2 away from the corresponding focal planes, respectively. These two principal planes may be thought of as the left and right sides of a thin lens that has a focal power of $1/f = 2/(f_1 + f_2)$. The transfer matrix from before the first to behind the second principal plane is thus identical to the transfer matrix of Eq. (1.6), i.e., of a thin lens of focal length f.

1.4 EXAMPLES FOR THE USE OF TRANSFER MATRICES

The value of transfer matrices is best demonstrated by discussing some general problems of light optics.

1.4.1 A Lens Doublet and Its Optical Properties

For a two-lens system, a so-called doublet, which is illustrated in Fig. 1.13, the transfer matrix between an object and an image profile plane, is given by

$$\begin{pmatrix} (x|x) & (x|a) \\ (a|x) & (a|a) \end{pmatrix} = \begin{pmatrix} 1 & w_2 \\ 0 & 1 \end{pmatrix} \begin{pmatrix} 0 & f_2 \\ -1/f_2 & 0 \end{pmatrix} \begin{pmatrix} 1 & d_f \\ 0 & 1 \end{pmatrix} \begin{pmatrix} 0 & f_1 \\ -1/f_1 & 0 \end{pmatrix} \begin{pmatrix} 1 & w_1 \\ 0 & 1 \end{pmatrix}$$

$$= \frac{1}{f_1 f_2} \begin{pmatrix} w_2 d_f - f_2^2 & d_f w_1 w_2 - w_1 f_2^2 - w_2 f_1^2 \\ d_f & w_1 d_f - f_1^2 \end{pmatrix}. \tag{1.27}$$

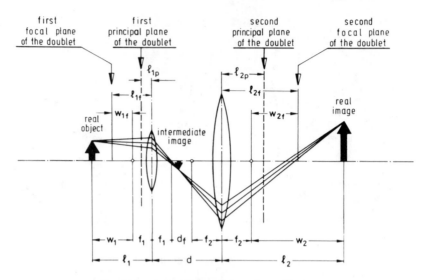

Fig. 1.13. A doublet consisting of two thin lenses.

Here it is assumed that the two focal lengths of the first lens are both f_1 and that the two focal lengths of the second lens are both f_2. The refractive power $1/f$ of the doublet described by Eq. (1.27) is $-(a|x)$ or

$$\frac{1}{f} = \frac{-d_f}{f_1 f_2} = \frac{f_1 + f_2 - d}{f_1 f_2},$$ (1.28)

with d defined in Fig. 1.13 as $f_1 + f_2 + d_f$.

According to Eq. (1.28), the overall focal length f can be varied by varying d. Note that in order for f to be positive, one or all three of the quantities f_1, f_2, d_f must be negative.

1.4.1.1 Focal Planes and Principal Planes of a Doublet

According to the ideas presented in Sections 1.3.2.3 and 1.3.2.4, the positions of the focal planes 1 and 2 of a doublet are found by setting $(x|x)$ and $(a|a)$ of Eq. (1.27) individually to zero. Equation (1.28) then yields

$$w_{1f} = f_1^2/d_f = -f(f_1/f_2),$$ (1.29a)

$$w_{2f} = f_2^2/d_f = -f(f_2/f_1).$$ (1.29b)

Here, w_{1f} describes the distance between focal plane 1 of the doublet and focal plane 1 of the first lens. Similarly, w_{2f} describes the distance between focal plane 2 of the doublet and focal plane 2 of the second lens as illustrated in Fig. 1.13.

It is often necessary to calculate the distances l_{1f} and l_{2f} between the focal planes of the doublet and the positions of the first and second lens (for simplification, both lenses are assumed to be thin lenses). These lengths are taken from Fig. 1.13 and Eqs. (1.28) and (1.29) to be

$$l_{1f} = w_{1f} + f_1 = f[1 - (d/f_2)], \tag{1.30a}$$

$$l_{2f} = w_{2f} + f_2 = f[1 - (d/f_1)]. \tag{1.30b}$$

The principal planes 1 and 2 must each be a distance f apart from the focal planes 1 and 2 of the doublet, respectively. Thus, the distance between the principal planes and the two corresponding lenses are found to be

$$l_{1p} = l_{1f} - f = -d(f/f_2), \tag{1.31a}$$

$$l_{2p} = l_{2f} - f = -d(f/f_1), \tag{1.31b}$$

whereas the distance d_p between the two principal planes is

$$d_p = l_{1p} + d + l_{2p} = -d^2 f/f_1 f_2. \tag{1.31c}$$

1.4.1.2 The Telescope: A Special Lens Doublet

A commonly used lens doublet is a telescope. Such a system transforms an incoming parallel beam into an outgoing parallel beam (Fig. 1.14). Since $(\alpha|x)$ must vanish according to Section 1.3.2.2 for a telescope, one finds $d_f = 0$ from Eq. (1.28), or $d = f_1 + f_2$. This is equivalent to stating that the second focal plane of the front lens and the first focal plane of the back lens coincide. If all rays enter the doublet as a parallel bundle under the angle α_1 relative to the optic axis, we find $a_2 = (a|a)a_1$, or $\alpha_2 \approx (a|a)\alpha_1$. Taking $(a|a)$ from Eq. (1.27), the angular magnification of a telescopic focusing doublet is found for $d_f = 0$ as

$$M_{ang} = a_4/a_1 = -f_1/f_2. \tag{1.32}$$

Note that according to Eq. (1.32), a high angular magnification requires a large f_1 and a small f_2.

Two focusing lenses, i.e., $f_1 > 0$ and $f_2 > 0$, characterize the so-called astronomical or Kepler telescope for which M_{ang} is negative. A focusing and a defocusing lens, i.e., $f_1 > 0$ and $f_2 < 0$, characterize the so-called Dutch or Galilean telescope, for which M_{ang} is positive. In both cases, the two lenses of the doublet should be well-designed lens multiplets instead of single thin lenses so that the image aberrations of the lenses do not limit the resolving power.

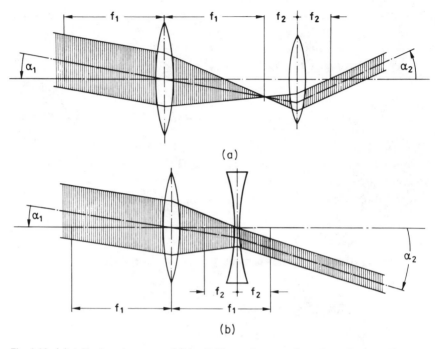

Fig. 1.14. (a) A Kepler telescope and (b) a Galilean telescope of equal angular magnification.

1.4.2 A Doublet Consisting of a Focusing and a Defocusing Thin Lens

For many doublet applications, we combine a defocusing and a focusing lens, so that $f_1 f_2 < 0$. We call such a defocusing–focusing doublet a DF doublet. In the case of $f > 0$ and $d > 0$, Eqs. (1.31a) and (1.31b) yield $l_{1p} > 0$ and $l_{2p} < 0$ for $f_1 > 0$ and $l_{1p} < 0$ and $l_{2p} > 0$ for $f_2 > 0$. This shows that both principal planes are shifted to the side of the focusing lens (Fig. 1.15). This property of a DF doublet can be used to construct lenses for which the principal and focal planes are shifted to suitable positions relative to the physical dimensions of the lenses.

1.4.3 Numerical Examples of Lens Doublets

1.4.3.1 Photographic Lenses of Long and Short Focal Lengths

In a photographic camera, the image of a very distant object ($w_1 \approx \infty$) is calculated from Eq. (1.13) to be situated at $w_2 = f^2/w_1 \approx 0$, i.e., immediately behind the second focal plane of a lens of focal length f. The size of

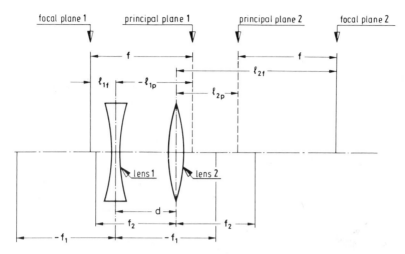

Fig. 1.15. A focusing doublet ($f > 0$) is shown consisting of one defocusing ($f_1 < 0$) and one focusing lens (f_2). Note that both principal planes are shifted to the side of the focusing lens. The doublet shown corresponds to $f_1 = -250$ mm, $f_2 = 200$ mm, and $d = 150$ mm, and thus, $f = 250$ mm, $l_{1p} = -187.5$ mm, and $l_{2p} = 150$ mm.

the image is the size of the object multiplied by the lateral magnification $M_x = f/w_1$. For a very distant object it is thus advantageous to use a lens of long focal length because the demagnification M_x then stays within limits, and details in the image remain distinguishable. On the other hand, since the image lies just behind the second focal plane of the lens, a camera is unwieldy if it employs a lens of long focal length.

An ideal camera for distant objects should have a lens of long focal length f, with the second focal plane of this lens only a short distance l_{2f} behind the first glass surface. Values of $f = 250$ mm snd $l_{2f} = 125$ mm, for instance, would be good choices. To achieve these values by a doublet, as shown in Fig. 1.16, we may attempt to vary f_1, f_2, and d. From Eqs. (1.28), and (1.30b), the dependences of f and l_{2f} on f_1, f_2, and d are known if the doublet consists of two thin lenses. These equations can be transformed to

$$l_{2f}/f_1 = (l_{2f}/d)[1 - (l_{2f}/f)], \qquad (1.33a)$$

$$l_{2f}/f_2 = 1 + (l_{2f}/d)[1 - (f/1_{2f})]. \qquad (1.33b)$$

Choosing $l_{2f} = 50$ mm, $f = 250$ mm and $d + l_{2f} = 125$ mm, we find that $f/l_{2f} = 5$ and $d/l_{2f} = 1.5$. Equations (1.33a) and (1.33b), yield in this case, $l_{2f}/f_1 \approx 0.533$, $l_{2f}/f_2 \approx -1.66$ or $f_1 \approx 93.8$ mm and $f_2 \approx -30$ mm. Thus, the first and second principal planes of the system are situated at $l_{1p} = 625$ mm and $l_{2p} - d = -125$ mm. That is, the distance d_p between the two principal planes is calculated from Eq. (1.31c) to be approximately 500 mm with both

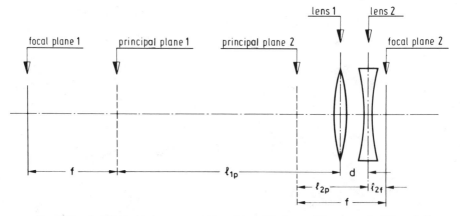

Fig. 1.16. A doublet is shown consisting of two thin lenses 75 mm apart. Although the doublet focal length is 250 mm, the focal plane 2 of the doublet is only 50 mm behind the second lens. The size of the total mechanical system is $d + l_{2f} = 125$ mm. Note that both principal planes are shifted to the side of the focusing lens.

principal planes found before the front lens. The geometry of this lens doublet is shown in Fig. 1.16.

Any tele lens used for photographic purposes consists of a design similar to that shown in Fig. 1.16. This can easily be checked by comparing the rated focal length with the physical length of such a lens including its mount. Photographic lenses used in satellite or aerial photography employ this principle in an even more elaborate manner, since there the required focal length of 10 and more meters greatly exceeds the size of the carrier.

A similar problem arises if one wants to use a lens of very short focal length (a wide-angle lens) on a camera that was designed to be used with a lens of longer focal length. Postulating for instance, $f = 20$ mm, $l_{2f} = 30$ mm, and $d = 30$ mm, Eqs. (1.33a) and (1.33b) yield $f_1 = -60$ mm and $f_2 = 22.5$ mm.

1.4.3.2   Huygens' Eyepiece

As a more difficult application of a doublet we can correct some image aberration of a lens doublet (Halbach, 1964). In a telescope or a microscope the first lens produces an intermediate real image. From this real image a magnified, virtual image is formed by the second lens, the eyepiece. The image formation of such an eyepiece is shown in Fig. 1.17 together with the eye that observes this image. If the eyepiece exhibits chromatic aberrations, the angle α in Fig. 1.17 varies as well as the distance l between the virtual image and the eye for light of different wavelengths.

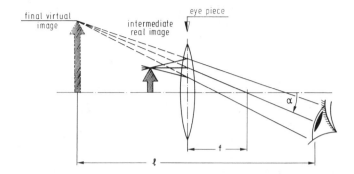

Fig. 1.17. Formation of a virtual image by an eyepiece of a Kepler telescope or a microscope.

What can be done to minimize these aberrations, at least in the neighborhood of one wavelength λ? The best choice would be to keep both l and α constant for some region of λ. Unfortunately, this is impossible when the eyepiece consists of only a simple lens arrangement. If l varies with λ, the top of the arrow in Fig. 1.17 will be sharply focused at the back of the eye for one color and be somewhat out of focus for the others. A variation of α on the other hand is identical to a variation in angular magnification of the telescope. Thus, the top of the arrow would be focused to different places at the back of the eye. The second option is clearly less desirable, since in that case any point would appear as a short line showing all colors of the spectrum. In the first case, however, the images produced by light of different colors at least fall on top of each other, although they may not be exact points but rather small circles of varying diameter. Thus, we should ask for a constant α, which is the same as asking for a constant f, which we may take from Fig. 1.17 or Eqs. (1.23a) and (1.23b).

The desired eyepiece may now be designed as a doublet consisting of two thin lenses of focal lengths f_1 and f_2 separated by a distnace d. These focal lengths depend on the wavelength $\lambda = \lambda_0(1 + \delta)$ of the light as $f_1 = f_{10}(1 + \delta + \cdots)$ and $f_2 = f_{20}(1 + \delta + \cdots)$, where f_{10} and f_{20} are the focal lengths for λ_0. Introducing these f_1 and f_2 into Eq. (1.28), the focal length f of the doublet is found as

$$\frac{1}{f} = \frac{f_{10}f_{20} - d}{f_{10}f_{20}} - \frac{f_{10}f_{20} - 2d}{f_{10}f_{20}}\delta + \cdots . \tag{1.34}$$

Choosing

$$d = \tfrac{1}{2}(f_{10} + f_{20}), \tag{1.35a}$$

the quantity f becomes independent of $\delta = (\lambda - \lambda_0)/\lambda_0$ to first order. With

d determined according to Eq. (1.35a), the Eqs. (1.34) and (1.30) yield

$$f = 2f_{10}f_{20}/(f_{10} + f_{20}), \tag{1.35b}$$

$$l_{1f} = (f_{10}f_{20} - f_{20}^2)/(f_{10} + f_{20}), \tag{1.35c}$$

$$l_{2f} = (f_{10}f_{20} - f_{10}^2)/(f_{10} + f_{20}). \tag{1.35d}$$

For $f_{10} > 0$ and $f_{20} > 0$, Eqs. (1.35c) and (1.35d) indicate that only one focal plane can be real because either l_{1f} or l_{2f} are negative. Since the eye should be close to the second focal plane, we advantageously choose $l_{2f} > 0$ and $l_{1f} < 0$. For $f_{10} < f_{20}$, this is achieved in Huygens' original design, which was characterized by $f_{10} = 3f_{20}$.

REFERENCES

Banford, A. P. (1966). "The Transport of Charged Particle Beams." E. & F. N. Spon Ltd., London.
Braams, C. M. (1956). Thesis, Utrecht.
Brown, K. L. Belbeoch, R., and Bounin, R. (1964). *Rev. Sci. Instrum.* **35**, 481.
Cotte, M. (1938). *Ann. Phys. (Paris)* **10**, 333.
Halbach, K. (1964). *Am. J. Phys.* **32**, 90.
Herzberger, M. (1958). "Modern Geometrical Optics." Wiley (Interscience), New York.
Penner, S. (1961). *Rev. Sci. Instrum.* **32**, 150.
Wollnik, H. (1967). Nucl. Instrum. Methods **52**, 250.

2

General Relations for the Motion of Charged Particles in Electromagnetic Fields

To determine the trajectories of charged particles within electromagnetic fields, it is necessary to know the spatial distribution of these fields as well as the particles' rigidity against electromagnetic deflections. This rigidity depends on the mass, charge, and velocity of the particle under investigation.

2.1 ENERGY, VELOCITY, AND MASS OF ACCELERATED PARTICLES

Assume a particle of positive charge (ze) to be at rest at zero electrostatic potential (here, e is the charge of a proton, and z is an integer). Under the action of an electrostatic field, such a particle of rest mass m_0 will move to a point of potential $-V$ and then have the velocity $v = |\mathbf{v}|$. Since the total particle energy $E = mc^2$ is the sum of m_0c^2 and the gained kinetic energy $K = -(ze)V$, we can express K and the increased mass m as

$$K = -(ze)V = (m - m_0)c^2, \tag{2.1a}$$

$$m = \frac{m_0}{\sqrt{1 - (v^2/c^2)}} = m_0\gamma \tag{2.1b}$$

or $v^2/c^2 = 1 - (1/\gamma^2)$. Here, $c = 299,792,458$ m/sec ≈ 300 m/μsec is the velocity of light, and

$$\gamma = 1 + 2\eta \quad \text{with} \quad \eta = K/(2m_0c^2). \tag{2.1c}$$

In Eqs. (2.1a) and (2.1b) the kinetic energy K and the mass m are expressed as functions of the particle velocity $|\mathbf{v}|$. Usually, however, the particle energy is known and the particle velocity unknown. Therefore it is useful (see for instance, Glaser, 1956) to rewrite Eqs. (2.1a) and (2.1b) as

$$m = m_0(1 + 2\eta), \tag{2.2a}$$

$$v = \frac{2c}{1 + 2\eta}\sqrt{\eta(1 + \eta)} = \frac{\sqrt{(1 + \eta)2K/m_0}}{1 + 2\eta}. \tag{2.2b}$$

For numerical calculations, it is advantageous to introduce dimensionless quantities $\bar{m}_0$, $\bar{K}$, and $\bar{v}$ with

$$K = \bar{K} \text{ MeV}, \quad m_0 = \bar{m}_0 \text{ u}, \quad v = \bar{v} \text{ m}/\mu\text{sec}.$$

Remember now that the energy equivalent m_0c^2 of one mass unit* $[u]$, is approximately 931.501 million electron volts† [MeV] so that the numerical value of η is found from Eq. (2.1c):

$$\eta = \bar{K}/1863.003\,\bar{m}_0. \tag{2.3}$$

Thus, m and v are calculated from Eqs. (2.2a) and (2.2b) as

$$\bar{m} = \bar{m}_0(1 + 2\eta), \tag{2.4a}$$

$$\bar{v} = 13.891\frac{\sqrt{\bar{K}(1 + \eta)/\bar{m}_0}}{1 + 2\eta}. \tag{2.4b}$$

For relativistically slow particles $[v \ll c$ or $(K/m_0) \ll 931$ MeV/u], we find $\eta \ll 1$ so that the Eqs. (2.4a) and (2.4b) simplify to

$$\bar{m} \approx \bar{m}_0, \tag{2.4as}$$

$$\bar{v} \approx 13.891332\sqrt{\bar{K}/\bar{m}_0}. \tag{2.4bs}$$

---

* One mass unit ($u \approx 1.660565 \times 10^{-24}$ g) is the mass of 1/12 of a ^{12}C atom.

† One electron volt (eV = 10^{-6} MeV $\approx 1.60219 \times 10^{-19}$ VAsec) is the energy that a particle of one charge unit gains by moving through a potential difference of 1 V.

For a singly charged ^{100}Mo ion ($m_0 \approx 99.908$ u) of 0.01 MeV, we find $\bar{K}/\bar{m}_0 \approx 1.0009 \times 10^{-4}$ or $\eta \approx 5 \times 10^{-8}$. Thus, Eqs. (2.4as) and (2.4bs) yield in a sufficient approximation $\bar{m} \approx \bar{m}_0$ and $\bar{v} \approx 0.139$ or $m \approx 99.9$ u and $v \approx 0.139$ m/μsec.

For relativistically fast particles [$v \approx c$ or $(K/m_0) \gg 931$ MeV/u], we find that $\eta \gg 1$, so that the Eqs. (2.4a) and (2.4b) simplify to

$$\bar{m} \approx \bar{K}/931.5, \tag{2.4af}$$

$$\bar{v} \approx 300. \tag{2.4bf}$$

For a proton ($m_0 \approx 1.0073$ u) of 30 GeV ($K = 30{,}000$ MeV), we find $\bar{K}/\bar{m}_0 \approx 29{,}783$ or $\eta \approx 16 \gg 1$. Thus, approximate Eqs. (2.4af) and (2.4bf) yield $\bar{m} \approx 32.21$ and $\bar{v} \approx 300$, whereas exact Eqs. (2.4a) and (2.4b) would have yielded $\bar{m} \approx 33.21$ and $\bar{v} \approx 299.6$ or $v \approx 0.9995c \approx 299.6$ m/μsec.

2.2 FORCES ON CHARGED PARTICLES IN MAGNETIC AND ELECTROSTATIC FIELDS

Assume a particle of mass m and charge (ze) that moves with the velocity **v** in either an electrostatic field **E** or a magnetic field **H**, where $\mathbf{B} = \mu\mu_0\mathbf{H}$ is the magnetic flux density.* In the electrostatic field, this particle experiences a Coulomb force $\mathbf{F}_{\mathscr{E}}$ and in the magnetic field, a Lorentz force $\mathbf{F}_{\mathscr{B}}$ with

$$\mathbf{F}_{\mathscr{B}} = d(m\mathbf{v})/dt = (ze)\mathbf{v} \times \mathbf{B}, \tag{2.5a}$$

$$\mathbf{F}_{\mathscr{E}} = d(m\mathbf{v})/dt = (ze)\mathbf{E}. \tag{2.5b}$$

The force $\mathbf{F}_{\mathscr{B}}$ is perpendicular to **v** and **B** and points in the direction indicated in Fig. 2.1 for positively charged particles. The force $\mathbf{F}_{\mathscr{E}}$ points in the direction of E; that is, from the positive to the negative potential.

According to Eqs. (2.5a) and (2.5b), the motion of charged particles in electrostatic and magnetic fields is described by the Lorentz equation

$$d(m\mathbf{v})/dt = (ze)\mathbf{E} + (ze)\mathbf{v} \times \mathbf{B}, \tag{2.6}$$

which is the basis for most calculations in the following chapters.

To describe particle trajectories in comparison to light rays, one must replace c, the velocity of light, by v, the velocity of the particle under consideration. Thus we can (see also Wollnik and Berz, 1985) redefine the

---

* The electrostatic field strength **E** is given in volts per meter, the magnetic flux density $\mathbf{B} = \mu\mu_0\mathbf{H}$ in teslas (1 T = 1 Vsec/m^2 = 10,000 G), and the magnetic field strength **H** in A/m (1 A/m = $4\pi \times 10^{-3}$ O). Here, μ_0 equals $4\pi \times 10^{-7}$ Vsec/Am, and μ is a material constant that is 1 for vacuum. The forces $\mathbf{F}_{\mathscr{B}}$ and $\mathbf{F}_{\mathscr{E}}$ are both in volt ampere seconds per meter.

Fig. 2.1. Spatial relation between the velocity **v** of a positively charged particle, the direction of the magnetic flux density **B**, and the Lorentz force $\mathbf{F}_{\mathscr{B}}$. In order to find the direction of $\mathbf{F}_{\mathscr{B}}$ in space, it is a trivial but useful rule to form a tripod with the thumb, the forefinger, and the middle finger of the right hand. If the thumb points in the direction of **v** and the forefinger in the direction of **B**, i.e., from a magnetic north to a magnetic south pole, then the middle finger points in the direction of $\mathbf{F}_{\mathscr{B}}$. Note here also that on a compass needle, a mark is applied at the magnetic north pole that points to the geographic north pole (a magnetic south pole). Thus, we call a magnet north pole also a north-seeking pole.

quantities a and b of Eqs. (1.15a) and (1.15b) as

$$a = \frac{mv_x}{mv_0} = \frac{p_x}{p_0} = \frac{(1+\delta_{\mathrm{p}})\tan\alpha}{\sqrt{1+\tan^2\alpha+\tan^2\beta}} = \frac{(1+\delta_{\mathrm{p}})\sin\alpha}{\sqrt{1+\tan^2\beta\,\cos^2\alpha}}, \qquad (2.7a)$$

$$b = \frac{mv_y}{mv_0} = \frac{p_y}{p_0} = \frac{(1+\delta_{\mathrm{p}})\tan\beta}{\sqrt{1+\tan^2\alpha+\tan^2\beta}} = \frac{(1+\delta_{\mathrm{p}})\sin\beta}{\sqrt{1+\tan^2\alpha\,\cos^2\beta}}. \qquad (2.7b)$$

Here, $p = p_0(1+\delta_{\mathrm{p}}) = \sqrt{p_x^2 + p_y^2 + p_z^2}$ is the momentum of the particle under consideration, and p_0 is the momentum of a reference particle at the same potential. Furthermore, one should make certain that a, b, p_x, p_0, α, and β are all taken at the same position. Note here that in the case of a planar motion $\beta = 0$ (Fig. 2.2) or $\alpha = 0$, the Eqs. (2.7a) and (2.7b) simplify to

$$a = \sin\alpha, \qquad b = 0, \qquad\qquad (2.8a)$$

or

$$a = 0, \qquad\qquad b = \sin\beta. \qquad\qquad (2.8b)$$

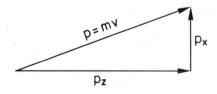

Fig. 2.2. For the case $\beta = 0$, the components of the momentum $p = ip_x + jp_z$ are shown for a particle of mass m and velocity $v = iv_x + jv_z$.

With the definition of a and b given in Eq. (2.7), we can now rewrite the equation of motion [Eq. (2.6)] and split it into its x, y components:

$$d(mv_x)/dt = d(ap_0)/dt = (ze)E_x + (ze)(\mathbf{v} \times \mathbf{B})_x, \qquad (2.9a)$$

$$d(mv_y)/dt = d(bp_0)/dt = (ze)E_y + (ze)(\mathbf{v} \times \mathbf{B})_y. \qquad (2.9b)$$

2.2.1 Magnetic Rigidity

Equation (2.5a) states that in a pure magnetic field, the Lorentz force $\mathbf{F}_{\mathscr{B}}$ is always perpendicular to the particle velocity $\mathbf{v}$. Thus, $\mathbf{v}$ changes its direction but not its magnitude. The angle $d\phi$ between the particle velocities $\mathbf{v}$ and $\mathbf{v} + d\mathbf{v}$ at two consecutive instants is $d\phi = dv/v$ with $v = |\mathbf{v}|$ (Fig. 2.3), and consequently $dv/dt = v\,d\phi/dt$. On the other hand, the magnitude of the particle velocity $\mathbf{v}$ equals $\rho(d\phi/dt)$, where ρ denotes the momentary radius of curvature of the trajectory so that we find

$$dv/dt = v\,d\phi/dt = v^2/\rho.$$

For a fast-moving particle, the mass m increases for v approaching c, the velocity of light. However, for a particle whose velocity v changes only in direction and not in magnitude, the mass m is constant. Thus, dm/dt is always zero for a particle moving in a pure magnetic field, and as far as the magnitudes of $v = |\mathbf{v}|$ and $B = |\mathbf{B}|$ are concerned, Eq. (2.5a) transforms to

$$m(dv/dt) = mv^2/\rho = (ze)vB.$$

Here, the vector $\mathbf{v} \times \mathbf{B}$ is replaced by vB, assuming that $\mathbf{v}$ is perpendicular to $\mathbf{B}$. This relation may be interpreted as describing the balance between the centrifugal force (mv^2/ρ) and the centripetal Lorentz force $(ze)vB$ and can be rewritten as

$$\chi_B = B\rho = mv/(ze) = p/(ze), \qquad (2.10)$$

where mv is the momentum p of the particle. The quantity χ_B we call the magnetic rigidity of the particle or its $B\rho$ value, both measured in tesla meters. According to Eq. (2.10), a particle of given $mv/(ze)$ moves in a

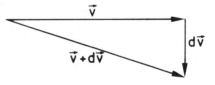

Fig. 2.3. The velocities $\mathbf{v}$ and $\mathbf{v} + d\mathbf{v}$ of a particle are shown at two consecutive instants. In a pure magnetic field, $d\mathbf{v}$ is always perpendicular to $\mathbf{v}$.

constant flux density B along a circle of constant radius ρ. Introducing Eqs. (2.2a) and (2.2b) into Eq. (2.10), we find in tesla meters (Tm),

$$\chi_B = \frac{m_0}{(ze)} \sqrt{\frac{2K}{m_0}(1 + \eta)}. \tag{2.11a}$$

Note that because of Eq. (2.2b) we can replace $\sqrt{2K(1 + \eta)/m_0}$ by $(1 + 2\eta)v/c = \gamma v/c$. In some cases, Eq. (2.11a) is also written by using Eq. (2.1c) as

$$\chi_B = \frac{K}{(ze)c} \sqrt{1 + \frac{2m_0c^2}{K}} \tag{2.11b}$$

so that χ_B is given in MeV/c, i.e., approximately $1/300$ Tm.

For numerical calculations, it is useful to introduce a dimensionless rigidity $\bar{\chi}_B$ or $\tilde{\chi}_B$ with

$$\chi_B = \bar{\chi}_B \ \text{Tm}, \qquad \chi_B = \tilde{\chi}_B \ \text{MeV}/c$$

where $\tilde{\chi}_B \approx 300\bar{\chi}_B$. Using $m_0c^2 \approx 931.5\bar{m}_0$ or $\eta \approx \bar{K}/(1863\bar{m}_0)$, according to Eq. (2.3) we thus find

$$\bar{\chi}_B \approx 0.14397467 \frac{\bar{m}_0}{(z)} \sqrt{\frac{\bar{K}}{\bar{m}_0}(1 + \eta)}, \qquad \tilde{\chi}_B \approx \frac{\bar{K}}{(z)} \sqrt{1 + \frac{1863.003\,\bar{m}_0}{\bar{K}}}. \tag{2.12}$$

For relativistically slow particles $[v \ll c$ or $(K/m_0) \ll 931$ MeV/u$]$, we find $\eta \ll 1$, so that the Eq. (2.12) simplifies to

$$\bar{\chi}_B \approx \frac{0.144}{(z)} \sqrt{\bar{K}\bar{m}_0}, \qquad \tilde{\chi}_B \approx \frac{43.16}{(z)} \sqrt{\bar{K}\bar{m}_0}. \tag{2.12s}$$

For a singly charged ^{100}Mo ion $(m_0 \approx 99.908$ u$)$ of 0.01 MeV, we find $\bar{K}/\bar{m}_0 \approx 10^{-4}$ or $\eta \approx 5 \times 10^{-8}$. Consequently, Eq. (2.12s) yields a magnetic rigidity $\chi_B \approx 0.144$ Tm ≈ 43.2 MeV/c with a deviation of less than 0.1% from the exact value determined by Eq. (2.12).

For relativistically fast particles $[v \approx c$ or $(K/m_0) \gg 931$ MeV/u$]$, we find $\eta \gg 1$,

$$\bar{\chi}_B \approx \bar{K}/300(z) \approx -\bar{V}/300, \qquad \tilde{\chi}_B \approx \bar{K}/(z) \approx -\bar{V} \tag{2.12f}$$

with the acceleration potential $V = \bar{V}$ MV. According to Eq. (2.12f), the magnetic rigidities for very energetic particles depend only on the magnitude of $-V$, i.e., on the energy per charge $K/(ze)$ of the particle but not on its rest mass m_0, which is assumed here to be negligibly small. For a proton $(m_0 \approx 1.0073$ u$)$ or for an electron $(m_0 \approx 0.54858 \times 10^{-3}$ u$)$ of 9000 MeV energy Eq. (2.12f) thus yields the same magnetic rigidity of 30 Tm or

9000 MeV/c, whereas the exact values would have been 33.0 Tm or 9894 MeV/c and 30.0 Tm or 9000 MeV/c, respectively, according to Eq. (2.12).

2.2.2 Electrostatic Rigidity

Equation (2.4b) states that a positively charged particle moving in an electrostatic field **E** experiences a Coulomb force $\mathbf{F}_{\mathscr{E}}$ in the direction of the field lines, i.e., from a positive to a negative charge. Thus, the magnitude of the particle velocity **v** varies if the particle moves in this direction. In a field in which the field lines are all perpendicular to the particle motion, only the direction of the particle velocity is varied and not its magnitude. This situation is approximated by charged particles moving through a cylindrical, spherical, or toroidal electric field as indicated in Fig. 2.4.

For the definition of electrostatic rigidity, we use here only the special case $E_{\perp}\mathbf{v}$, which is illustrated in Fig. 2.4. As in Section 2.2.1, the magnitude of **v** thus does not change with time, so that the mass of the particle does not vary either. Consequently, dm/dt vanishes, and Eq. (2.5b) yields

$$m(dv/dt) = mv^2/\rho = (ze)E,$$

where dv/dt is replaced by v^2/ρ, as derived in Section 2.2.1 for dv and E perpendicular to **v**. This relation may be interpreted analogously to Eq. (2.10) as describing the balance between the centrifugal force mv^2/ρ and the centripetal Coulomb force $(ze)\mathbf{E}$. It may be rewritten as

$$\chi_E = E\rho = mv^2/(ze) = pv/(ze), \tag{2.13}$$

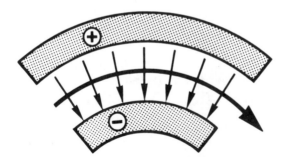

Fig. 2.4. The electrostatic force on a positively charged particle is in the direction of the electrostatic field, i.e., from the positive to the negative potential. Thus, the potentials $\pm V_0$ at two concentric electrodes can be chosen such that a particle of energy K_0 and charge $(z_0 e)$ can move along a circle in the middle between the two electrodes. In this case, the electrostatic force is always directed toward the center of curvature of the electrodes and is thus perpendicular to the circular particle trajectory. Consequently, along this circle, charged particles move with constant velocity.

or with Eqs. (2.10), (2.11a), and (2.2a),

$$\chi_E = \chi_B^2 \frac{(ze)}{m} = \frac{2K(1 + \eta)}{(ze)(1 + 2\eta)} = -\frac{2V(1 + \eta)}{1 + 2\eta}.$$ (2.14)

We call this quantity the electrostatic rigidity, or the $E\rho$ value of a particle. Here, χ_E is in megavolts if $K/(ze) = -V$ is given in mega-electron-volts per charge with η calculated from Eq. (2.3).

For relativistically slow particles [$v \ll c$ or $(K/m_0) \ll 931$ MeV/u], we find $\eta \ll 1$, so that Eq. (2.14) simplifies to

$$\chi_E = -2V.$$ (2.14s)

For relativistically fast particles [$v \approx c$ or $(K/m_0) \gg 931$ MeV/u], we find $\eta \gg 1$, so that Eq. (2.14) simplifies to

$$\chi_E = -V.$$ (2.14f)

In both limiting cases, the electrostatic rigidity depends only on the accelerating potential V or, in other words, on the gained energy per charge of the particle and not on the particle rest mass. Only in the region between the two extrema does the electrostatic rigidity vary with the particle rest mass. In both limiting cases, two particles of equal energy to charge ratios $K/(ze)$ are thus deflected equally by an electrostatic field independent of the masses of the two particles.

As an example, where the electrostatic rigidity must be calculated from the more complex Eq. (2.14) in which χ_E depends on the particle rest mass, we may determine this quantity for an electron ($m_0 \approx 0.54858 \times 10^{-3}$ u) of 0.5-MeV energy. In this case, $\bar{K}/\bar{m}_0$ equals 911.444, so that we find $\eta = 0.4892$, which is neither large nor small compared to 1. Thus, Eqs. (2.14s) and (2.14f) would have yielded 1 and 0.5 MV, respectively, whereas the exact result is obtained from Eq. (2.14) as $\chi_E \approx 0.7527$ MV.

2.3 DESCRIPTION OF A CHROMATIC PARTICLE BUNDLE

Whatever process we choose to generate ions, we always find that besides the desired ion species, ions of other masses are formed as well, at least in small quantities. The rest mass m_0 for an individual particle can thus deviate from the rest mass m_{00} of a reference particle. The most relevant quantity for the motion of charged particles in electromagnetic fields, however, is not the mass but the mass–charge ratio. If the charge for any individual particle and for a reference particle are (ze) and (z_0e), we can define

$$m_0/(ze) = [m_{00}/(z_0e)](1 + \delta_m)_{K=\text{const}}.$$ (2.15)

To accelerate a cloud of charged particles, it is necessary to create at least a small electrostatic field within the cloud in order to move the particles from the point where they have been generated. Consequently, at least small potential differences $\pm \Delta V$ must exist within the cloud of particles. Assume now a reference particle of charge $(z_0 e)$ generated at potential zero and an arbitrary particle of charge (ze) generated at potential ΔV. When both particles move to a point of reference potential V_0, the reference particle gains the energy $K_0 = -(z_0 e)V_0$ and the arbitrary particle the energy $K_r = -(ze)(V_0 - \Delta V)$. Thus, we can write

$$K_0/(z_0 e) = -V_0, \tag{2.16a}$$

$$K_r/(ze) = [K_0/(z_0 e)](1 + \delta_K)_{m=\text{const}}, \tag{2.16b}$$

where δ_K is an abbreviation for the small quantity $-\Delta V/V_0$. Moving the same arbitrary particle to a point at potential $V_0 + \hat{V}$, the particle gains the energy $K = -(ze)(V_0 + \hat{V} - \Delta V)$,

$$\frac{K}{(ze)} = \frac{K_0}{(z_0 e)}(1 + \delta_K) + \hat{V} = \frac{K_0}{(z_0 e)}(1 + \delta_K + \hat{\delta}_K), \tag{2.16c}$$

where $\hat{\delta}_K$ is an abbreviation for the small quantity $\hat{V}/V_0$.

In order to finally determine the motion of an arbitrary particle ($\delta_m \neq 0$, $\delta_K \neq 0$, $\hat{\delta}_K \neq 0$) relative to the trajectory of a reference particle ($\delta_m = \delta_K = 0$) moving along the optic axis ($\hat{\delta}_K = 0$), it is necessary to know the rigidities and the velocities of both particles. A reference particle of rest mass m_{00} and charge $(z_0 e)$ generated at potential zero moves with the velocity v_0 when it comes to a point of reference potential V_0 and has the magnetic and electrostatic rigidities χ_{B0} and χ_{E0}. By using Eqs. (2.2b), (2.11a), (2.14) we can write, with $\eta_0 = K_0/2m_{00}c^2$

$$v_0 = \frac{1}{1 + 2\eta_0} \sqrt{\frac{2K_0}{m_{00}}(1 + \eta_0)}, \tag{2.17a}$$

$$(z_0 e)\chi_{B0} = m_{00}\sqrt{\frac{2K_0}{m_{00}}(1 + \eta_0)}, \tag{2.17b}$$

$$(z_0 e)\chi_{E0} = \frac{2K_0(1 + \eta_0)}{1 + 2\eta_0}. \tag{2.17c}$$

An arbitrary particle of rest mass $m_0 = m_{00}(1 + \delta_m)$ and charge (ze), which would be at rest at a potential ΔV, moves with the velocity v_r when it comes to a point of reference potential V_0 and with the velocity v at a point of arbitrary potential $V = V_0 + \hat{V}$, where it has the magnetic and electrostatic rigidities χ_B and χ_E. In detail, we thus find with $\eta = K/2m_0 c^2$

or $\eta = \eta_0(1 + \delta_K + \hat{\delta}_K)/(1 + \delta_m)$,

$$v = v_0 \frac{1 + 2\eta_0}{1 + 2\eta} \sqrt{\frac{(1 + \delta_K + \hat{\delta}_K)(1 + \eta)}{(1 + \delta_m)(1 + \eta_0)}},$$

$$\chi_B = \chi_{B0}\sqrt{(1 + \delta_m)(1 + \delta_K + \hat{\delta}_K)(1 + \eta)/(1 + \eta_0)},$$

$$\chi_E = \chi_{E0}(1 + \delta_K + \hat{\delta}_K)\frac{(1 + \eta)(1 + 2\eta_0)}{(1 + 2\eta)(1 + \eta_0)}.$$

Writing these equations explicitly, we obtain

$$v = \frac{v_0}{1 + [\delta_m + 2\eta_0(\delta_K + \hat{\delta}_K)]/(1 + 2\eta_0)}$$

$$\times \sqrt{(1 + \delta_K + \hat{\delta}_K)\left[1 + \frac{\delta_m + \eta_0(\delta_K + \hat{\delta}_K)}{1 + \eta_0}\right]}, \tag{2.18a}$$

$$\chi_B = \chi_{B0}\sqrt{(1 + \delta_K + \hat{\delta}_K)\left[1 + \frac{\delta_m + \eta_0(\delta_K + \hat{\delta}_K)}{1 + \eta_0}\right]}, \tag{2.18b}$$

$$\chi_E = \chi_{E0}(1 + \delta_K + \hat{\delta}_K)\frac{1 + [\delta_m + \eta_0(\delta_K + \hat{\delta}_K)]/(1 + \eta_0)}{1 + [\delta_m + 2\eta_0(\delta_K + \hat{\delta}_K)]/(1 + 2\eta_0)}. \tag{2.18c}$$

Combining Eqs. (2.18b) and (2.18c), we can write

$$\chi = \chi_0\left\{\frac{h(1 + \delta_K + \hat{\delta}_K)\{1 + [\delta_m + \eta_0(\delta_K + \hat{\delta}_K)]/(1 + \eta_0)\}}{1 + [\delta_m + 2\eta_0(\delta_K + \hat{\delta}_K)]/(1 + 2\eta_0)}\right.$$

$$\left. + (1 - h)\sqrt{(1 + \delta_K + \hat{\delta}_K)\left[1 + \frac{\delta_m + \eta_0(\delta_K + \hat{\delta}_K)}{1 + \eta_0}\right]}\right\}, \tag{2.18d}$$

where h equals 0 for a magnetic and 1 for an electrostatic field; i.e.,

$$h_{\text{magnetic}} = 0, \tag{2.19a}$$

$$h_{\text{electrostatic}} = 1. \tag{2.19b}$$

From Eqs. (2.18a) and (2.18c), we also find $\lambda = (v^2/\chi_E) = (v_0^2/\chi_{E0})(1 + 2\eta_0)/[1 + \delta_m + 2\eta_0(1 + \delta_K + \hat{\delta}_K)]$. Expanding Eqs. (2.18a)–(2.18d) in a power series in δ_m, δ_K, and $\hat{\delta}_K$, we find

$$v = v_0\left[1 + \frac{\delta_K + \hat{\delta}_K - \delta_m}{2(1 + \eta_0)(1 + 2\eta_0)} + \frac{A}{8(1 + \eta_0)^2(1 + 2\eta_0)^2} + \cdots\right], \tag{2.20a}$$

with

$$A = (3 - 4\eta_0^2)\delta_m^2 - 2(1 - 6\eta_0 - 8\eta_0^2)\delta_m(\delta_K + \hat{\delta}_K)$$

$$- (1 + 12\eta_0 + 12\eta_0^2)(\delta_K + \hat{\delta}_K)^2$$

$$\chi_B = \chi_{B0}(1 + \Delta_B) = \chi_{B0}\left[1 + \frac{\delta_m + (1 + 2\eta_0)(\delta_K + \hat{\delta}_K)}{2(1 + \eta_0)}\right.$$

$$\left. - \frac{(\delta_K + \hat{\delta}_K - \delta_m)^2}{8(1 + \eta_0)^2} + \cdots\right], \tag{2.20b}$$

$$\chi_E = \chi_{E0}(1 + \Delta_E) = \chi_{E0}\left[1 + \frac{\eta_0\delta_m + (\delta_K + \hat{\delta}_K)[1 + 2\eta_0(1 + \eta_0)]}{(1 + \eta_0)(1 + 2\eta_0)}\right.$$

$$\left. - \frac{\eta_0(\delta_K + \hat{\delta}_K - \delta_m)^2}{(1 + \eta_0)(1 + 2\eta_0)^2} + \cdots\right], \tag{2.20c}$$

$$\chi = \chi_0(1 + \Delta) = \chi_0\left[1 + \frac{[(1 + 2\eta_0)^2 + h](\delta_K + \hat{\delta}_K) + (1 + 2\eta_0 - h)\delta_m}{2(1 + \eta_0)(1 + 2\eta_0)}\right.$$

$$+ \frac{h[1 - 4\eta_0(1 + \eta_0)] - (1 + 2\eta_0)^2}{8(1 + \eta_0)^2(1 + 2\eta_0)^2}$$

$$\left. \times (\delta_K + \hat{\delta}_K - \delta_m)^2 + \cdots\right]. \tag{2.20d}$$

Note that in a pure magnetic field, $\hat{V}$ vanishes as does $\hat{\delta}_K$.

For relativistically slow particles $[v_0 \ll c$ and $(K_0/m_{00}) \ll 931$ MeV/u], we find $\eta_0 \ll 1$, so that the Eqs. (2.20) simplify to

$$v = v_0[1 + \tfrac{1}{2}(\delta_K + \hat{\delta}_K - \delta_m) - \tfrac{3}{8}\delta_m^2 - \tfrac{1}{4}\delta_m(\delta_K + \hat{\delta}_K) - \tfrac{1}{8}(\delta_K + \hat{\delta}_K)^2 + \cdots], \tag{2.20as}$$

$$\chi_B = \chi_{B0}\sqrt{(1 + \delta_m)(1 + \delta_K + \hat{\delta}_K)}$$
$$= \chi_{B0}[1 + \tfrac{1}{2}(\delta_K + \hat{\delta}_K + \delta_m) - \tfrac{1}{8}(\delta_K + \hat{\delta}_K - \delta_m)^2 + \cdots], \tag{2.20bs}$$

$$\chi_E = \chi_{E0}(1 + \delta_K + \hat{\delta}_K) \tag{2.20cs}$$

$$= \chi_0\left[1 + \frac{1 + h}{2}(\delta_K + \hat{\delta}_K) + \frac{1 - h}{2}\delta_m - \frac{1 - h}{8}(\delta_K + \hat{\delta}_K - \delta_m)^2 + \cdots\right], \tag{2.20ds}$$

with v_0, χ_{B0}, and χ_{E0} given numerically as $\bar{v}_0 \approx 13.9\sqrt{\bar{K}_0/\bar{m}_{00}}$, $\bar{\chi}_{B0} \approx 0.144\sqrt{\bar{K}_0\bar{m}_{00}}/(z_0) \approx \chi_{B0}/300$ and $\chi_{E0} = -2V_0$ according to Eqs. (2.4bs), (2.12s), and (2.14s). For relativistically fast particles $[v_0 \approx c$ and $(K_0/m_{00}) \gg 931$ MeV/u], we find $\eta_0 \gg 1$, so that the Eqs. (2.20b) and (2.20c) simplify to

$$\chi_B = \chi_{B0}(1 + \delta_K + \hat{\delta}_K), \tag{2.20af}$$

$$\chi_E = \chi_{E0}(1 + \delta_K + \hat{\delta}_K), \tag{2.20bf}$$

with χ_{B0} given numerically as $\tilde{\chi}_{B0} = K_0/(z_0) = -V_0 \approx 300\tilde{\chi}_{B0}$ and $\chi_{E0} = -V_0$ according to Eqs. (2.12f) and (2.14f).

2.4 REFRACTIVE INDEX OF THE ELECTROMAGNETIC FIELD

Although the motion of charged particles in electromagnetic fields is fully described by the Lorentz equation [Eq. (2.6)], we can try to develop alternative, mathematically elegant descriptions in which the functional form must be modified such that the final result is identical to Eq. (2.6). One of those descriptions is Fermat's principle, which states that among all conceivable ways to move from point P_1 to point P_2, the actual trajectory of a particle is characterized by

$$\delta \int_{P_1}^{P_2} n \, ds = 0, \qquad (2.21a)$$

[see also Eq. (2.37b)]. Here, s is the path coordinate, and n is the locally varying so-called refractive index of the electromagnetic field, which must be defined such that the final equation of motion is identical to Eq. (2.6). To determine n we may compare Eq. (2.21a) to Hamilton's principle $\delta \int (\mathbf{p} \mathbf{s_u}) \, ds = 0$ of Eq. (2.37a), another mathematical procedure that finally yields Eq. (2.6). In this description [see also Eq. (2.34)], the quantity $\mathbf{p}$ is the generalized momentum $[m\mathbf{v} + (ze)\mathbf{A}]$ and $\mathbf{s_u}$ is a unit vector in the direction of the trajectory. This comparison yields (Glaser, 1956)

$$n = k[mv + (ze)(\mathbf{A}\mathbf{s_u})], \qquad (2.21b)$$

where v equals $\mathbf{v}\mathbf{s_u}$, and k is a constant that is currently undetermined. Replacing $mv/(ze)$ by the magnetic rigidity χ_B of Eq. (2.10) at a point of potential $V = V_0 + \hat{V}$, Eq. (2.21b) yields

$$n = (ze)k[\chi_B + \mathbf{A}\mathbf{s_u}].$$

We may now choose $(ze)k$ to be equal to $1/\chi_{Br}$, where (ze) characterizes the charge and χ_{Br} the magnetic rigidity of the particle under investigation at a reference potential V_r. Thus, we can define a dimensionless refractive index n_i at position z_i with

$$n_i \chi_{Br} = \chi_{Bi} + \mathbf{A}_i \mathbf{s_u}. \qquad (2.21c)$$

According to Eq. (2.10), the rigidity χ_B equals $p/(ze)$, so that Eq. (2.21c) can be rewritten as

$$n_i p_r = p_i + (ze)\mathbf{A}_i \mathbf{s_u}. \qquad (2.21d)$$

With $p_r = p_0(1 + \delta_p)/n_i$ describing the momentum of the particle under consideration at a reference potential V_{0i} at z_i, we can rewrite Eqs. (2.7a) and (2.7b) at positions z_j, where the magnetic vector potential $\mathbf{A}_j$ vanishes as

$$a_j = \frac{p_x(1+\delta_p)}{n_j p_r} = \frac{(1+\delta_p)\sin\alpha_j}{\sqrt{1+\tan^2\beta_j\cos^2\alpha_j}}, \tag{2.22a}$$

$$b_j = \frac{p_y(1+\delta_p)}{n_j p_r} = \frac{(1+\delta_p)\sin\beta_j}{\sqrt{1+\tan^2\alpha_j\cos^2\beta_j}}. \tag{2.22b}$$

To see the analogy to the refractive index in light optics, one may observe a particle moving across a dipole sheet, as indicated in Fig. 2.5. This dipole sheet shall be assumed to separate a region of potential $V_{01} = -K_1/(ze)$, corresponding to the refractive index n_1, from a region of potential $V_{02} = -K_2/(ze)$, corresponding to the refractive index n_2. Such a dipole sheet exerts forces only in the z direction, so that one finds the components p_{1x} and p_{2x} of the momentum p of the particle under consideration to be equally large on both sides of the dipole sheet. Consequently, the Eqs. (2.22a) and (2.22b) postulate for $j = 1, 2$,

$$a_2 n_2 = a_1 n_1, \qquad b_2 n_2 = b_1 n_1,$$

so that the transformation across the dipole sheet reads $\mathbf{X}_2 = T\mathbf{X}_1$, or explicitly,

$$\begin{pmatrix} x_2 \\ a_2 \end{pmatrix} = \begin{pmatrix} 1 & 0 \\ 0 & n_1/n_2 \end{pmatrix} \begin{pmatrix} x_1 \\ a_1 \end{pmatrix}.$$

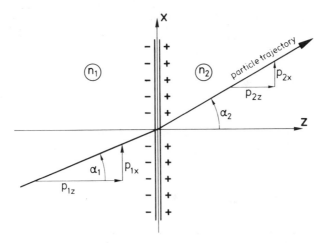

Fig. 2.5. A charged particle is shown moving obliquely across a dipole sheet. According to Eqs. (2.24), we find $n_1 \sin\alpha_1 = n_2 \sin\alpha_2$, since the x component of the particle momentum is equal on both sides of the dipole sheet, whereas the z component is varied.

The determinant of this transfer matrix is n_1/n_2 and not 1 as in Eqs. (1.19a) and (1.19b). The difference comes from the accelerations, which were not taken into account in Chapter 1. As shall be proven later in Section 8.3.2, the statement $|T| = n_1/n_2$ derived here is not only correct for the transfer matrix of a dipole sheet, but for any transfer matrix [Eq. (8.27a)]. Thus, we can write

$$(x|x)(a|a) - (x|a)(a|x) = n_1/n_2, \tag{2.23a}$$

and understand Eq. (1.19b) as the special case of Eq. (2.23a) in which n_1 equals n_2. By the same arguments as for Eq. (2.23a), we find for the transverse y direction,

$$(y|y)(b|b) - (y|b)(b|y) = n_1/n_2. \tag{2.23b}$$

Evaluating Eq. (2.22a) for a planar particle motion, i.e., $\beta_j = 0$, across a dipole sheet we find for $j = 1, 2$,

$$\frac{\sin \alpha_1}{\sin \alpha_2} = \frac{n_2}{n_1} = \frac{p_2}{p_1} = \sqrt{\frac{K_2(1 + \eta_2)}{K_1(1 + \eta_1)}}, \tag{2.24}$$

with $\eta_1 = K_1/2m_0c^2$ and $\eta_2 = K_2/2m_0c^2$.

For relativistically slow particles $[v_1, v_2 \ll c$ or $(K_1/m_0), (K_2/m_0) \ll 931$ MeV/u$]$, we find $\eta_1 \ll 1$ and $\eta_2 \ll 1$. Thus, Eq. (2.24) reads

$$\sqrt{K_2} \sin \alpha_2 = \sqrt{K_1} \sin \alpha_1, \tag{2.24s}$$

expressing the well-known fact (see, for instance, Helmer, 1966) that for slow particles, the product of the square root of the particle energy and of the sines of the angles of inclination of a trajectory is a constant. For relativistically fast particles $[v_1, v_2 \approx c$ or $(K_1/m_0), (K_2/m_0) \gg 931$ MeV/u$]$, we find $\eta_1 \gg 1$ and $\eta_2 \gg 1$, so that $(1 + \eta_2)/(1 + \eta_1) \approx \eta_2/\eta_1 = K_2/K_1$, which transforms Eq. (2.24) to

$$K_2 \sin \alpha_2 = K_1 \sin \alpha_1. \tag{2.24f}$$

Note here that a practical example of such a dipole sheet is discussed in Section 7.2.2.

2.5 EULER–LAGRANGE EQUATIONS

The Lorentz equation [Eq. (2.6)] describes the motion of charged particles in magnetic and electrostatic fields; however, in some cases it is advantageous (Glaser, 1956) to derive generally applicable equations of motion in curvilinear coordinates x, y, z, as are illustrated in Fig. 2.6. In this

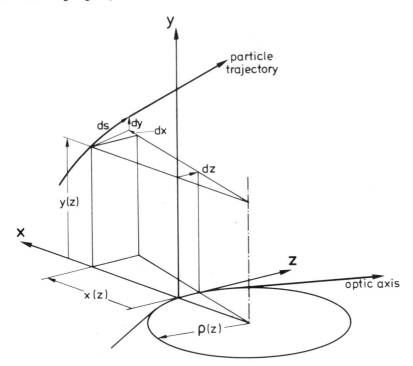

Fig. 2.6. A curvilinear x, y, z-coordinate system is indicated. This coordinate system is Cartesian for $\rho(z) = \infty$ and cylindrical for $\rho(z) = \rho_0$. Note that from simple geometry, we find $ds^2 = dx^2 + dy^2 + (1 + x/\rho_0)^2\, dz^2$.

coordinate system the variational principle of Eq. (2.21a) may be expressed as

$$\delta \int_{P_1}^{P_2} n\, ds = \delta \int_{z_1}^{z_2} \bar{n}\, dz = 0, \qquad (2.25)$$

where instead of integrating along an actual flight path s, we proceed along the optic axis z of Fig. 2.6 curved with a radius $\rho_0(z)$. The path length-element ds is thus given by $\mathbf{i}\, dx + \mathbf{j}\, dy + \mathbf{k}\, dz(1 + x/\rho_0)$, so that its magnitude is

$$ds = \sqrt{dx^2 + dy^2 + (1 + x/\rho_0)^2\, dz^2}.$$

With $x' = dx/dz$, $y' = dy/dz$, and $dz = \rho_0\, d\phi$, this relation transforms to

$$ds = dz\sqrt{(1 + x/\rho_0)^2 + x'^2 + y'^2}.$$

Introducing this ds as well as the n of Eq. (2.21c) into $\bar{n} = n\, ds/dz$, as found from Eq. (2.25) yields

$$\bar{n} = \frac{\chi_B}{\chi_{Br}} \sqrt{\left(1 + \frac{x}{\rho_0}\right)^2 + x'^2 + y'^2} + \frac{1}{\chi_{Br}}\left[A_z\left(1 + \frac{x}{\rho_0}\right) + A_x x' + A_y y'\right],$$

(2.26)

where A_x, A_y, A_z are the three components of the magnetic vector potential $\mathbf{A}(x, y, z)$. Knowing $\bar{n}$, we can then derive from the variational principle of Eq. (2.25) the equations of motion

$$\frac{\partial \bar{n}}{\partial x} - \frac{d}{dt}\frac{\partial \bar{n}}{\partial x'} = 0,$$

(2.27a)

$$\frac{\partial \bar{n}}{\partial y} - \frac{d}{dt}\frac{\partial \bar{n}}{\partial y'} = 0,$$

(2.27b)

analogous to the derivation of the Lagrange equations in Section 2.A1.

APPENDIX

2.A1 Hamilton's Variational Principle and the Lagrange Equations

Principally, the motion of charged particles in electromagnetic fields is fully described by the Lorentz equation [Eq. (2.6)]. However, there are alternative descriptions which in many cases simplify the choice of an appropriate coordinate system. One such alternative is Hamilton's variational principle. It states that the trajectory along which a particle moves from point P_1 to point P_2 is distinguished from all other possible paths by the fact that the line integral

$$W = \int L(x, y, z, \dot{x}, \dot{y}, \dot{z}, t)\, dt = \int L(\mathbf{r}, \mathbf{v}, t)\, dt$$

(2.28)

has a minimal value (see, for instance, Goldstein, 1980). The L of Eq. (2.28) is the so-called Lagrange function, which must be chosen such that the final result is identical to Eq. (2.6). This L is a function of the particle position $\mathbf{r}$ and velocity $\mathbf{v}$ at the time t. Here, both $\mathbf{r}$ and $\mathbf{v}$ are denoted in Cartesian coordinates as

$$\mathbf{r}(t) = \mathbf{i}x(t) + \mathbf{j}y(t) + \mathbf{k}z(t), \qquad \mathbf{v}(t) = \mathbf{i}\dot{x}(t) + \mathbf{j}\dot{y}(t) + \mathbf{k}\dot{z}(t)$$

with $\dot{x} = dx/dt$, $\dot{y} = dy/dt$, and $\dot{z} = dz/dt$. The integration of Eq. (2.28) must be performed between the times t_1 and t_2 at which the particle passes the points $P_1(\mathbf{r}_1)$ and $P_2(\mathbf{r}_2)$, respectively. For this purpose, the Lagrange function L of Eq. (2.28) must be evaluated for all times t at the corresponding position and velocity coordinates $\mathbf{r}(t)$ and $\mathbf{v}(t)$. For an infinitesimal variation in the flight path,

$$\mathbf{r}(t) \rightarrow \mathbf{r}(t) + \Delta\mathbf{r}(t), \qquad \mathbf{v}(t) \rightarrow \mathbf{v}(t) + \Delta\mathbf{v}(t)$$

the variation δW of Eq. (2.28) is

$$\delta W = \int_{t_1}^{t_2} L\, dt = \int_{t_1}^{t_2} \left(\frac{\partial L}{\partial \mathbf{r}} \delta\mathbf{r} + \frac{\partial L}{\partial \mathbf{v}} \delta\mathbf{v} \right) dt, \tag{2.29}$$

where terms of second and higher order in $\Delta\mathbf{r}$ and $\Delta\mathbf{v}$ are omitted and where the following abbreviations are used:

$$\frac{\partial L}{\partial \mathbf{r}} = \mathbf{i}\frac{\partial L}{\partial x} + \mathbf{j}\frac{\partial L}{\partial y} + \mathbf{k}\frac{\partial L}{\partial z}, \qquad \frac{\partial L}{\partial \mathbf{v}} = \mathbf{i}\frac{\partial L}{\partial \dot{x}} + \mathbf{j}\frac{\partial L}{\partial \dot{y}} + \mathbf{k}\frac{\partial L}{\partial \dot{z}}.$$

By partial integration of the second part of Eq. (2.29) we find because of $\mathbf{v} = (d\mathbf{r}/dt) = (d/dt)\,\delta\mathbf{r}$,

$$\delta W = \frac{\partial L}{\partial \mathbf{v}} \delta\mathbf{r} \int_{t_1}^{t_2} (\cdots) + \int_{t_1}^{t_2} \left[\frac{\partial L}{\partial \mathbf{r}} - \frac{d}{dt}\frac{\partial L}{\partial \mathbf{v}} \right] \mathbf{r}\, dt. \tag{2.30}$$

Since it was postulated before that all flight paths must pass through the points P_1 and P_2 at the times t_1 and t_2, respectively, the quantities $\delta\mathbf{r}(t_1)$ and $\delta\mathbf{r}(t_2)$ vanish, and so does the first term of Eq. (2.30).

Let us postulate now $\delta W = 0$ so that among all possible trajectories that pass through the points P_1 and P_2, the real trajectory is that one for which W takes up a minimal value. This condition can be satisfied only if the integrand of Eq. (2.30) vanishes; that is, if the relation

$$\frac{\partial L}{\partial \mathbf{r}} - \frac{d}{dt}\frac{\partial L}{\partial \mathbf{v}} = 0 \tag{2.31a}$$

is valid, which reads explicitly

$$\mathbf{i}\left[\frac{\partial L}{\partial x} - \frac{d}{dt}\frac{\partial L}{\partial \dot{x}}\right] + \mathbf{j}\left[\frac{\partial L}{\partial y} - \frac{d}{dt}\frac{\partial L}{\partial \dot{y}}\right] + \mathbf{k}\left[\frac{\partial L}{\partial z} - \frac{d}{dt}\frac{\partial L}{\partial \dot{z}}\right] = 0. \tag{2.31b}$$

To fulfill Eq. (2.31b) for all x, y, z, and t, each of the three expressions in brackets must vanish individually. This postulate yields the three so-called Lagrange equations in Cartesian coordinates for which the Lagrange function L still must be determined.

The simplest procedure to obtain L is to assume

$$L = m_0 c^2[1 - \sqrt{1 - (v/c)^2}] - (ze)V + (ze)\mathbf{v}\mathbf{A} \tag{2.32}$$

and prove that this assumption causes Eqs. (2.31a) to be identical to the Lorentz equation [Eq. (2.6)]. The quantity c in Eq. (2.32) again denotes the velocity of light, (ze) the charge of the particle under consideration, and $m = m_0(1 + 2\eta) = m_0/\sqrt{1 - (v/c)^2}$ the mass of this particle of rest mass m_0. Furthermore,

$$v = |\mathbf{v}| = \frac{2c}{1 + 2\eta}\sqrt{\eta(1 + \eta)} = \sqrt{\dot{x}^2 + \dot{y}^2 + \dot{z}^2}$$

describes the velocity of the particle with $\eta = K/2m_0c^2$. Finally, V is the electrostatic potential and $\mathbf{A}$ the magnetic vector potential at the point of investigation. From this vector potential the magnetic flux density is calculated as

$$\mathbf{B} = \text{curl } \mathbf{A},$$

which determines the magnetic field strength H as $B/\mu\mu_0$. The corresponding electrostatic field strength $\mathbf{E}$ is calculated from the electrostatic potential V and the variation of the magnetic vector potential $\mathbf{A}$ as

$$\mathbf{E} = -\text{grad } V - d\mathbf{A}/dt,$$

according to Maxwell's equations. Introducing the assumed Lagrange function L of Eq. (2.32) into Eq. (2.31a), we find

$$-(ze)\,\text{grad } V + (ze)\mathbf{v}\,\text{curl } \mathbf{A} = \frac{d}{dt}\frac{m_0\mathbf{v}}{\sqrt{1 - (v/c)^2}} + (ze)\mathbf{A},$$

or using Eq. (2.1b), $d(m\mathbf{v})/dt = (ze)\mathbf{E} + (ze)\mathbf{v} \times \mathbf{B}$. This is identical to Eq. (2.6), so that it may be accepted that the L of Eq. (2.32) is correct.

2.A2 Motion of Charged Particles in Time-Independent Magnetic and Electrostatic Fields

Forming the derivative of L with respect to t yields

$$\frac{dL}{dt} = \frac{\partial L}{\partial t} + \frac{\partial L}{\partial \mathbf{r}}\frac{\partial \mathbf{r}}{\partial t} + \frac{\partial L}{\partial \mathbf{v}}\frac{\partial \mathbf{v}}{\partial t}.$$

With $\partial L/\partial \mathbf{r}$ obtained from Eq. (2.31a), we thus find

$$\frac{d}{dt}\left[\mathbf{v}\left(\frac{\partial L}{\partial \mathbf{v}}\right) - L\right] = -\frac{\partial L}{\partial t}.$$

In all cases in which the Lagrange function L of Eq. (2.32) does not explicitly depend on the time t, we can thus write

$$\mathbf{v}\left(\frac{\partial L}{\partial \mathbf{v}}\right) - L = E = \text{const.} \tag{2.33}$$

Inserting the Lagrange function L of Eq. (2.32) into Eq. (2.33), we find that E is the sum of the kinetic energy $(m - m_0)c^2 = m_0 c^2 [1/\sqrt{1 - (v/c)^2} - 1]$, and the potential energy $(ze)V$:

$$E = (m - m_0)c^2 + (ze)V. \qquad (2.34)$$

Defining a so-called generalized momentum

$$\mathbf{p} = \frac{\partial L}{\partial \mathbf{v}} = m\mathbf{v} + (ze)\mathbf{A}, \qquad (2.35)$$

we can express L of Eq. (2.32) as $-E + \mathbf{pv}$. Introducing this L into Eq. (2.30) yields

$$\delta W = E(t_2) - E(t_1) + \int_{t_1}^{t_2} \mathbf{pv}\, dt = E(t_2) - E(t_1) + \int_{r_1}^{r_2} \mathbf{p}\, d\mathbf{r}, \quad (2.36)$$

with $\mathbf{v} = d\mathbf{r}/dt$. Here $\mathbf{r}_1$ and $\mathbf{r}_2$ describe the coordinates of the points P_1 and P_2 at which the particle is located at the times t_1 and t_2.

One may now vary the flight path but observe only such paths that have the same kinetic and potential energies at P_1 and P_2, so that $E(t_1) = E(t_2)$. In this case, the W of Eq. (2.28) becomes minimal for

$$\delta \int_{r_1}^{r_2} \mathbf{p}\, d\mathbf{r} = \delta \int_{s_1}^{s_2} (\mathbf{ps}_u)\, ds = 0, \qquad (2.37a)$$

where ds is an element of the path coordinate along a trajectory and where $\mathbf{s}_u$ denotes a unit vector in the direction of this trajectory. This Eq. (2.37a) is known as Hamilton's principle. It transforms into Fermat's principle by replacing $\mathbf{ps}_u$ by the so-called refractive index n:

$$\delta \int_{s_1}^{s_2} n\, ds = 0. \qquad (2.37b)$$

This relation can also be written [see also Eqs. (2.25) and (2.26)] as

$$\int_{z_1}^{z_2} \bar{n}\, dz = 0, \qquad (2.37c)$$

where z is the coordinate along a reference trajectory, the so-called optic axis. Similarly, as the Lagrange equations [Eq. (2.31)] were derived from Eq. (2.29), we now could derive Eqs. (2.27a) and (2.27b) from Eq. (2.37c) where the independent time variable is changed to the z position so that all d/dt and $\partial/\partial t$ operators are exchanged to d/dz and $\partial/\partial z$.

2.A3 Ion Optical Refractive Index for a Chromatic Particle Bundle

In the case of a pure magnetic field, the magnetic rigidities χ_B and $\chi_{Br} = \chi_B(\hat{\delta}_K = 0)$ in Eq. (2.26) are equal for any particle of charge (ze) and rest mass m_0, since the electrostatic potential in the magnet equals V_0 everywhere so that $\hat{V}$ vanishes as does $\hat{\delta}_K$ in Eqs. (2.16c) and (2.18b). Consequently, the refractive index of Eq. (2.26) reads

$$\bar{n}_B = \sqrt{(1 + x/\rho_0)^2 + x'^2 + y'^2} + 1/\chi_{Br}[A_z(1 + x/\rho_0) + A_x x' + A_y y'],$$
$$(2.38a)$$

or to third order in x/ρ_0, x', y' for the common case $A_x = A_y = 0$, $A_z \neq 0$,

$$\bar{n}_B = \left(1 + \frac{x}{\rho_0}\right)\left(1 + \frac{A_z}{\chi_{Br}}\right) + \frac{1}{2}\left(1 + \frac{2x}{\rho_0}\right)(x'^2 + y'^2) + \cdots. \quad (2.38b)$$

In the case of a pure electrostatic field, the magnetic vector potential vanishes everywhere, i.e., $A_x = A_y = A_z = 0$. Assuming a particle of charge (ze) and rest mass m_0, which has the kinetic energy $K_r = K_0(1 + \delta_K)$ at the optic axis and the kinetic energy $K = K_0(1 + \delta_K + \hat{\delta}_K) = K_r + \hat{K} = -(ze)[(V_0 - \Delta V) + \hat{V}]$ at an arbitrary point of potential $V_0 + \hat{V}$ we can determine the refractive index of Eq. (2.26). From Eq. (2.11a) we find at these two points $(ze)^2 \chi_B^2 = 2(K_r + \hat{K})(1 + \eta)m_0$ and $(ze)^2 \chi_{Br}^2 = 2K_r(1 + \eta_r)m_0$ with $\eta_r = K_r/2m_0c^2$ and $\eta = (K_r + \hat{K})/2m_0c^2 = \eta_r(1 + \hat{K}/K_r)$. Thus, Eq. (2.26) yields

$$\bar{n}_E^2 = \left[1 + \frac{\hat{K}\eta_r}{K_r(1 + \eta_r)}\right]\left(1 + \frac{\hat{K}}{K_r}\right)\left[\left(1 + \frac{x}{\rho_0}\right)^2 + x'^2 + y'^2\right]. \quad (2.39)$$

Using the relation $(ze)\chi_{Er} = 2K_r(1 + \eta_r)/(1 + 2\eta_r)$ of Eq. (2.14), one can transform Eq. (2.39) to

$$\bar{n}_E^2 = \left\{1 + \frac{2\hat{V}}{\chi_{Er}}\left[1 + \frac{\hat{V}}{2\chi_{Er}}\left(\frac{4\eta_r(1 + \eta_r)}{(1 + 2\eta_r)^2}\right)\right]\right\}\left[\left(1 + \frac{x}{\rho_0}\right)^2 + x'^2 + y'^2\right].$$
$$(2.40)$$

Note here that Eq. (2.40) can still be simplified with $4\eta_r(1 + \eta_r)/(1 + 2\eta_r)^2 = v_r^2/c^2$ according to Eq. (2.2b).

At the optic axis the electrostatic rigidity is $\chi_{E0} = [2K_0/(z_0e)](1 + \eta_0)/(1 + 2\eta_0)$ for reference particles and for arbitrary particles $\chi_{Er} = \chi_{E0}(1 + \Delta_E) = [2K_0/(z_0e)](1 + \delta_K)(1 + \eta_r)/(1 + 2\eta_r)$ with $\eta_r = \eta_0(1 + \delta_K)/(1 + \delta_m)$. Thus, Eq. (2.40) becomes

$$\bar{n}_E^2 = \left\{1 + \frac{2\hat{V}}{\chi_{E0}(1 + \Delta_E)}\left(1 + \frac{\lambda\hat{V}}{2\chi_{E0}}\right)\right\}\left[\left(1 + \frac{x}{\rho_0}\right)^2 + x'^2 + y'^2\right], \quad (2.41)$$

with $\lambda = (v_r/c)^2/(1 + \Delta_E)$ and v_r as well as $(1 + \Delta_E)$ are given in Eqs. (2.18) and (2.20). Thus, (Wollnik, 1967) λ reads $(v_0/c)^2(1 + 2\eta_0)/[1 + \delta_m + 2\eta_0(1 + \delta_K)] = 2K_0(1 + \eta_0)/\{m_0(1 + 2\eta_0)[1 + \delta_m + 2\eta_0(1 + \delta_K)]\}$. Note here that for relativistically slow particles, λ becomes $(v_0/c)^2/(1 + \delta_m) = 2K_0/[m_0(1 + \delta_m)]$, whereas for relativistically fast particles, λ equals $1/(1 + \delta_K)$.

REFERENCES

Glaser, W. (1956). "Handbuch der Physik," Vol. XXXIII. Springer Verlag, Berlin.

Goldstein, H. (1980). "Classical Mechanics," 2nd ed. Addison-Wesley Publ., Reading, Massachusetts.

Helmer, J. C. (1966). *Am. J. Phys.* **34**, 222.

Wollnik, H. (1967). In "Focusing of Charged Particles," A. Septier (ed.) p. 161. Academic, New York.

Wollnik, H., and Berz, M. (1985). *Nucl. Instrum. Methods* **238**, 127.

3

Quadrupole Lenses

A lens is defined in Section 1.1.2 as an element in which a charged particle traversing that element experiences a bend toward or away from the optic axis, the angle of bend being proportional to the distance of this particle from the optic axis. Searching for a lens, we could look for a system in which the x and y components of the magnetic flux density $\mathbf{B}$ or the electrostatic field strength $\mathbf{E}$ increase linearly with the distance from a straight optic axis. One such system is a quadrupole lens (Melkirch, 1947; Courant *et al.* 1952). It consists of four hyperbolically shaped pole faces or electrodes (Fig. 3.1). Quadrupole lenses are focusing in one plane and defocusing in the perpendicular one. Thus, several such lenses must normally be combined for a useful lens system. Since hyperbolic pole faces or electrodes are difficult to fabricate, we often approximate the hyperbolas by circles or even by steps, as indicated in Fig. 3.2.

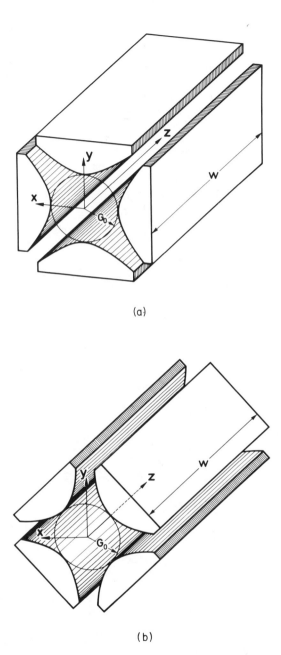

(a)

(b)

Fig. 3.1.   An electrostatic quadrupole (a) and a magnetic quadrupole (b). Note the rotation by 45° in going from one system to the other. Both systems are defocusing in y direction if they are focusing in x direction.

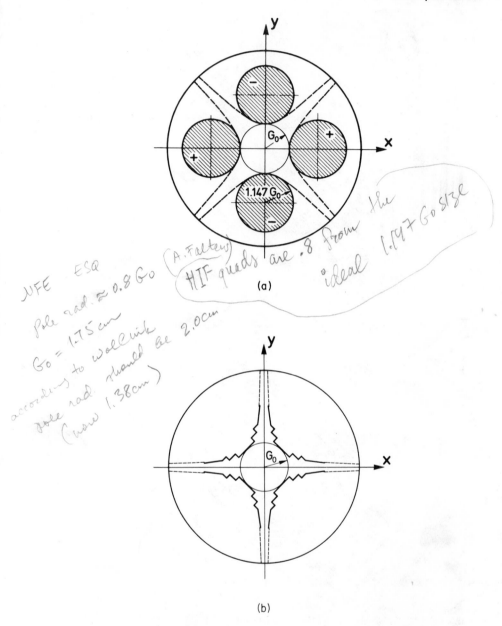

Handwritten annotations:

NFE   ESQ
Pole rad. ≈ 0.8 G₀ (A. Faltens)
G₀ = 1.75 cm
according to wall link
pole rad. should be 2.0 cm
(now 1.38 cm)

HIF quads are .8 from the
ideal 1.147 G₀ size

(a)

(b)

Fig. 3.2. Hyperbolic electrodes are often approximated by (a) circles and (b) hyperbolic pole faces by steps. In case of cylindrical electrodes or pole faces, it is advisable to choose the cylinder radius to be 1.1468 times the distance from the electrode of pole tip to the optic axis (Dayton *et al.*, 1954; Denison, 1971).

3.1. PARTICLE TRAJECTORIES IN QUADRUPOLE LENSES

To determine particle trajectories in a quadrupole, we must know the electrostatic potential $\hat{V}(x, y, z)$ relative to the optic axis and the magnetic flux density $\mathbf{B}(x, y, z)$ throughout the region between the hyperbolically shaped electrodes or polefaces. From the geometry of Fig. 3.1, it is evident that except for the entrance and exit regions, both $\hat{V}$ and $\mathbf{B}$ depend only on x and y but not on z.

For an electrostatic quadrupole of diameter $2G_0$ (Figs. 3.1a and 3.2a) and electrode potentials $\pm V_T$, the electrostatic potential distribution relative to the optic axis is found from Eq. (3.39a) in the appendix to this chapter as

$$\hat{V}(x, y) = [(x^2 - y^2)/G_0^2] V_T.$$

With $\mathbf{E} = -\text{grad }\hat{V}$, the components of the electrostatic field strength $\mathbf{E}$ are then obtained as

$$E_x(x, y) = -\partial \hat{V}/\partial x = -(2 V_T/G_0^2)x = -g_E x, \tag{3.1a}$$

$$E_y(x, y) = -\partial \hat{V}/\partial y = (2 V_T/G_0^2)y = g_E y, \tag{3.1b}$$

$$E_z(x, y) = 0, \tag{3.1c}$$

yielding constant field strength gradients along the x and y axes. For particles of mass m and charge (ze), the equations of motion are found from Eq. (2.6) as $m\ddot{x} = (ze)E_x$ and $m\ddot{y} = (ze)E_y$, yielding

$$m\ddot{x} = -(ze)g_E x, \qquad m\ddot{y} = (ze)g_E y. \tag{3.2a}$$

For a magnetic quadrupole of aperture $2G_0$ (Figs. 3.1b and 3.2b), the scalar magnetic potential is found from Eq. (3.39b) as

$$V_B(x, y) = g_B xy.$$

With $\mathbf{B} = -\text{grad }V_B$, the components of the magnetic flux density $\mathbf{B}$ then are obtained as

$$B_x(x, y) = -\partial V_B/\partial x = -g_B y, \tag{3.3a}$$

$$B_y(x, y) = -\partial V_B/\partial y = -g_B x, \tag{3.3b}$$

$$B_z(x, y) = 0, \tag{3.3c}$$

yielding constant flux density gradients along the x and y axes. At the pole tips, that is, at $x_T = \pm y_T = \pm G_0/\sqrt{2}$ inFigs. 3.1b and 3.2b, the magnetic flux density is $B_T = \sqrt{B_x^2 + B_y^2} = \pm G_0 g_m$. Thus, one can describe the g_B of Eqs. (3.3a) and (3.3b) as

$$g_B = -B_T/G_0.$$

With $B_z = 0$, the equations of motion are found from Eq. (2.6) for particles of mass m, charge (ze), and velocity $\mathbf{v}$, where v_z is the z component of $\mathbf{v}$, as $m\ddot{x} = -(ze)v_z B_y$ and $m\ddot{y} = (ze)v_z B_x$, yielding

$$m\ddot{x} = (ze)v_z g_B x, \qquad m\ddot{y} = (ze)v_z g_B y. \tag{3.2b}$$

With $\ddot{x} = (d^2x/dz^2)(dz/dt)^2 = x''v_z^2$ and $\ddot{y} = (d^2y/dz^2)(dz/dt)^2 = y''v_z^2$, Eqs. (3.2a) and (3.2b) can be combined [see also Eq. (1.16)]:

$$x'' = -k^2 x, \tag{3.4a}$$

$$y'' = +k^2 y. \tag{3.4b}$$

For the electrostatic and magnetic cases, respectively, k^2 then equals $k_E^2 = g_E(ze)/mv_z^2$ and $k_B^2 = g_B(ze)/mv_z$, or explicitly,

$$k_E^2 = \frac{2 V_T(ze)}{G_0^2 mv_z^2} \approx \frac{2 V_T}{G_0^2 \chi_E}, \qquad k_B^2 = \frac{B_T(ze)}{G_0 mv_z} \approx \frac{B_T}{G_0 \chi_B}.$$

Here, $(ze)\chi_E = mv^2 \approx mv_z^2$ and $(ze)\chi_B = mv \approx mv_z$ describe the rigidities of the particles under investigation as defined in Eqs. (2.10) and (2.13).

3.1.1 Equations of Motion

To first order the particle trajectories in quadrupole lenses are described by Eqs. (3.4a) and (3.4b). For $k^2 > 0$, these differential equations are satisfied by

$$x(z) = c_1 \cos(kz) + d_1 \sin(kz), \qquad y(z) = c_2 \cosh(kz) + d_2 \sinh(kz),$$

with undetermined coefficients c_1, c_2, d_1, d_2. This solution is periodic in the xz plane and diverges in the yz plane. The angles of inclination $\alpha(z), \beta(z)$ shown in Fig. 1.1 can be found from the relations $\tan \alpha(z) = dx/dz$ and $\tan \beta(z) = dy/dz$:

$$\tan \alpha(z) = -c_1 k \sin(kz) + d_1 k \cos(kz),$$

$$\tan \beta(z) = c_2 k \sinh(kz) + d_2 k \cosh(kz).$$

Knowing the position where and the angles of inclination with which a particle entered a quadrupole at $z = z_1$ in Fig. 3.3,

$$x(z_1) = x_1, \qquad y(z_1) = y_1, \qquad \alpha(z_1) = \alpha_1, \qquad \beta(z_1) = \beta_1,$$

we can determine the coefficients c_1, c_2, d_1, d_2. Thus, the particle trajectories are given as

$$x(z) = x_1 \cos(kz) + (\tan \alpha_1/k) \sin(kz), \tag{3.5a}$$

$$\tan \alpha(z) = -x_1 k \sin(kz) + \tan \alpha_1 \cos(kz), \tag{3.5b}$$

$$y(z) = y_1 \cosh(kz) + \tan \beta_1/k) \sinh(kz), \tag{3.5c}$$

$$\tan \beta(z) = y_1 k \sinh(kz) + \tan \beta_1 \cosh(kz). \tag{3.5d}$$

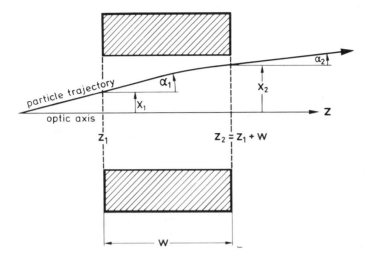

Fig. 3.3. Trajectory of a charged particle moving through a quadrupole lens of length w. Note the initial values of $x_1 = x_1(z_1)$, $\alpha_1 = \alpha(z_1)$ and the final values of $x_2 = x(z_1 + w)$, $\alpha_2 = \alpha(z_1 + w)$.

Note here that the focusing (F) action (i.e., the bend toward the optic axis) is in the xz plane, and the defocusing (D) action (i.e., the bend away from the optic axis) is in the yz plane. These focusing properties may be illustrated by Fig. 3.4, where we assume a glass lens that is ground so as to be concave in the yz plane and convex in the xz plane.

Choosing the quantity z to be the length w of a quadrupole, as indicated in Figs. 3.1, 3.3, and 3.4, Eqs. (3.5) can be written in the form of transfer matrices as $\mathbf{X}_2 = T_x\mathbf{X}_1$ and $\mathbf{Y}_2 = T_y\mathbf{Y}_1$, or

$$\begin{pmatrix} x_2 \\ a_2 \end{pmatrix} = \begin{pmatrix} (x|x) & (x|a) \\ (a|x) & (a|a) \end{pmatrix} \begin{pmatrix} x_1 \\ a_1 \end{pmatrix} = \begin{pmatrix} \cos(kw) & k^{-1}\sin(kw) \\ -k\sin(kw) & \cos(kw) \end{pmatrix} \begin{pmatrix} x_1 \\ a_1 \end{pmatrix},$$

$$\begin{pmatrix} y_2 \\ b_2 \end{pmatrix} = \begin{pmatrix} (y|y) & (y|b) \\ (b|y) & (b|b) \end{pmatrix} \begin{pmatrix} y_1 \\ b_1 \end{pmatrix} = \begin{pmatrix} \cosh(kw) & k^{-1}\sinh(kw) \\ k\sinh(kw) & \cosh(kw) \end{pmatrix} \begin{pmatrix} y_1 \\ b_1 \end{pmatrix}.$$

Similarly, as in Eqs. (1.14), $\tan \alpha$ and $\tan \beta$ are replaced here by a and b, with

$$a = \frac{\sin \alpha}{\sqrt{1 + \tan^2 \beta \cos^2 \alpha}}, \qquad b = \frac{\sin \beta}{\sqrt{1 + \tan^2 \alpha \cos^2 \beta}},$$

according to Eqs. (2.7a) and (2.7b). This replacement is correct in a first-order theory as used in this chapter. Actually, it is also correct in a second-order theory if both a and b are multiplied by $(1 + \delta_p)$; for higher order calculations, it must be interpreted as a shorthand notation.

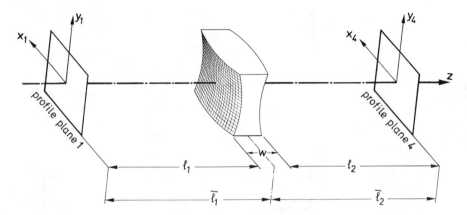

Fig. 3.4. Principal arrangement of a quadrupole singlet (illustrated by an especially ground glass lens), which has its focusing action in the xz plane and its defocusing action in the yz plane. The drift distances l_1 and l_2 denote lengths relative to the front and back sides of the quadrupole and the distances $\bar{l}_1$ and $\bar{l}_2$ denote lengths relative to the center plane of the quadrupole.

With $k_x = \sqrt{k^2}$ and $k_y = \sqrt{-k^2}$, the previous relations can be abbreviated as

$$\begin{pmatrix} x_2 \\ a_2 \end{pmatrix} = \begin{pmatrix} (x|x) & (x|a) \\ (a|x) & (a|a) \end{pmatrix} \begin{pmatrix} x_1 \\ a_1 \end{pmatrix} = \begin{pmatrix} c_x & s_x \\ -k_x^2 s_x & c_x \end{pmatrix} \begin{pmatrix} x_1 \\ a_1 \end{pmatrix}, \tag{3.6a}$$

$$\begin{pmatrix} y_2 \\ b_2 \end{pmatrix} = \begin{pmatrix} (y|y) & (y|b) \\ (b|y) & (b|b) \end{pmatrix} \begin{pmatrix} y_1 \\ b_1 \end{pmatrix} = \begin{pmatrix} c_y & s_y \\ -k_y^2 s_y & c_y \end{pmatrix} \begin{pmatrix} y_1 \\ b_1 \end{pmatrix}, \tag{3.6b}$$

$$c_x = \cos(k_x w), \qquad s_x = k_x^{-1} \sin(k_x w),$$

$$c_y = \cos(k_y w), \qquad s_y = k_y^{-1} \sin(k_y w).$$

For negative values of k_x^2 or k_y^2, the arguments of the sin and cos functions become imaginary. Thus, these functions transform into sinh and cosh with $k_x = \sqrt{-k_x^2}$ or $k_y = \sqrt{-k_y^2}$. The necessary k_x and k_y values are taken from Eq. (3.4a) and (3.4b) as

$$k_{xE}^2 = -k_{yE}^2 = \pm 2V_T / G_0^2 \chi_E, \tag{3.7a}$$

$$k_{xB}^2 = -k_{yB}^2 = \pm B_T / G_0 \chi_B, \tag{3.7b}$$

for the magnetic and the electrostatic cases, respectively. The quantities B_T and V_T denote pole-tip flux densities in teslas and electrode potentials in megavolts relative to the optic axis; $2G_0$ describes the quadrupole apertures in meters; and χ_B and χ_E are magnetic and electrostatic rigidities of reference particles at the optic axis in tesla meters and megavolts [see also Eqs. (2.11) and (2.14)]. Note that the Eqs. (3.7a) and (3.7b) hold only if the field strength gradients in the middle of the quadrupole are independent of the quadrupole length w, that is to say the quadrupole aperture $2G_0$ stays smaller than approximately $w/2$.

The position vectors $\mathbf{X}_2 = (x_2, a_2)$ and $\mathbf{Y}_2 = (y_2, b_2)$ at the profile plane $(z = z_2)$ are determined by Eqs. (3.6a) and (3.6b), provided that the position vectors $\mathbf{X}_1 = (x_1, a_1)$ and $\mathbf{Y}_1 = (y_1, b_1)$ at the profile plane $(z = z_1)$ are known. In Eqs. (3.5a)–(3.5d) it was assumed that the quadrupole lens had its focusing action (F) in the xz plane and its defocusing action (D) in the yz plane (see also Fig. 3.4). This definition corresponds to the upper signs in Eqs. (3.7a) and (3.7b). In order to exchange the focusing and the defocusing direction of the quadrupole, we can either change the polarity of the electrostatic potentials or of the magnetic flux densities or we can mechanically rotate the quadrupole by 90°.

Note also that Eqs. (3.6a) and (3.6b) disregard the fringing field action so that the field strength is assumed to rise and fall abruptly at the field boundaries. As outlined in Section 6, this approximation is correct for the first- and second-order description of particle trajectories in quadrupole lenses.

3.1.1.1 Focusing Properties of Quadrupole Lenses

Using the transfer matrices of Eqs. (3.6a) and (3.6b) particle trajectories may be traced from a profile plane 1 to a profile plane 4, distances l_1 and l_2 before and behind a quadrupole lens, respectively:

$$\begin{pmatrix} (x|x) & (x|a) \\ (a|x) & (a|a) \end{pmatrix} = \begin{pmatrix} 1 & l_2 \\ 0 & 1 \end{pmatrix} \begin{pmatrix} c_x & s_x \\ -k_x^2 s_x & c_x \end{pmatrix} \begin{pmatrix} 1 & l_1 \\ 0 & 1 \end{pmatrix}, \tag{3.8a}$$

$$\begin{pmatrix} (y|y) & (y|b) \\ (b|y) & (b|b) \end{pmatrix} = \begin{pmatrix} 1 & l_2 \\ 0 & 1 \end{pmatrix} \begin{pmatrix} c_y & s_y \\ -k_y^2 s_y & c_y \end{pmatrix} \begin{pmatrix} 1 & l_1 \\ 0 & 1 \end{pmatrix}. \tag{3.8b}$$

For $k_x^2 = -k_y^2 = k^2 > 0$, i.e., for a quadrupole that is focusing in the x direction and defocusing the y direction, we read from Eqs. (3.6a) and (3.6b)

$$c_x = \cos(kw), \qquad s_x = k^{-1}\sin(kw),$$

$$c_y = \cosh(kw), \qquad s_y = k^{-1}\sinh(kw).$$

The refractive powers $1/f_x = -(a|x)$ in the xz plane and $1/f_y = -(b|y)$ in the yz plane are read from Eq. (3.8a) and (3.8b) as:

$$1/f_x = k\sin(kw) = k_x^2 s_x, \tag{3.9a}$$

$$1/f_y = -k \sinh(kw) = k_y^2 s_y. \tag{3.9b}$$

The distances l_{fx1} and l_{fx2} between the first and second focal planes and the quadrupole entrance and exit are found from Eq. (3.8a) and $(x|x) = 0$ and $(a|a) = 0$, as outlined in Sections 1.3.1.3 and 1.3.2.4. The corresponding distances l_{fy1} and l_{fy2} are found from Eq. (3.8b) for $(y|y) = 0$ and $(b|b) = 0$. Thus, we obtain $l_{fx1} = l_{fx2} = l_{fx}$ and $l_{fy1} = l_{fy2} = l_{fy}$ as

$$l_{fx} = \cotan(kw)/k = c_x/k_x^2 s_x, \tag{3.10a}$$

$$l_{fy} = -\cotanh(kw)/k = c_y/k_y^2 s_y \tag{3.10b}$$

(Fig. 3.5). The principal planes 1 and 2 must be distances f_x and f_y apart from the focal planes 1 and 2. Their distances l_{px} and l_{py} from the entrance and exit boundaries of the quadrupole are thus found as

$$l_{px} = l_{fx} - f_x = -k^{-1} \tan(kw/2), \tag{3.11a}$$

$$l_{py} = l_{fy} - f_y = -k^{-1} \tanh(kw/2) \tag{3.11b}$$

(see again Fig. 3.5).

Replacing both l_1 and l_2 in Eq. (3.8a) by l_{fx} and in Eq. (3.8b) by l_{fy}, we find the transfer matrices between the first and second focal planes of a quadrupole to be

$$\begin{pmatrix} 1 & c_x/k_x^2 s_x \\ 0 & 1 \end{pmatrix} \begin{pmatrix} c_x & s_x \\ -k_x^2 s_x & c_x \end{pmatrix} \begin{pmatrix} 1 & c_x/k_x^2 s_x \\ 0 & 1 \end{pmatrix} = \begin{pmatrix} 0 & f_x \\ -1/f_x & 0 \end{pmatrix} \tag{3.12a}$$

$$\begin{pmatrix} 1 & c_y k_y^2 s_y \\ 0 & 1 \end{pmatrix} \begin{pmatrix} c_y & s_y \\ -k_y^2 s_y & c_y \end{pmatrix} \begin{pmatrix} 1 & c_y/k_y^2 s_y \\ 0 & 1 \end{pmatrix} = \begin{pmatrix} 0 & f_y \\ -1/f_y & 0 \end{pmatrix}, \tag{3.12b}$$

with f_x and f_y given by Eqs. (3.9a) and (3.9b) for $k_x^2 = -k_y^2 = k^2 > 0$. As we might expect, the result is identical to Eq. (1.24) for the xz plane as well as for the yz plane. Note, however, that the focal planes for the xz and the yz planes do not coincide, as illustrated in Fig. 3.5.

Quadrupole lenses in any complex optical system can be described equally well by Eqs. (3.6) or (3.12). For numerical calculations with pencil and paper Eqs. (3.12) are normally preferred to Eqs. (3.6). For computer calculations, however, Eqs. (3.6a) and (3.6b) are used almost exclusively.

3.1.1.2 Focusing Properties of Thin-Lens Quadrupoles

As long as kw remains small compared to 1, which is often the case, Eqs., (3.9)–(3.11) can be expanded in a power series around $kw = 0$. To

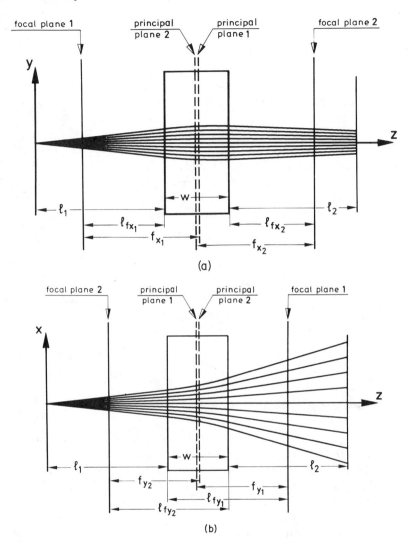

Fig. 3.5. Particle trajectories in the focusing xz plane (a) and in the defocusing yz plane (b) of a quadrupole of length w. Note the geometric relations between f_x, l_{fx1}, l_{fx2}, as well as between f_y, l_{fy1}, l_{fy2}.

zeroth and first order in kw these expansions yield for $k_x^2 = -k_y^2 = k^2 > 0$,

$$f_x = (1/k^2w) + (w/6) + \cdots, \qquad f_y = (1/k^2w) - (w/6) + \cdots.$$
$$l_{fx} = (1/k^2w) - (w/3) + \cdots, \qquad l_{fy} = (1/k^2w) - (w/3) + \cdots.$$
$$l_{px} = (-w/2) + \cdots, \qquad l_{py} = (-w/2) + \cdots.$$

Neglecting the terms $w/6$ and replacing $w/3$ by $w/2$, these relations approximate a real quadrupole by a thin lens of focal lengths

$$1/f_x \approx -1/f_y \approx k^2 w \qquad (3.13)$$

located at the center of the quadrupole, i.e., at $-w/2$ seen from the exit side. Assuming that for a given problem a solution has been found with a thin-lens quadrupole of focal length $f_x = -f_y = f$, we can expect that the properties of an extended real quadrupole are similar if its $k^2 w$ value is chosen to be $1/f$. In any case, such an approximation is good as an initial value for an iterative numerical solution of the real problem. Having determined k_B or k_E from Eq. (3.13), we find from Eqs. (3.7a) and (3.7b) approximately the necessary pole-tip flux density B_T or the electrode potential V_T of a lens of length w and aperture $2G_0$ (see Fig. 3.1) from

$$2V_T/G_0 = (\chi_E/f)(G_0/w), \qquad (3.13\text{el})$$

$$B_T = (\chi_B/f)(G_0/w), \qquad (3.13\text{ma})$$

with χ_B and χ_E determined by Eqs. (2.12) and (2.14).

Note again that for relativistically slow particles, χ_E equals twice the particle acceleration potential V_0 [Eq. (2.14s)] and that χ_B equals approximately $0.144\sqrt{Km_0}/(ze)$ in tesla meters, with K in mega-electron-volts and m in units u [Eq. (2.12s)].

3.2 DESIGN OF QUADRUPOLE MULTIPLETS

Optical systems normally consist of two or more quadrupole lenses for which we will define as standard that the first quadrupole has its defocusing action (D) in the xz plane and its focusing action (F) in the yz plane as in Figs. 3.6 and 3.13. The overall transfer matrix of such a quadrupole multiplet is determined by the product of several transfer matrices of drift lengths and quadrupoles. An example could be the product matrix of two quadrupoles and three drift distances:

$$\begin{pmatrix} (x|x)(x|a) \\ (a|x)(a|a) \end{pmatrix} = \begin{pmatrix} 1 & l_2 \\ 0 & 1 \end{pmatrix} \begin{pmatrix} c_{x2} & s_{x2} \\ -k_{x2}^2 s_{x2}^2 & c_{x2} \end{pmatrix} \begin{pmatrix} 1 & d_1 \\ 0 & 1 \end{pmatrix} \begin{pmatrix} c_{x1} & s_{x1} \\ -k_{x1}^2 s_{x1}^2 & c_{x1} \end{pmatrix} \begin{pmatrix} 1 & l_1 \\ 0 & 1 \end{pmatrix},$$

$$(3.14\text{a})$$

$$\begin{pmatrix} (y|y)(y|b) \\ (b|y)(b|b) \end{pmatrix} = \begin{pmatrix} 1 & l_2 \\ 0 & 1 \end{pmatrix} \begin{pmatrix} c_{y2} & s_{y2} \\ -k_{y2}^2 s_{y2} & c_{y2} \end{pmatrix} \begin{pmatrix} 1 & d_1 \\ 0 & 1 \end{pmatrix} \begin{pmatrix} c_{y1} & s_{y1} \\ -k_{y1}^2 s_{y1} & c_{y1} \end{pmatrix} \begin{pmatrix} 1 & l_1 \\ 0 & 1 \end{pmatrix}.$$

$$(3.14\text{b})$$

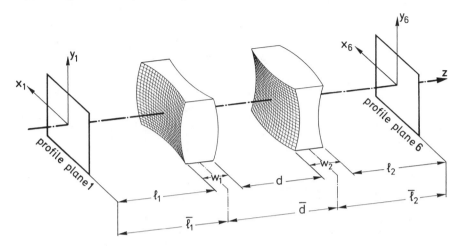

Fig. 3.6. Principal arrangement of a quadrupole doublet (illustrated by two specially ground glass lenses) the first lens of which is defocusing in x direction, and the second lens is focusing in x direction. Such a doublet is called a DF system.

For a system of n-quadrupole lenses separated by $n-1$ distances, the corresponding transfer matrices would read

$$\begin{pmatrix} (x|x)(x|a) \\ (a|x)(a|a) \end{pmatrix} = \begin{pmatrix} 1 & l_2 \\ 0 & 1 \end{pmatrix} \prod_{i=1}^{n} \begin{pmatrix} c_{xi} & s_{xi} \\ -k_{xi}^2 s_{xi} & c_{xi} \end{pmatrix} \begin{pmatrix} 1 & d_i \\ 0 & 1 \end{pmatrix}, \qquad (3.15a)$$

$$\begin{pmatrix} (y|y)(y|b) \\ (b|y)(b|b) \end{pmatrix} = \begin{pmatrix} 1 & l_2 \\ 0 & 1 \end{pmatrix} \prod_{i=1}^{n} \begin{pmatrix} c_{yi} & s_{yi} \\ -k_{yi}^2 s_{yi} & c_{yi} \end{pmatrix} \begin{pmatrix} 1 & d_i \\ 0 & 1 \end{pmatrix}, \qquad (3.15b)$$

Here d_1 is the object distance l_1 of Eq. (3.14a) and (3.14b) whereas d_i describes the distance between the quadrupoles i and $i+1$ for $i>1$.

For all optical designs, we assume an object of size $2x_{10}2y_{10}$, each point $P(x_1, y_1)$ of which emits charged particles. The corresponding trajectories are inclined relative to the optic axis under angles $\alpha_1 < \alpha_{10}$ and $\beta_1 < \beta_{10}$ with analogous quantities $a_1 < a_{10}$ and $b_1 < b_{10}$. Such a particle beam fills an upright parallelogram-like phase space area as defined in Section 5.2.

3.2.1 Point-to-Parallel Focusing

A point-to-parallel focusing system illustrated in Fig. 1.10d postulates $(a|a) = 0$ and/or $(b|b) = 0$ in Eqs. (3.14) or (3.15) so that particle trajectories diverging from a point of an object are finally parallel in the x and/or y direction. If the conditions $(a|a) = 0$ and $(b|b) = 0$ are fulfilled simultaneously, the first x and y focal planes of the overall lens system coincide at the position of the object, as is the case in the quadrupole doublet of

Fig. 3.7 or the quadrupole triplet of Fig. 3.14. The distances $l_{2x} = l_2 - l_{fx2}$ and $l_{2y} = l_2 - l_{fy2}$ from the final profile plane to the second focal planes of the overall system are open to choice. Thus, we find

$$T_x = \begin{pmatrix} (x|x) & (x|a) \\ (a|x) & 0 \end{pmatrix} = \begin{pmatrix} 1 & l_{2x} \\ 0 & 1 \end{pmatrix} \begin{pmatrix} 0 & f_x \\ -1/f_x & 0 \end{pmatrix}, \tag{3.16a}$$

$$T_y = \begin{pmatrix} (y|y) & (y|b) \\ (b|y) & 0 \end{pmatrix} = \begin{pmatrix} 1 & l_{2y} \\ 0 & 1 \end{pmatrix} \begin{pmatrix} 0 & f_y \\ -1/f_y & 0 \end{pmatrix}. \tag{3.16b}$$

The quantities $1/f_x$ and $1/f_y$ here describe the refractive powers of the complete quadrupole system for the xz and yz planes, respectively.

For particles that start from a point $P_1(x_1, y_1)$ with $x_1 < x_{10}, y_1 < y_{10}$, and angles of inclination $\alpha_1 \approx a_1$, $\beta_1 \approx b_1$, where $a_1 < a_{10}$, $b_1 < b_{10}$, we find the maximal angles of inclination $\pm\tan \alpha_m \approx \pm S_x = \pm a_m$ and $\pm\tan \beta_m \approx \pm S_y = \pm b_m$, as well as the beam widths $\pm x_m = \pm R_x$, $\pm y_m = \pm R_y$ behind such a quadrupole system from Eqs. (3.16a) and (3.16b) as

$$S_x \approx \pm\tan \alpha_m = \pm x_{10}/f_x, \tag{3.17a}$$

$$S_y \approx \pm\tan \beta_m = \pm y_{10}/f_y, \tag{3.17b}$$

$$R_x \approx \pm a_{10}f_x \pm l_{2x}(x_{10}/f_x), \tag{3.17c}$$

$$R_y \approx \pm b_{10}f_y \pm l_{2y}(y_{10}/f_y). \tag{3.17d}$$

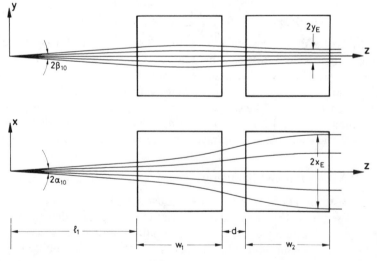

Fig. 3.7. A point-to-parallel focusing quadrupole doublet showing particle trajectories that all start at the center of the object ($x_1 = y_1 = 0$). Note that this doublet is a DF system.

Note that for a point-to-parallel focusing system, all particles, that originated from a certain point $P(x_1, y_1)$ at the object with angles α_1, β_1, are finally inclined by x_1/f_x, y_1/f_y behind a point-to-parallel focusing system independent of α_1, β_1. Note further that for a reversed parallel-to-point focusing system, all particles that originally were parallel are finally focused to a point.

3.2.2 Point-to-Point Focusing

A point-to-point focusing system (Section 1.3.2.1 and Fig. 1.10a) requires $(x|a) = 0$ and/or $(y|b) = 0$ in Eqs. (3.14) or (3.15) so that particle trajectories that diverge from a point of an object are focused to a point of the image. In this case, an object–image relation exists between the first and the last profile plane of the optical system. If the conditions $(x|a) = 0$ and $(y|b) = 0$ are fulfilled simultaneously as in the quadrupole doublet of Fig. 3.8 or in the quadrupole sextet of Fig. 3.17, we speak of a stigmatically focusing system for which one point of the object corresponds to one point of the image. Should either $(x|a)$ or $(y|b)$ vanish alone, we speak of an astigmatically focusing system. Analogous to Eq. (1.24) we find

$$\begin{pmatrix} (x|x) & 0 \\ (a|x) & (a|a) \end{pmatrix} = \begin{pmatrix} M_x & 0 \\ -f_x^{-1} & M_x^{-1} \end{pmatrix}, \tag{3.18a}$$

$$\begin{pmatrix} (y|y) & 0 \\ (b|y) & (b|b) \end{pmatrix} = \begin{pmatrix} M_y & 0 \\ -f_y^{-1} & M_y^{-1} \end{pmatrix}. \tag{3.18b}$$

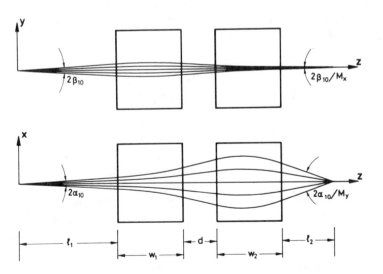

Fig. 3.8. A stigmatic focusing quadrupole doublet showing particle trajectories that all start at the center of the object ($x_1 = y_1 = 0$). Note that this doublet is a DF system.

The quantities $1/f_x$ and $1/f_y$ here describe the overall refractive powers of the quadrupole multiplet for the xz and yz planes, respectively.

Note that in Eqs. (3.18a) and (3.19b) the lateral overall magnifications $(x|x) = M_x$ and $(y|y) = M_y$ are inversely proportional to the overall angular magnifications $(a|a) = 1/M_x$ and $(b|b) = 1/M_y$. This result could also have been found from $|T_x| = 1$ and $|T_y| = 1$ according to Eq. (1.19), where T_x and T_y denote the transfer matrices of Eqs. (3.18a) and (3.18b).

3.2.3   Point-to-Parallel-to-Point Focusing

In a stigmatic focusing lens system, all particles are focused to one point in the image plane if they originated from one point in the object plane. Consequently, in some profile plane within the system the corresponding particle trajectories must be parallel relative to the xz plane, and in some other profile plane they must be parallel relative to the yz plane. If those two profile planes coincide, we find a spatially parallel beam in that intermediate profile plane, as is the case in the system of Fig. 3.17. At this location we can thus separate the lens systems from one another by some field-free region without disturbing the overall stigmatic focusing properties.

Conversely we can combine one point-to-parallel and one parallel-to-point focusing multiplet of n quadrupoles each to one point-to-point focusing multiplet of $2n$ quadrupoles. The first of these quadrupole multiplets is characterized by its first focal plane coinciding with the object, and the second quadrupole multiplet is characterized by its second focal plane coinciding with an image. Thus, such a system is described by the following transfer matrices:

$$\begin{pmatrix} (x|x) & 0 \\ (a|x) & (a|a) \end{pmatrix} = \begin{pmatrix} 0 & f''_x \\ -1/f''_x & 0 \end{pmatrix} \begin{pmatrix} 1 & d_{fx} \\ 0 & 1 \end{pmatrix} \begin{pmatrix} 0 & f'_x \\ -1/f'_x & 0 \end{pmatrix}, \qquad (3.19a)$$

$$\begin{pmatrix} (y|y) & 0 \\ (b|y) & (b|b) \end{pmatrix} = \begin{pmatrix} 0 & f''_y \\ -1/f''_y & 0 \end{pmatrix} \begin{pmatrix} 1 & d_{fy} \\ 0 & 1 \end{pmatrix} \begin{pmatrix} 0 & f'_y \\ -1/f'_y & 0 \end{pmatrix}. \qquad (3.19b)$$

Here, d_{fx} and d_{fy} are the distances between the downstream focal planes of the front-end lens system and the upstream focal planes of the back-end lens system in the xz and yz planes, respectively. The quantities f'_x, f'_y and f''_x, f''_y further describe the overall focal lengths of the two-lens systems. The distances between the second principal plane of the front-end lens and the first principal plane of the back-end lens are thus $d_{fx} + f'_x + f''_x$ or $d_{fy} + f'_y + f''_y$ for the x and the y directions, respectively.

Note that the coefficients $(x|a)$ and $(y|b)$ of the product matrices in Eqs. (3.19a) and (3.19b) do vanish as postulated by Eqs. (3.18a) and (3.18b) and that the lateral magnifications are found as

$$M_x = (x|x) = -f''_x/f'_x, \qquad M_y = (y|y) = -f''_y/f'_y. \tag{3.20}$$

3.2.4 Systems for Which $(x|a)$ and $(b|b)$ Vanish Simultaneously

Sometimes it is desirable to postulate an image $[(x|a) = 0]$ in the x direction at the final profile plane and simultaneously a parallel beam $[(b|b) = 0]$ in the y direction for particles leaving one point of the object (Fig. 3.9). In this case, Eqs. (3.16a) and (3.18b) are both fulfilled simultaneously.

3.2.5 Telescopic Focusing Systems

A telescopic focusing system achieves simultaneous point-to-point as well as parallel-to-parallel focusing (Sections 1.3.2.1 and 1.3.2.2). The corresponding transfer matrices thus read,

$$\begin{pmatrix} (x|x) & 0 \\ 0 & (a|a) \end{pmatrix} = \begin{pmatrix} M_x & 0 \\ 0 & M_x^{-1} \end{pmatrix}, \tag{3.21a}$$

$$\begin{pmatrix} (y|y) & 0 \\ 0 & (b|b) \end{pmatrix} = \begin{pmatrix} M_y & 0 \\ 0 & M_y^{-1} \end{pmatrix}. \tag{3.21b}$$

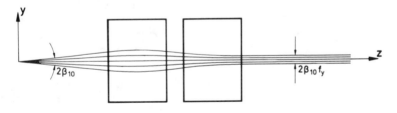

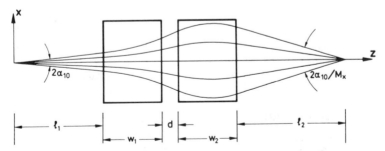

Fig. 3.9. A quadrupole doublet for which $(x|a)$ and $(b|b)$ vanish simultaneously in Eqs. (3.22a) and (3.22b). Analogously to Figs. 3.7 and 3.8, only particle trajectories are shown that start at the center of the object ($x_1 = y_1 = 0$). Note that this doublet is a DF system.

These transfer matrices are identical to those of Eqs. (3.18a) and (3.18b) with $f_x = f_y = \pm\infty$. Note that the magnifications M_x and M_y in Eqs. (3.18) and (3.21) normally are not equal.

A telescopic system can always be considered as a two-lens system for which the second focal plane of the first lens coincides with the first focal plane of the second lens. Only in this case can a parallel beam be transformed into a second parallel beam.

3.3   PROPERTIES OF THIN-LENS QUADRUPOLE MULTIPLETS

Normally, the choice of parameters for an optical system consisting of j quadrupoles and $j + 1$ drift distances such that the conditions of Eqs. (3.16), (3.18), or (3.21) are satisfied for the product transfer matrix is very complex. In most cases, it is not even possible to formulate these conditions algebraically and then solve the resulting equations by proper choices of drift distances and inhomogeneity parameters of individual quadrupoles. Thus, the problem must be formulated as in Eqs. (3.14) and (3.15) and solved by an iterative numerical procedure. As a starting point, I propose to use the solution to the problem which is found by approximating all quadrupoles by thin lenses of focal lengths $f_x = 1/(k^2 w) = -f_y$, as in Eq. (3.13). From this approximate solution we can normally also judge whether the real system is technically feasible.

Note that the solution of a given problem in a thin-lens approximation can be formulated in such a way that all quantities of interest are determined directly from the optical postulates. Note further that the derived formulas are so simple that any pocket calculator is adequate for this evaluation. However, be aware that this approximate solution can be quite different from a realistic solution if the quantity

$$kw \approx \sqrt{|w/f|}$$

in Eqs. (3.7) and (3.13) is not smaller than 1. On the other hand, the approximate solution is mostly adequate as a starting point for a numerical iterative procedure if the quantity kw stays less than 0.8 or 0.9.

3.3.1   Quadrupole Doublets

With two quadrupoles characterized by lengths w_1 and w_2 and inhomogeneity parameters $k_{x1}^2 = -k_{y1}^2$ and $k_{x2}^2 = -k_{y2}^2$ defined in Eq. (3.7a) or (3.7b), the overall x and y transfer matrices of a quadrupole doublet are given by Eqs. (3.14a) and (3.14b). These are two products of five transfer matrices transferring x- and y-position vectors in profile plane 1 to x- and y-position vectors in profile plane 6. These transfer matrices describe a first drift distance l_1, the first quadrupole of length w_1, a second drift distance $d = d_1$, the second quadrupole of length w_2, and a third drift distance l_2. In a thin-lens approximation these transfer matrices read

$$\begin{pmatrix} (x|x) & (x|a) \\ (a|x) & (a|a) \end{pmatrix} = \begin{pmatrix} 1 & \bar{l}_2 \\ 0 & 1 \end{pmatrix} \begin{pmatrix} 1 & 0 \\ -\bar{f}_2^{-1} & 1 \end{pmatrix} \begin{pmatrix} 1 & \bar{d} \\ 0 & 1 \end{pmatrix} \begin{pmatrix} 1 & 0 \\ -\bar{f}_1^{-1} & 1 \end{pmatrix} \begin{pmatrix} 1 & \bar{l}_1 \\ 0 & 1 \end{pmatrix}, \tag{3.22a}$$

$$\begin{pmatrix} (y|y) & (y|b) \\ (b|y) & (b|b) \end{pmatrix} = \begin{pmatrix} 1 & \bar{l}_2 \\ 0 & 1 \end{pmatrix} \begin{pmatrix} 1 & 0 \\ \bar{f}_2^{-1} & 1 \end{pmatrix} \begin{pmatrix} 1 & \bar{d} \\ 0 & 1 \end{pmatrix} \begin{pmatrix} 1 & 0 \\ \bar{f}_1^{-1} & 1 \end{pmatrix} \begin{pmatrix} 1 & \bar{l}_1 \\ 0 & 1 \end{pmatrix}, \tag{3.22b}$$

where $\bar{l}_1 = l_1 + w_1/2$, $\bar{d} = d + (w_1 + w_2)/2$, and $\bar{l}_2 = l_2 + w_2/2$, and according to Eq. (3.13), $\bar{f}_1^{-1} = k_{x1}^2 w_1$, and $\bar{f}_2^{-1} = k_{x2}^2 w_2$. The coefficients of Eqs. (3.22a) and (3.22b) read explicitly,

$$(a|x) = -\frac{1}{\bar{f}_x} = \frac{\bar{d}}{\bar{f}_1 \bar{f}_2} - \frac{1}{\bar{f}_1} - \frac{1}{\bar{f}_2}, \tag{3.23a}$$

$$(b|y) = -\frac{1}{\bar{f}_y} = \frac{\bar{d}}{\bar{f}_1 \bar{f}_2} + \frac{1}{\bar{f}_1} + \frac{1}{\bar{f}_2}, \tag{3.23b}$$

$$(a|a) = \left(1 - \frac{\bar{l}_1}{\bar{f}_1}\right)\left(1 - \frac{\bar{d}}{\bar{f}_2}\right) - \frac{\bar{l}_1}{\bar{f}_2}, \tag{3.23c}$$

$$(b|b) = \left(1 + \frac{\bar{l}_1}{\bar{f}_1}\right)\left(1 + \frac{\bar{d}}{\bar{f}_2}\right) + \frac{\bar{l}_1}{\bar{f}_2}, \tag{3.23d}$$

$$(x|a) = \bar{l}_1 + \bar{d}\left(1 - \frac{\bar{l}_1}{\bar{f}_1}\right) + \bar{l}_2(a|a), \tag{3.23e}$$

$$(y|b) = \bar{l}_1 + \bar{d}\left(1 + \frac{\bar{l}_1}{\bar{f}_1}\right) + \bar{l}_2(b|b), \tag{3.23f}$$

$$(x|x) = 1 - \frac{\bar{d}}{\bar{f}_1} + \bar{l}_2(a|x), \tag{3.23g}$$

$$(y|y) = 1 + \frac{\bar{d}}{\bar{f}_1} + \bar{l}_2(b|y). \tag{3.23h}$$

3.3.1.1 Doubly Point-to-Parallel Focusing Quadrupole Doublets

A quadrupole doublet that is point-to-parallel focusing in both the xz and yz planes, as shown in Fig. 3.7 requires $(a|a)$ and $(b|b)$ in Eqs. (3.14a) and (3.14b) to vanish simultaneously as in Eqs. (3.16a) and (3.16b). An approximate solution for this problem is found by postulating $(a|a) = (b|b) = 0$ in Eqs. (3.22a) and (3.22b) for a thin-lens quadrupole doublet. With Eqs. (3.23a)–(3.23d), these postulates yield

$$-\bar{l}_1/\bar{f}_1 = \bar{f}_2/\bar{d} = \pm\sqrt{1 + \bar{l}_1/\bar{d}}, \tag{3.24a}$$

$$\bar{l}_1/\bar{f}_x = 1 \mp \sqrt{(1 + \bar{l}_1/\bar{d})^{-1}}, \tag{3.24b}$$

$$\bar{l}_1/\bar{f}_y = 1 \pm \sqrt{(1 + \bar{l}_1/\bar{d})^{-1}}. \tag{3.24c}$$

In the case of the upper signs, the first quadrupole is defocusing ($\bar{f}_1 < 0$), and the second quadrupole is focusing ($\bar{f}_2 > 0$) insofar as the x direction is concerned (Figs. 3.6 and 3.7). This case will be referred to as a DF-arrangement. According to Section 1.4.2, the principal planes of a DF or an FD lens doublet are always shifted toward the focusing lens (see also Fig. 1.15). For a negative $\bar{f}_1$ and a positive $\bar{f}_2$, we thus always find $\bar{f}_x > \bar{f}_y$ with both $\bar{f}_x$ and $\bar{f}_y$ positive. For most quadrupole doublets, which must be point-to-parallel focusing, we normally require the focal length $\bar{f}_x$ to have a certain value. Thus, it is worthwhile to rewrite Eqs. (3.24a)–(3.24c) as

$$-\bar{l}_1/\bar{f}_1 = \bar{f}_2/\bar{d} = (1 - \bar{l}_1/\bar{f}_x)^{-1} \tag{3.25a}$$

$$\bar{l}_1/\bar{d} = (1 - \bar{l}_1/\bar{f}_x)^{-2} - 1, \tag{3.25b}$$

$$\bar{l}_1/\bar{f}_y = 2 - \bar{l}_1/\bar{f}_x. \tag{3.25c}$$

To get some idea about the possible thin-lens solutions and at the same time obtain initial values for an iterative numerical procedure to fulfill the condition $(a|a) = (b|b) = 0$ in Eqs. (3.14a) and (3.14b), the quantities $\bar{l}_1/\bar{f}_1$, $\bar{l}_1/\bar{f}_2$, $\bar{l}_1/\bar{f}_x$, and $\bar{l}_1/\bar{f}_y$ are plotted in Fig. 3.10 for various values of $\bar{d}/\bar{l}_1$.

3.3.1.2 Stigmatic Focusing Quadrupole Doublets

A stigmatic focusing quadrupole doublet as shown in Fig. 3.8 requires the matrix elements $(x|a)$ and $(y|b)$ in Eqs. (3.14a) and (3.14b) to vanish simultaneously. As in the case of Eqs. (3.18a) and (3.18b), all particles that leave a point $P(x_1, y_1)$ of the object are focused to a point $P(x_6, y_6)$ of the image. An approximate solution for this problem is found by postulating $(x|a) = (y|b) = 0$ in Eqs. (3.22a) and (3.22b), for a thin-lens quadrupole doublet. With Eqs. (3.23a), (3.23b), (3.23g), and (3.23h), these postulates yield

$$\frac{\bar{l}_1}{\bar{f}_1} = \mp\sqrt{\left(\frac{\bar{l}_1 + \bar{d} + \bar{l}_2}{\bar{d}}\right)\frac{\bar{l}_1 + \bar{d}}{\bar{l}_2 + \bar{d}}}, \tag{3.26a}$$

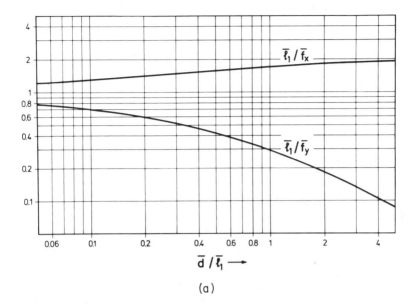

(a)

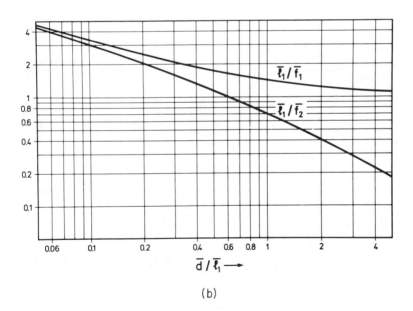

(b)

Fig. 3.10. For a point-to-parallel focusing thin-lens quadrupole doublet $[(a|a) = (b|b) = 0$ in Eqs. (3.22a) and (3.22b)], the quantities $\bar{l}_1/\bar{f}_1$, $\bar{l}_1/\bar{f}_2$ and $\bar{l}_1/\bar{f}_x$, $\bar{l}_1/\bar{f}_y$ are plotted for various values of $\bar{d}/\bar{l}_1$, according to Eqs. (3.24a)–(3.24c).

$$\frac{\bar{l}_2}{\bar{f}_2} = \pm \sqrt{\left(\frac{\bar{l}_1 + \bar{d} + \bar{l}_2}{\bar{d}}\right) \frac{\bar{l}_2 + \bar{d}}{\bar{l}_1 + \bar{d}}}, \tag{3.26b}$$

$$\bar{l}_1 \bar{M}_x = -(\bar{d} + \bar{l}_2) \pm \sqrt{\bar{d}(\bar{l}_1 + \bar{d} + \bar{l}_2) \frac{\bar{l}_2 + \bar{d}}{\bar{l}_1 + \bar{d}}}, \tag{3.26c}$$

$$\bar{l}_1 \bar{M}_y = -(\bar{d} + \bar{l}_2) \mp \sqrt{\bar{d}(\bar{l}_1 + \bar{d} + \bar{l}_2) \frac{\bar{l}_2 + \bar{d}}{\bar{l}_1 + \bar{d}}}, \tag{3.26d}$$

where $(x|x) = \bar{M}_x$ and $(y|y) = \bar{M}_y$ are the lateral x and y magnifications. Here again the upper signs are valid for the case in which the first quadrupole is defocusing ($\bar{f}_1 < 0$) and the second quadrupole is focusing ($\bar{f}_2 > 0$) in the x direction as in Figs. 3.6 and 3.8. In this case we also speak of a DF arrangement. From Eqs. (3.26c) and (3.26d) we find

$$\bar{l}_1 \bar{M}_y = \bar{l}_1 \bar{M}_x \mp 2\sqrt{\bar{d}(\bar{l}_1 + \bar{d} + \bar{l}_2) \frac{\bar{l}_2 + \bar{d}}{\bar{l}_1 + \bar{d}}},$$

which proves that equal x and y magnifications ($\bar{M}_x = \bar{M}_y$) are not possible in a stigmatic focusing quadrupole doublet. This is easily understood, since for positive $\bar{f}_2$ and negative $\bar{f}_1$, the xz principal planes are shifted toward the image, and the y principal planes are shifted toward the object as in Fig. 1.15.

For most stigmatic focusing quadrupole doublets, the magnitudes of the x magnification are required to have certain values. Thus, it is often worthwhile to determine $\bar{l}_2$ from Eq. (3.26c) as

$$\bar{l}_2 = \bar{d}/2 - \bar{M}_x(\bar{l}_1 + \bar{d}) \pm \sqrt{\bar{d}[\bar{d}/4 + \bar{M}_x(\bar{M}_x - 1)(\bar{l}_1 + \bar{d})]}, \tag{3.27}$$

and introduce this quantity into Eqs. (3.26a) and (3.26b) to obtain $\bar{l}_1/\bar{f}_1$ and $\bar{l}_2/\bar{f}_2$ as functions of $\bar{M}_x$ and $\bar{d}/\bar{l}_1$.

The maximal beam cross section in any DF quadrupole doublet occurs in the second quadrupole in the x direction as can be seen in Figs. 3.7 and 3.8. For a point source ($x_{10} = y_{10} = 0$) this maximal beam cross section is found as

$$2R_x = 2G_x \tan \alpha_{10} \approx 2a_{10}G_x.$$

Here, G_x is an abbreviation for the coefficient $(x|a)$ of Eq. (3.23c) with $\bar{l}_2 = 0$, which reads explicitly,

$$G_x = \bar{l}_1 + \bar{d} - (\bar{d}\bar{l}_1/\bar{f}_1), \tag{3.28}$$

where $\bar{l}_1/\bar{f}_1$ must be determined from Eq. (3.26a).

To get some idea of the possible thin-lens solutions of Eqs. (3.26a)–(3.26b) and at the same time obtain a starting point for an iterative numerical

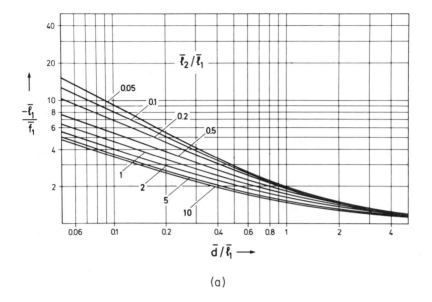

(a)

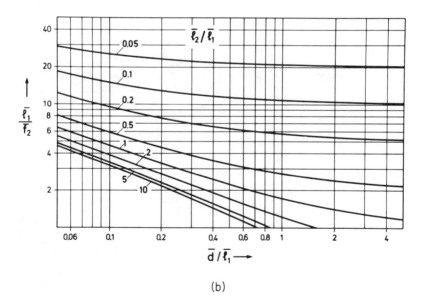

(b)

Fig. 3.11. For a stigmatic focusing thin-lens quadrupole doublet $[(x|a) = (y|b) = 0$ in Eqs. (3.22a) and (3.22b)], the quantities $\bar{l}_1/\bar{f}_1$, $\bar{l}_1/\bar{f}_2$, $\bar{M}_x$, and $\bar{M}_y$ are plotted for various values of $\bar{d}/\bar{l}_1$ and $\bar{l}_2/\bar{l}_1$ according to Eqs. (3.26a)–(3.26d).

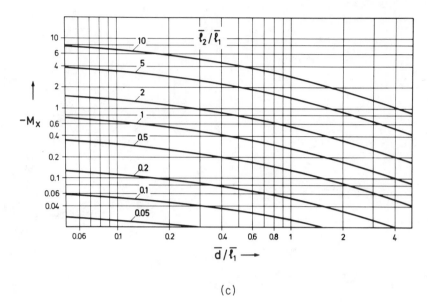

(c)

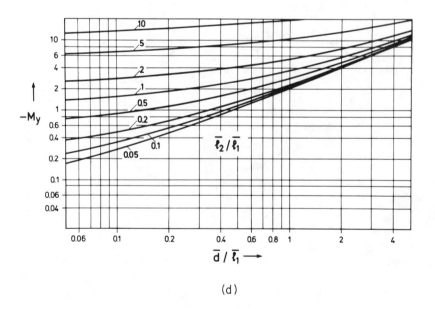

(d)

Fig. 3.11 (continued)

procedure to fulfill $(x|a) = (y|b) = 0$ in Eqs. (3.14a) and (3.14b), the quantities $\bar{l}_1/\bar{f}_1$, $\bar{l}_1/\bar{f}_2$, and $\bar{M}_x$, $\bar{M}_y$ are plotted in Fig. 3.11 for various values of $\bar{d}/\bar{l}_1$ and $\bar{l}_2/\bar{l}_1$.

3.3.1.3 Quadrupole Doublets for Which (x|a) and (b|b) Vanish Simultaneously

A third case of interest concerns those quadrupole doublets which are point-to-point focusing in the xz plane and point-to-parallel focusing in the yz plane, as indicated in Fig. 3.9, causing both $(x|a)$ and $(b|b)$ to vanish in Eqs. (3.14a) and (3.14b). An approximate solution for this problem is found for a thin-lens quadrupole doublet by postulating $(x|a) = (b|b) = 0$ in Eqs. (3.22a) and (3.22b). With Eqs. (3.23b), (3.23d), (3.23e), and (3.23g), these postulates yield

$$\frac{\bar{l}_1}{\bar{f}_1} = \mp\sqrt{\left(\frac{\bar{l}_1 + \bar{d} + 2\bar{l}_2}{\bar{d}}\right)\frac{\bar{l}_1 + \bar{d}}{2\bar{l}_2 + \bar{d}}}, \tag{3.29a}$$

$$\frac{2\bar{l}_2}{\bar{f}_2} = 1 \pm \sqrt{\left(\frac{\bar{l}_1 + \bar{d} + 2\bar{l}_2}{\bar{d}}\right)\frac{2\bar{l}_2 + \bar{d}}{\bar{l}_1 + \bar{d}}}, \tag{3.29b}$$

$$2\bar{l}_1\bar{M}_x = -(2\bar{l}_2 + \bar{d}) \pm \sqrt{\bar{d}(\bar{l}_1 + \bar{d} + 2\bar{l}_2)\frac{2\bar{l}_2 + \bar{d}}{\bar{l}_1 + \bar{d}}}, \tag{3.29c}$$

$$\frac{2\bar{l}_2\bar{l}_1}{\bar{f}_y} = -(\bar{l}_2 + \bar{d}) \mp \sqrt{\bar{d}(\bar{l}_1 + \bar{d} + 2\bar{l}_2)\frac{2\bar{l}_2 + \bar{d}}{\bar{l}_1 + \bar{d}}}. \tag{3.29d}$$

Here again, the upper signs are valid for the case in which the first quadrupole is defocusing ($\bar{f}_1 < 0$) and the second quadrupole is focusing ($\bar{f}_2 > 0$) insofar as the x direction is concerned. The overall focal length $\bar{f}_y$ for the yz plane determines the beam cross section in y direction at the position of the x image according to Eq. (3.17d):

$$R_y = \pm[b_{10}\bar{f}_y + y_{10}(\bar{l}_2/\bar{f}_y)], \tag{3.29e}$$

with the corresponding maximal inclination determined from Eq. (3.17b) as

$$S_y \approx \tan\beta_r = \pm y_{10}/\bar{f}_y. \tag{3.29f}$$

In most cases, $\bar{M}_x$ the magnitude of the x magnification is required to attain a certain value. Thus, it is often worthwhile to rewrite Eq. (3.29c) such that $\bar{l}_2$, which was assumed to be known in Eqs. (3.29), is calculated as

$$\bar{l}_2 = -\tfrac{1}{4}\bar{d} - \bar{M}_x(\bar{l}_1 + \bar{d}) \pm \sqrt{\tfrac{1}{16}\bar{d}^2 + \bar{M}_x\bar{d}(\bar{l}_1 + \bar{d})(\bar{M}_x - \tfrac{1}{2})}. \tag{3.30}$$

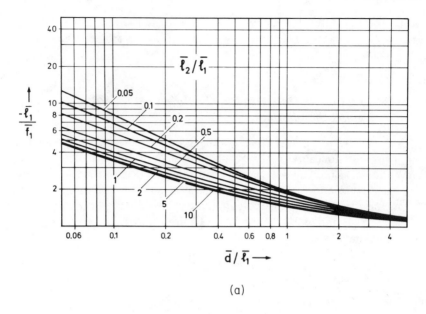

(a)

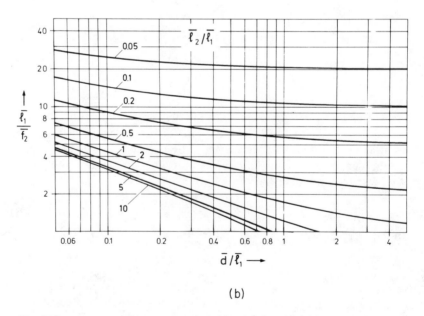

(b)

Fig. 3.12.   For a quadrupole doublet for which $(x|a)$ and $(b|b)$ vanish simultaneously, the quantities $\bar{l}_1/\bar{f}_1$, $\bar{l}_1/\bar{f}_2$, $\bar{M}_x$, and $\bar{l}_1/\bar{f}_y$ are plotted for various values of $\bar{d}_1/\bar{l}_1$ and $\bar{l}_2/\bar{l}_1$ according to Eqs. (3.29a)-(3.29d).

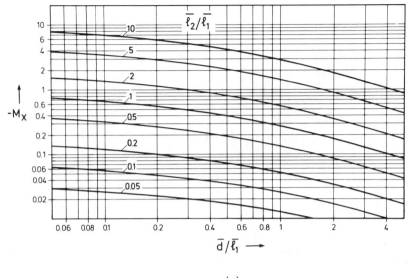

(c)

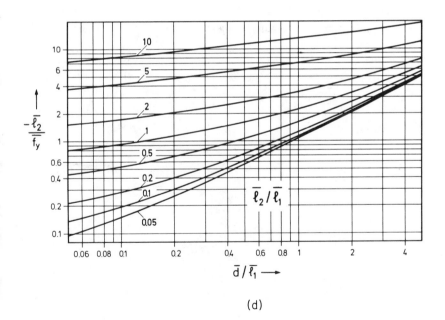

(d)

Fig. 3.12 (continued)

The maximal beam cross section again occurs in the second quadrupole for the x direction. In case of a point source ($x_{10} = y_{10} = 0$), this maximal cross section $2R_x = 2a_{10}G_x$ is determined for a stigmatic focusing quadrupole doublet with G_x again being given by Eq. (3.28). Here, however, $d\bar{l}_1/\bar{f}_1$ must be calculated from Eq. (3.29a).

To get some idea of the possible solutions of Eq. (3.29) and at the same time obtain initial conditions for an iterative numerical procedure to fulfill $(x|a) = (b|b) = 0$ in Eqs. (3.14a) and (3.14b) the quantities $\bar{l}_1/\bar{f}_1$, $\bar{l}_1/\bar{f}_2$ as well as $\bar{M}_x$ and $\bar{l}_1/\bar{f}_y$ are plotted in Fig. 3.12 for various values of $\bar{d}/\bar{l}_1$ and $\bar{l}_2/\bar{l}_1$.

3.3.2 Quadrupole Triplets

Instead of a quadrupole doublet, a quadrupole triplet as indicated in Fig. 3.13 can also be used to shape a particle beam. The advantage here is that the overall beam usually stays rounder. With three quadrupoles characterized by lengths w_1, w_2, w_3 and inhomogeneity parameters $k_{x1}^2 = -k_{y1}^2$, $k_{x2}^2 = -k_{y2}^2$, $k_{x3}^2 = -k_{y3}^2$ defined in Eqs. (3.7a) and (3.7b), the overall transfer matrices are found from Eqs. (3.15a) and (3.15b) with $n = 3$ as two products of seven transfer matrices transferring x- and y-position vectors in profile plane 1 to x- and y-position vectors in profile plane 8. These transfer matrices describe a first drift distance l_1, the first quadrupole; a second drift distance d_1, the second quadrupole; a third drift distance d_2, the third

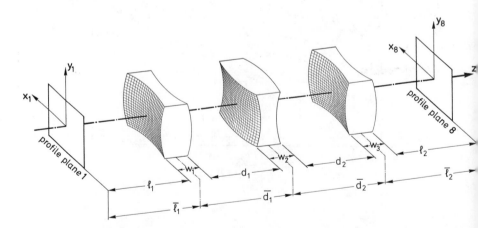

Fig. 3.13. Principal arrangement of a quadrupole triplet, the first lens of which is defocusing in the x direction, the second lens is focusing, and the third lens is defocusing. Such a triplet is called a DFD system.

quadrupole; and a fourth drift distance l_2. In a thin-lens approximation, these transfer matrices are written as

$$
\begin{pmatrix} (x|x) & (x|a) \\ (a|x) & (a|a) \end{pmatrix} = \begin{pmatrix} 1 & \bar{l}_2 \\ 0 & 1 \end{pmatrix} \begin{pmatrix} 1 & 0 \\ -\bar{f}_3^{-1} & 1 \end{pmatrix} \begin{pmatrix} 1 & \bar{d}_2 \\ 0 & 1 \end{pmatrix} \begin{pmatrix} 1 & 0 \\ -\bar{f}_2^{-1} & 1 \end{pmatrix} \begin{pmatrix} 1 & \bar{d}_1 \\ 0 & 1 \end{pmatrix}
$$

$$
\times \begin{pmatrix} 1 & 0 \\ -\bar{f}_1^{-1} & 1 \end{pmatrix} \begin{pmatrix} 1 & \bar{l}_1 \\ 0 & 1 \end{pmatrix}, \tag{3.31a}
$$

$$
\begin{pmatrix} (y|y) & (b|y) \\ (b|y) & (b|b) \end{pmatrix} = \begin{pmatrix} 1 & \bar{l}_2 \\ 0 & 1 \end{pmatrix} \begin{pmatrix} 1 & 0 \\ \bar{f}_3^{-1} & 1 \end{pmatrix} \begin{pmatrix} 1 & \bar{d}_2 \\ 0 & 1 \end{pmatrix} \begin{pmatrix} 1 & 0 \\ \bar{f}_2^{-1} & 1 \end{pmatrix} \begin{pmatrix} 1 & \bar{d}_1 \\ 0 & 1 \end{pmatrix}
$$

$$
\times \begin{pmatrix} 1 & 0 \\ \bar{f}_1^{-1} & 1 \end{pmatrix} \begin{pmatrix} 1 & \bar{l}_1 \\ 0 & 1 \end{pmatrix}, \tag{3.31b}
$$

with $\bar{l}_1 = l_1 + w_1/2$, $\bar{d}_1 = d_1 + (w_1 + w_2)/2$, $\bar{d}_2 = d_2 + (w_2 + w_3)/2$, $\bar{l}_2 = l_2 + w_3/2$, and $\bar{f}_1^{-1} \approx k_{x1}^2 w_1$, $\bar{f}_2^{-2} \approx k_{x2}^2 w_2$, $\bar{f}_3^{-1} \approx k_{x3}^2 w_3$. For $\bar{d} = \bar{d}_1 = \bar{d}_2$, the coefficients of Eqs. (3.31a) and (3.31b) read explicitly;

$$
(a|x) = \frac{-1}{\bar{f}_x} = \left(\frac{\bar{d}}{\bar{f}_1} - 1 \right) \frac{1}{\bar{f}_3} + \left(\frac{\bar{d}}{\bar{f}_3} - 1 \right) \left(\frac{1}{\bar{f}_1} + \frac{1}{\bar{f}_2} - \frac{\bar{d}}{\bar{f}_1 \bar{f}_2} \right), \tag{3.32a}
$$

$$
(b|y) = \frac{-1}{\bar{f}_y} = \left(\frac{\bar{d}}{\bar{f}_1} + 1 \right) \frac{1}{\bar{f}_3} + \left(\frac{\bar{d}}{\bar{f}_3} + 1 \right) \left(\frac{1}{\bar{f}_1} + \frac{1}{\bar{f}_2} + \frac{\bar{d}}{\bar{f}_1 \bar{f}_2} \right), \tag{3.32b}
$$

$$
(a|a) = \left(\frac{\bar{d}}{\bar{f}_2 \bar{f}_3} - \frac{1}{\bar{f}_3} - \frac{1}{\bar{f}_2} \right) \left(\bar{l}_1 + \bar{d} \frac{\bar{l}_1}{\bar{f}_1} \right) + \left(1 - \frac{\bar{d}}{\bar{f}_3} \right) \left(1 - \frac{\bar{l}_1}{\bar{f}_1} \right), \tag{3.32c}
$$

$$
(b|b) = \left(\frac{d}{\bar{f}_2 \bar{f}_3} + \frac{1}{\bar{f}_3} + \frac{1}{\bar{f}_2} \right) \left(\bar{l}_1 + \bar{d} + \frac{\bar{d}\bar{l}_1}{\bar{f}_1} \right) + \left(1 + \frac{\bar{d}}{\bar{f}_3} \right) \left(1 + \frac{\bar{l}_1}{\bar{f}_1} \right), \tag{3.32d}
$$

$$
(x|a) = \bar{l}_1 + \bar{d} - \bar{l}_1 \bar{d} \left(\frac{1}{\bar{f}_1} + \frac{1}{\bar{f}_2} \right) + \bar{d} \left(1 - \frac{\bar{l}_1}{\bar{f}_1} \right) \left(1 - \frac{\bar{d}}{\bar{f}_2} \right) + \bar{l}_2 (a|a), \tag{3.32e}
$$

$$
(y|b) = \bar{l}_1 + \bar{d} + \bar{l}_1 \bar{d} \left(\frac{1}{\bar{f}_1} + \frac{1}{\bar{f}_2} \right) + \bar{d} \left(1 + \frac{\bar{l}_1}{\bar{f}_1} \right) \left(1 + \frac{\bar{d}}{\bar{f}_2} \right) + \bar{l}_2 (b|b), \tag{3.32f}
$$

$$
(x|x) = \left(1 - \frac{\bar{d}}{\bar{f}_1} \right) \left(1 - \frac{\bar{d}}{\bar{f}_2} \right) - \frac{\bar{d}}{\bar{f}_1} + \bar{l}_2 (a|x), \tag{3.32g}
$$

$$
(y|y) = \left(1 + \frac{\bar{d}}{\bar{f}_1} \right) \left(1 + \frac{\bar{d}}{\bar{f}_2} \right) + \frac{\bar{d}}{\bar{f}_1} + \bar{l}_2 (b|y). \tag{3.32h}
$$

3.3.2.1   Doubly Point-to-Parallel Focusing Quadrupole Triplets

A quadrupole triplet, which is point-to-parallel focusing as indicated in Fig. 3.14, in both the xz and yz planes, requires $(a|a)$ and $(b|b)$ in Eqs. (3.15a) and (3.15b) to vanish simultaneously as in Eqs. (3.16a) and (3.16b). Unlike the quadrupole doublet in Section 3.2.1.1, here not only the overall focal length $\bar{f}_x$ but also the overall focal length $\bar{f}_y$ of the triplet can be postulated to attain a certain value. An approximate solution for this problem is found by postulating $(a|a) = (b|b) = 0$ in Eqs. (3.32c) and (3.32d) for a thin-lens quadrupole triplet. For the simple case $\bar{f}_x = \bar{f}_y$, which is shown in Fig. 3.14, Eqs. (3.32a)–(3.32d) yield

$$\bar{f}_x = \bar{f}_y = \bar{l}_1 + \bar{d}, \tag{3.33a}$$

$$-\bar{f}_1/l_1 = \bar{d}/\bar{f}_2 = -2\bar{d}/\bar{f}_3 = \pm\sqrt{2\bar{d}/(\bar{l}_1 + \bar{d})}, \tag{3.33b}$$

where again the upper sign is valid if the first quadrupole is defocusing ($\bar{f}_1 < 0$) in the x direction.

The maximal beam cross section in such a quadrupole triplet occurs in the center quadrupole. For a point source ($x_{10} = y_{10} = 0$), this maximal cross section is $2R_x = 2a_{10}G_x$ as in the case of a stigmatic focusing doublet, with G_x given by Eq. (3.28) as $\bar{l}_1 + \bar{d} - \bar{d}\bar{l}_1/\bar{f}_1$. Here, however, $\bar{d}\bar{l}_1/\bar{f}_1$ must be calculated from Eq. (3.33b). To get some idea of the possible thin-lens solutions as determined from Eqs. (3.33a) and (3.33b) and at the same time obtain initial values for an iterative numerical procedure to solve, Eqs.

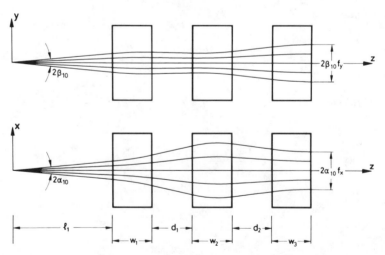

Fig. 3.14.   A point-to-parallel focusing quadrupole triplet showing particle trajectories that all start at the center of the object ($x_1 = y_1 = 0$). Note that this triplet is a DFD system.

(3.15a) and (3.15b), the quantities $\bar{l}_1/\bar{f}_1$, $\bar{l}_1/\bar{f}_2 = -2\bar{l}_1/\bar{f}_3$ are plotted in Fig. 3.15 for various values of $\bar{d}/\bar{l}_1 = \bar{f}_x/\bar{l}_1 - 1 = \bar{f}_y/\bar{l}_1 - 1$. If $\bar{f}_y$ should attain a value other than $\bar{f}_x$, the Eqs. (3.33a) and (3.33b) normally still yield good initial values for an iterative numerical procedure to solve Eqs. (3.15a) and (3.15b) if the quantity $\bar{d} + \bar{l}_1$ is chosen to be $(\bar{f}_x + \bar{f}_y)/2$.

3.3.2.2 Stigmatic Focusing Quadrupole Triplets

A stigmatic focusing quadrupole triplet requires the matrix elements $(x|a)$ and $(y|b)$ in Eqs. (3.15a) and (3.15b) to vanish simultaneously such that particles leaving a point $P(x_1, y_1)$ of the object are all focused to a point $P(x_8, y_8)$ of the image. The advantage of such a stigmatic focusing triplet over a corresponding doublet, as discussed in Section 3.2.2, is the fact that the lateral magnifications $M_x = (x|x)$ and $M_y = (y|y)$ can now be made equal.

An approximate solution for such a stigmatic focusing triplet can be found from a point-to-parallel-to-point focusing thin-lens quadrupole quadruplet discussed in Section 3.3.3 with the added condition that the second and the third quadrupoles are combined to a center quadrupole. The lateral magnifications M_x and M_y in this case are both negative, as we will find.

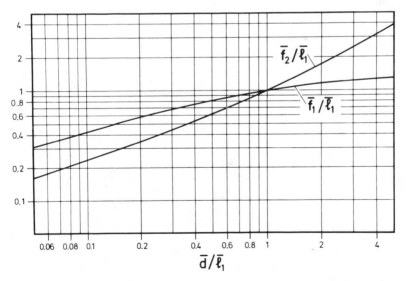

Fig. 3.15. For a point-to-parallel focusing thin-lens quadrupole triplet $[(a|a) = (b|b) = 0$ in Eqs. (3.31a) and (3.31b)], the quantities $\bar{l}_1/\bar{f}_1$ and $\bar{l}_1/\bar{f}_2 = 2\bar{l}_1/\bar{f}_3$ are plotted for various values of $\bar{d}/\bar{l}_1$, according to Eqs. (3.33a) and (3.33b).

3.3.3   Stigmatic Focusing Quadrupole Quadruplets

By combining a point-to-parallel and a parallel-to-point focusing quadrupole doublet, discussed in Sections 3.2.1 and 3.3.1.1, we obtain a point-to-parallel-to-point focusing quadrupole quadruplet for which the matrix elements $(x|a)$ and $(y|b)$ vanish simultaneously in Eqs. (3.15a) and (3.15b). In this case, the lateral magnifications M_x and M_y are both negative. In such a system we usually assume a DFFD arrangement in the xz plane as is illustrated in Fig. 3.16a, however, a DFDF arrangement as is illustrated in Fig. 3.16b is also feasible. An approximate solution for this problem is found from two point-to-parallel focusing thin-lens quadrupole doublets, as discussed in Section 3.3.1.1. The lateral magnifications of such an ideal-

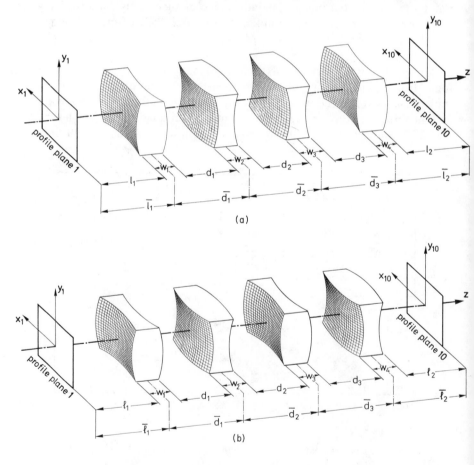

Fig. 3.16.   (a) A DFFD quadrupole quadruplet and (b) a DFDF quadrupole quadruplet.

ized system are read from Eqs. (3.20) and (3.24) as

$$\bar{M}_x = \frac{-f''_x}{\bar{f}'_x} = \left(\frac{-\bar{l}_2}{\bar{l}_1}\right) \frac{1 - 1/\sqrt{1 + \bar{l}_1/\bar{d}_1}}{1 \mp 1/\sqrt{1 + \bar{l}_2/\bar{d}_2}}, \tag{3.34a}$$

$$\bar{M}_y = \frac{-\bar{f}''_y}{\bar{f}'_y} = \left(\frac{-\bar{l}_2}{\bar{l}_1}\right) \frac{1 + 1/\sqrt{1 + \bar{l}_1/\bar{d}_1}}{1 \pm 1/\sqrt{1 + \bar{l}_2/\bar{d}_2}}. \tag{3.34b}$$

The upper and lower signs here correspond to a DFFD and to a DFDF arrangement of quadrupoles, respectively. For a DFDF arrangement, usually $|\bar{M}_x|$ is much smaller than $|\bar{M}_y|$, whereas for a DFFD arrangement, both magnifications are approximately equal. For $\bar{l}_1/\bar{d}_1 = \bar{l}_2/\bar{d}_2$, the Eqs. (3.34a) and (3.34b) yield for a DFFD arrangement,

$$\bar{M}_x = \bar{M}_y = -\bar{l}_2/\bar{l}_1 = -\bar{d}_2/\bar{d}_1.$$

Note that in case the initial maximal inclinations of a trajectory a_{10} and b_{10} are approximately equal, a point-to-parallel-to-point focusing thin-lens quadrupole quadruplet produces a parallel beam between the two doublets that is grossly asymmetric ($R_x \approx \pm a_{10}\bar{f}'_x$) and ($R_y \approx \pm b_{10}\bar{f}'_y$), with $\bar{f}'_x$ and $\bar{f}'_y$ calculated from Eqs. (3.24b) and (3.24c).

As an alternative to placing two point-to-parallel focusing quadrupole doublets back to back, we also could have combined two point-to-point-focusing quadrupole doublets, discussed in Section 3.3.1.2, to achieve an overall stigmatic focusing with lateral magnifications $\bar{M}_x > 0$ and $\bar{M}_y > 0$. As another alternative, we could also use two back-to-back arranged quadrupole doublets for each of which the matrix elements $(x|a)$ and $(b|b)$ vanish. In this case, we would have obtained an overall stigmatic focus with $\bar{M}_x > 0$ and $\bar{M}_y < 0$.

3.3.4 Point-to-Parallel-to-Point Focusing Sextets

Combining two point-to-parallel focusing triplets, discussed in Section 3.3.2.1, we obtain a point-to-parallel-to-point focusing sextet, as shown in Fig. 3.17. The advantage of this sextet over a quadruplet is that if $\alpha_{10} \approx a_{10}$ and $\beta_{10} \approx b_{10}$ are about equally large, a round parallel beam can be obtained between the two triplets. The beam radii here are $R_x = a_{10}\bar{f}'_x$ and $R_y = b_{10}\bar{f}'_y$, with $\bar{f}'_x$ and $\bar{f}'_y$ calculated from Eqs. (3.32a) and (3.32b). Furthermore, some additional flexibility exists in the choice of quadrupole parameters, if different magnifications $\bar{M}_x = \bar{M}_y$ are required.

3.3.5 A "Beam Rotator"

In some cases it is of interest to rotate a particle beam around its optic axis so that x and y are exchanged and, for instance, a vertical dispersion

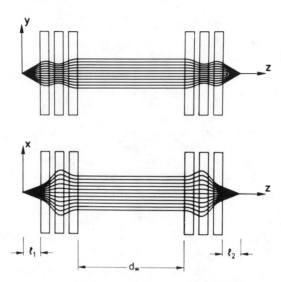

Fig. 3.17. A point-to-parallel-to-point focusing quadrupole sextet showing particle trajectories that all start at the center of the object. Note that this sextet is a DFDDFD system.

of some sector field is turned into a horizontal dispersion. This goal can certainly be achieved by passing a particle beam through the bore of a solenoid; however, it also can be realized by a quadrupole multiplet, as was shown by Kowalski and Enge (1972).

The simplest approach is to require the quadrupole system under consideration to fulfill an object–image relation $(x|a) = (y|b) = 0$ between two profile planes and postulate $(x|x) = -(y|y)$. In this case, points A_1 and B_1 at the object are translated to points A_n and B_n, Fig. 3.18, which is not really a rotation but rather a mirror action with respect to the $(x = y)$ plane of the quadrupole multiplet.

Assume A_1 and B_1 to be points of the x_1, y_1 axes of Fig. 3.18, rotated by an angle $-\Omega$ relative to the x, y coordinates of the quadrupole. Then, the points A_n and B_n are points on the x_n, y_n axes. Note that the x_n axis is rotated by an angle 2Ω, and the y_n axis is rotated by an angle $2\Omega + \pi$ relative to the x_1, y_1 coordinate system. For $\Omega = \pi/4$, consequently, the x and y axes are interchanged.

The more general approach requires not only the x, y coordinates to be "rotated," but the a, b values as well, in which case the quadrupole multiplet does not have to satisfy an object–image relation between the two profile planes in question. This problem can always be solved by a combination of five or more quadrupole lenses, which are all rotated by an angle $\pi/4$ (Kowalski and Enge, 1972).

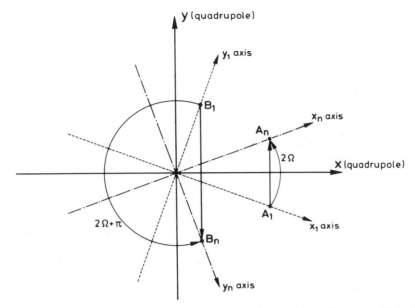

Fig. 3.18. An illustration of the action of a "beam rotator." A quadrupole multiplet is assumed to form an image with an intermediate x image but no intermediate y image, so that $M_x = +1$ and $M_y = -1$. Thus, the points A_1, B_1 are transformed to A_n, B_n and consequently, the x_1, y_1-coordinate system transforms into the x_n, y_n-coordinate system.

3.4 HOW TO CALCULATE QUADRUPOLE MULTIPLETS NUMERICALLY

In order to find a quadrupole multiplet of specified properties, we should set up the problem with numerical transfer matrices of quadrupoles and drift distances, multiply these transfer matrices as in Eqs. (3.15a) and (3.15b), and try to achieve iteratively all postulates by modifying the quadrupole field strengths and possibly the lengths of quadrupoles and drift distances. This iterative procedure should start from an approximate thin-lens solution that is consistent with the physical and technical constraints.

Before starting detailed ion optical calculations, we must know in each case

(1) the magnetic or electrostatic rigidities χ_B or χ_E of the particle beam defined in Eqs. (2.11) and (2.13);

(2) the source width $\pm x_{10}$ and $\pm y_{10}$; and

(3) the maximum possible inclinations $\pm\tan a_{10} \approx \pm a_{10}$ and $\pm\tan \beta_{10} \approx \pm b_{10}$ with which particles can start from the center of the source $x_1 = y_1 = 0$.

3.4.1   Examples of Point-to-Parallel Focusing Quadrupole Systems

To obtain a useful solution for doubly point-to-parallel focusing quadrupole systems, we must first state to what extent the particle beam shall finally be parallel; i.e., how small α_m and β_m of Eqs. (3.17a) and (3.17b) must be or what magnitudes f_x and f_y should have. Note that according to Eqs. (3.17c) and (3.17d) this postulate also determines the beam cross section behind the quadrupole system.

As an example, we may consider an electrostatic quadrupole system for an atomic collision experiment, as sketched in Fig. 3.19. Assuming ions of 1 keV energy for this experiment, the corresponding electrostatic rigidity is $\chi_E = 2000$ V. These particles may emerge from an orifice of diameter $2x_{10} = 2y_{10} = 0.2$ mm with maximum angles of inclination $\pm\alpha_{10} = \beta_{10} \approx \pm15$ mrad or $\pm a_{10} = b_{10} \approx \pm0.015$. In order that small scattering angles can be detected and the full ion beam still be used, the incoming particle beam is best transformed by a point-to-parallel focusing quadrupole system, so as to be widely parallel and also have a large cross section at the scattering site. The

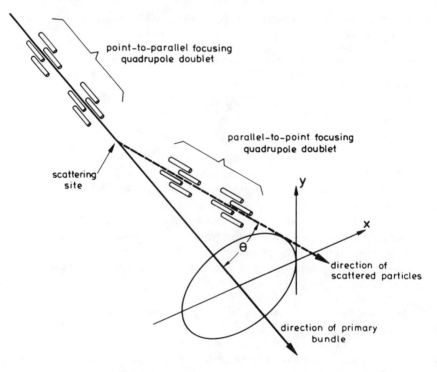

Fig. 3.19.   Outlay of an atomic collision experiment.

scattered ions are then optimally detected behind another parallel-to-point focusing quadrupole system so that the position of detection is a measure of the ion scattering angle [see also below Eq. (3.17d)]. If the energy loss in the scattering process is small, the two quadrupole multiplets can be identical in construction. Normally, the axes of the two quadrupole systems will coincide, however, it can be advantageous to incline the axis of the second system with respect to the first. Note that this inclination does not disturb the optical arrangement. The only difference is that in case of such an inclined arrangement, a different region of scattering angles is investigated.

Postulating that the scattering angle θ in Fig. 3.19 must be measured with an accuracy of $\alpha = \pm 1$ mrad or $S_x = \pm 0.001$, Eq. (3.17a) states that f_x must equal $x_{10}/S_x = 0.1/0.001 = 100$ mm. The magnitude of $f_x = 100$ mm automatically determines the maximal diameter of the beam at the scattering site as $2R_x = 2 \times a_{10} \times 100 = 3$ mm if the second term in Eq. (3.17c) is neglected as being small. Thus quadrupoles with an aperture of $2G_0 = 6$ mm should be quite sufficient, which suggests quadrupole lengths above 2 or 3 times 6 mm or values of $w_1 = w_2 > 15$ mm.

3.4.1.1 A Quadrupole Triplet

A solution for the problem of Fig. 3.19 can be found with a quadrupole triplet for which the Eq. (3.15) yield $(a|a) = (b|b) = 0$. Before starting exact numerical calculations with Eqs. (3.15a) and (3.15b), we might search for a solution with a thin-lens quadrupole triplet by using Eqs. (3.33a) and (3.33b). As a first attempt we choose l_1 to be 50 mm for quadrupoles of $w = 20$ mm in length, and find with $\bar{l}_1 = l_1 + w/2 = 60$ mm, from Eq. (3.33a) $\bar{d} = \bar{f}_x - \bar{l}_1 = 40$ mm. This $\bar{d}$ causes distances of $d_1 = d_2 = \bar{d} - w = 20$ mm between the quadrupoles. With this $\bar{l}_1$ and $\bar{d}$, the Eqs. (3.33a) and (3.33b) yield $\bar{f}_1 = -53.7$ mm and $\bar{f}_2 = -\bar{f}_3/2 = 44.7$ mm. For ions of 1000 eV energy, the corresponding voltages on the quadrupoles of diameter $2G_0 = 6$ mm are found from Eq. (3.13el) to be

$$\bar{V}_{T1} \approx -8.38 \text{ V}, \qquad \bar{V}_{T2} \approx 10.1 \text{ V}, \qquad \bar{V}_{T3} \approx -5.03 \text{ V}.$$

The maximum beam diameter in the center quadrupole is calculated from Eq. (3.28) to be $2R_x \approx 2a_{10}G_x \approx 4.3$ mm, which is well below $2G_0 = 6$ mm. With these voltages, the exact Eqs. (3.15a) and (3.15b) yield numerically $(a|a) \approx 0.65$, $(b|b) \approx 0.22$, $f_x \approx 492$ mm, and $f_y \approx 117$ mm. By iterative numerical methods we find that it is advantageous to change the quadrupole voltages slightly to $V_{T1} \approx -9.91$ V, $V_{T2} \approx 11.9$ V, $V_{T3} \approx -5.8$ V, which yields $(a|a) \approx (b|b) < 0.01$, $f_x \approx 99.4$ mm, and $f_y \approx 99.8$ mm, which, for most purposes should be adequate.

3.4.1.2   A Quadrupole Doublet

A solution for the previous problem can also be found with a quadrupole doublet for which the Eqs. (3.14a) and (3.14b) yield $(a|a) = (b|b) = 0$. Before doing exact numerical calculations we might again search for a solution with a thin-lens quadrupole doublet by using Eqs. (3.24), (3.25), or Fig. 3.10. We choose l_1 to be 25 mm, and find with quadrupoles of $w = 20$ mm in length, $\bar{l}_1 = l_1 + w/2 = 35$ mm and $\bar{d} = 25.6$ mm from Eq. (3.25b) for $\bar{f}_x = 100$ mm. This would still allow a distance $d \approx 5.6$ mm between the quadrupoles. With this $\bar{l}_1$ and $\bar{d}$, we find from Eqs. (3.24a) or (3.25a); $\bar{f}_1 \approx -22.7$ mm and $\bar{f}_2 \approx 39.4$ mm. For ions of 1000 eV energy, the corresponding voltages on the quadrupoles of diameter $2G_0 = 6$ mm are calculated from Eq. (3.13el) to be

$$\bar{V}_{T1} \approx -19.8 \text{ V}, \qquad \bar{V}_{T2} \approx 11.4 \text{ V}.$$

With these voltages, Eqs. (3.14a) and (3.14b) yield numerically $(a|a) \approx 0.33$, $(b|b) \approx 0.38$, $f_x \approx 411$ mm, and $f_y \approx 27$ mm. By iterative numerical methods, we find that changing the quadrupole voltages to $V_{T1} \approx -25$ V and $V_{T2} \approx 14$ V results in $(a|a) \approx (b|b) < 0.01$ as well as $f_x \approx 116$ mm and $f_y \approx 19$ mm, which for many purposes is adequate. Note, however, that in the case of a quadrupole doublet, f_x and f_y are quite different.

3.4.2   Examples of Point-to-Point Focusing Quadrupole Systems

In order to have a quadrupole system that is stigmatic focusing we must postulate $(x|a) = (y|b) = 0$ in Eqs. (3.15a) and (3.15b). Additionally, we normally postulate the overall x and y magnifications M_x and M_y to take up certain values that automatically determine the angular magnifications of Eqs. (3.18a) and (3.18b) as M_x^{-1} and M_y^{-1}.

3.4.2.1   A Stigmatic Focusing Quadrupole Doublet

As an example we assume that a beam of 3-keV ions must be focused to the entrance slit $(2x_6 = 0.1$ mm) of a mass spectrometer by a quadrupole doublet. These particles of an electrostatic rigidity, calculated from Eq. (2.14s), of 6000 V will be assumed to emerge from a virtual emission area of $2x_{10} = 2y_{10} = 0.5$ mm with maximum angles $\pm\alpha_{10} = \beta_{10} = \pm 15$ mrad or $\pm a_{10} = b_{10} = \pm 0.015$. The x magnification of a corresponding quadrupole system must be consequently postulated to be $M_x = -0.2$.

Before starting exact numerical calculations with Eqs. (3.15a) and (3.15b) for such a quadrupole doublet, we shall again investigate a solution in the thin-lens approximation of Eqs. (3.26a)–(3.26d). For $\bar{M}_x = -0.2$, $\bar{l}_1 = 20$ mm, and $\bar{d} = 15$ mm, Eq. (3.27) yields $\bar{l}_2 \approx 13$ mm, so that for quadrupoles of lengths $w = 10$ mm we find $l_1 = 15$ mm, $d = 5$ mm, and $l_2 = 8$ mm. With $\bar{l}_1$, $\bar{d}$, and $\bar{l}_2$, Eqs. (3.26a) and (3.26b) yield $\bar{f}_1 \approx -10$ mm and $\bar{f}_2 \approx 8.3$ mm. For ions of 3-keV energy, the corresponding voltages on quadrupoles of diameter $2G_0 = 5$ mm are found from Eq. (3.13el) to be

$$V_{T1} \approx -187 \text{ V}, \qquad V_{T2} \approx 226 \text{ V}.$$

The maximum beam diameter in the second quadrupole is calculated from Eq. (3.28) to be $2R_x \approx 2a_{10}G_x \approx 4$ mm, so that quadrupoles with apertures of $2G_0 = 5$ mm and lengths of $w = 10$ mm should be adequate.

Keeping these voltages fixed, the exact Eqs. (3.14a) and (3.14b) yield $(x|a) \approx (y|b) \approx 0.02$, $M_x \approx -0.68$, and $M_y \approx -1.8$. By iterative numerical procedures, we find that changing the quadrupole voltages to $V_{T1} \approx -240$ V and $V_{T2} \approx 307$ V results in $(x|a) \approx (y|b) \approx 0$, $M_x \approx -0.2$, and $M_y \approx -3.2$. The geometry of this quadrupole doublet is shown in Fig. 3.9, together with some particle trajectories starting at the center of the object.

3.4.2.2   A Point-to-Parallel-to-Point Focusing Sextet with $M_x = M_y$

As another example, we assume that an electron beam of 60 MeV shall be transported through a narrow, round hole in a 3-m-thick concrete shielding. These particles of a magnetic rigidity of 0.2 Tm, calculated from Eq. (2.11f), must pass through an aperture of $x_{10} = y_{10} = \pm 5$ mm with maximum angles of $a_{10} = b_{10} = \pm 5$ mrad. This aperture, placed 1.5 m before the shielding wall, shall be imaged 1.5 m behind the shielding wall to a target of $2x_{10} = 2y_{10} = 5$ mm. For this purpose we could possibly use a point-to-parallel focusing quadrupole system in front of and a parallel-to-point focusing quadrupole system behind the shielding wall.

Since the hole in the shielding wall is round, quadrupole triplets should be preferred on both sides (see Section 3.3.4). For simplicity, let us choose two identical triplets each of which consists of three quadrupoles of $w = 0.1$ m lengths having apertures of $2G_0 = 0.04$ m. These quadrupoles may be separated by distances of 0.1 m so that each lens triplet has an overall length of 0.50 m. Placing the quadrupole triplets both 0.3 m away from the concrete wall, both l_1 and l_2 must equal 0.7 m so that the target and aperture are 1.5 m away from the shielding wall.

Before starting exact numerical calculations with Eqs. (3.15a) and (3.15b), we shall again investigate a solution in the thin-lens approximation of Eqs. (3.33a) and (3.33b). With the values of $\bar{l}_1 = l_1 + w/2 = (0.7 + 0.05)$ m and $\bar{d} = d + w = (0.1 + 0.1)$ m already fixed, Eq. (3.33a) yields $\bar{f}_x \approx \bar{f}_y \approx 0.95$ m, and Eq. (3.32b) yields $\bar{f}_1 \approx -0.487$ m, $\bar{f}_2 \approx 0.308$ m, and $\bar{f}_3 \approx -0.616$ m. According to Eq. (3.13ma), the above required focal lengths correspond to pole-tip flux densities of

$$B_{T1} \approx -0.082 \text{ T}, \qquad B_{T2} \approx 0.13 \text{ T}, \qquad B_{T3} \approx -0.065 \text{ T}.$$

The maximal beam cross section in this triplet occurs in the center quadrupole, with its magnitude found from Eq. (3.26e) to be $2R_x = 2a_{10}G_x$ so that quadrupoles with an aperture $2G_0 \approx 0.04$ m should be appropriate.

Using these three magnetic flux densities, Eqs. (3.15a) and (3.15b) yield for each quadrupole triplet $(a|a) \approx 0.38$, $(b|b) \approx 0.18$, $f_x \approx 1.6$, and $f_y \approx 1.13$. By iterative numerical procedures we find that changing the magnetic flux densities to $B_{T1} \approx -0.104$ T, $B_{T2} \approx 0.153$ T, and $B_{T3} \approx -0.0767$ T results in $(a|a) \approx (b|b) \approx 0$, $f_x \approx 0.994$, and $f_y \approx 0.876$ with the full beam still passing through the 0.04 m quadrupole apertures.

APPENDIX: POTENTIAL DISTRIBUTION BETWEEN HYPERBOLIC ELECTRODES

The cross section of electrodes or pole faces of a quadrupole, as indicated in the z planes of Figs. 3.20 and 3.21, can be described as

$$\pm G_{e0}^2 = x_0^2 - y_0^2, \tag{3.35a}$$

$$\pm G_{m0}^2 = 2x_0y_0. \tag{3.35b}$$

Mapping this z plane $(z = x + iy)$ conformally to a w plane $(w = u + iv)$ by the function $w = z^2$ (see also Figs. 3.20 and 3.21), we find

$$u = x^2 - y^2, \tag{3.36a}$$

$$v = 2xy. \tag{3.36b}$$

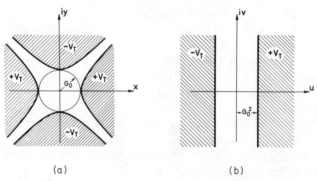

(a)   (b)

Fig. 3.20. Cross section of an electrostatic quadrupole lens (z plane) plus the result of a conformal mapping procedure $w = z^2$: (a) z plane; (b) w plane.

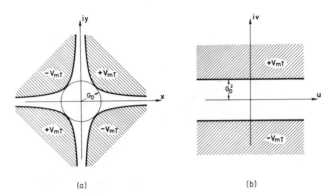

Fig. 3.21. Cross section of a magnetic quadrupole lens (z plane) plus the result of a conformal mapping procedure $w = z^2$: (a) z plane; (b) w plane.

Here, both u and v have the dimensions of the squares of a length. Each point of the hyperbolas of Eqs. (3.35a) and (3.35b) is thus transformed to

$$\pm G_{E0}^2 = u_0, \tag{3.37a}$$

$$\pm G_{B0}^2 = v_0. \tag{3.37b}$$

With V_{ET} and V_{BT} the electrostatic and magnetic scalar potentials at these electrodes, we find the potential distributions V_E and V_B in the region between the electrodes of the parallel plate condensors in the w planes of Figs. 3.20 and 3.21:

$$V_E(u, v) = (V_{ET}/G_{E0}^2)u, \tag{3.38a}$$

$$V_B(u, v) = (V_{BT}/G_{B0}^2)v, \tag{3.38b}$$

or with Eqs. (3.36a) and (3.36b) for quadrupole electrodes in the z planes of Figs. 3.20 and 3.21,

$$V_E(x, y) = (V_{ET}/G_{E0}^2)(x^2 - y^2), \tag{3.39a}$$

$$V_B(x, y) = (V_{BT}/G_{B0}^2)(2xy). \tag{3.39b}$$

Instead of using Cartesian coordinates x, y and u, v we could also have used polar coordinates $z = re^{i\theta}$ and $w = z^2 = r^2 e^{2i\theta}$ with $x = r\cos\theta$, $y = r\sin\theta$, $u = r^2 \cos 2\theta$, and $v = r^2 \sin 2\theta$, which transform Eqs. (3.38a) and (3.38b) to

$$V_E(r, \theta) = (V_{ET}/G_{E0}^2)r^2 \cos 2\theta \tag{3.40a}$$

$$V_B(r, \theta) = (V_{BT}/G_{B0}^2)r^2 \sin 2\theta. \tag{3.40b}$$

Note that according to Eqs. (3.40a) and (3.40b), the electrostatic or the scalar magnetic potential along a circle of radius r varies in a quadrupole as $\cos 2\theta$ or $\sin 2\theta$.

REFERENCES

Bennewitz, H. G., and Paul, W. (1954). *Z. Phys.* **139**, 489.
Courant, E. D., Livingston M. S., and Snyder, H. S. (1952). *Phys. Rev.* **88**, 1190.
Dayton, I. E., Shoemaker, F. C., and Mozley, R. F. (1954). *Rev. Sci. Instrum.* **25**, 485.
Denison, R. D. (1971). *J. Vac. Sci. Technol.* **8**, 266.
Kowalski, S. K., and Enge H. (1972). *Proc. Int. Conf. Magnet. Technol.* (*Brookhaven*). p. 181.
Melkirch, A. (1947). *Sitzungsber. Öster. Akad. Wiss. Wien* **155**, 393.

4

Sector Field Lenses

In magnetic and electrostatic sector fields (Fig. 4.1), particles of mass m, charge ze, and kinetic energy K or velocity v are deflected according to their magnetic rigidities $\chi_B = mv/(ze)$ and their electrostatic rigidities $\chi_E = mv^2/(ze)$, as described in Eqs. (2.10) and (2.13). Several such sector fields can be combined to beam-deflecting systems, which often employ additional quadrupole lenses. Depending on their application, such systems are called *particle separators*, *particle spectrometers*, *particle spectrographs*, or simply, *beam guidance systems*.

Particle separators are systems of nonvarying fields that separate particles according to their rigidities. A separator normally has one exit slit through which particles of one specific rigidity can pass. In some cases, however, several exit slits are used so that particles of specified rigidities can be collected simultaneously.

Particle spectrometers or particle spectrographs are arrangements of magnetic or electrostatic sector fields that record the rigidity distribution in a beam of particles. In a spectrograph, constant fields are employed with all particles within a limited range of rigidities being recorded by some

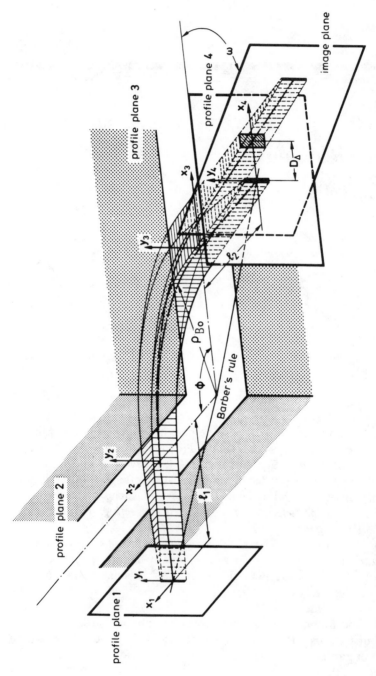

Fig. 4.1. The separation of particles of two rigidities in a homogeneous magnetic sector field. Note that there are no focusing forces in the y direction and that the focusing forces in the x direction cause the object (in profile plane 1), the image (in profile plane 4), and the center of curvature of the optic axis to all lie on a straight line (Barber, 1953). Particles of different rigidities are not all focused in profile plane 4 but in an inclined image plane. This is due to an image aberration of second order, as outlined in Chapter 8 [see also Eq. (8.13a)].

position-sensitive detector, for instance, a photographic plate or a multiwire avalanche counter. In a spectrometer, the strengths of the deflecting fields are varied with time so that the rigidity distribution of a particle beam is obtained as the time-dependent particle intensity behind one exit slit.

4.1 HOMOGENEOUS MAGNETIC SECTOR FIELDS

The most common magnetic sector field magnet uses parallel pole faces that create a constant flux density $B_y = B_0 \neq 0$, $B_x = B_z = 0$. In such a field, all charged particles move along trajectories for which the projections on the plane of symmetry ($y = 0$) are circles, as described in Eq. (2.10). Such a field exhibits focusing properties in the direction of x but not in the direction of y (Fig. 4.1).

4.1.1 Particle Trajectories in Homogeneous Magnetic Sector Fields

A reference particle of momentum $p_0 = m_{00}v_{00}$ and charge (z_0e) moves in a flux density B_0 on a circle of radius ρ_0, according to Eq. (2.9). This trajectory we call the circular *optic axis*. In the u, v coordinate system of Fig. 4.2, this optic axis is given as $u^2 + v^2 = \rho_0^2$. The trajectories of a beam of all other particles are then described relative to this optic axis (see, for instance, Ewald and Hintenberger, 1953).

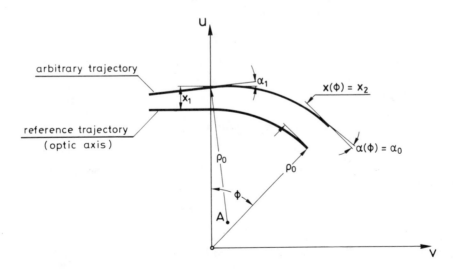

Fig. 4.2. The circular optic axis and an arbitrary trajectory of equal radius.

4.1.1.1   Trajectories in the Plane of Symmetry (y = 0) for Particles of Equal Magnetic Rigidities

We assume a reference particle of the magnetic rigidity $\chi_0 = m_{00}v_0/(z_0e) = B_0\rho_0$, which does not move along the optic axis but along a paraxial trajectory of equal radius of curvature. Assuming that in a u, v-coordinate system, the center of this circle is at a point $A(u, v) = A[x_1 + \rho_0(1 - \cos \alpha_1), \rho_0 \sin \alpha_1]$ in Fig. 4.2, we find the coordinates of a point on this trajectory to be

$$[u - x_1 - \rho_0(1 - \cos \alpha_1)]^2 + [v - \rho_0 \sin \alpha_1]^2 = \rho_0^2. \tag{4.1a}$$

Here, x_1 and α_1 denote the lateral deviation and the angle of inclination of the paraxial trajectory under consideration relative to the optic axis at $u = 0$. With $x_1 \ll \rho_0$ and $\alpha_1 \ll 1$, all terms higher than first order in x_1/ρ_0 and α_1 may be dropped in a power series expansion of Eq. (4.1a), yielding

$$u^2 - 2x_1u + v^2 - 2\rho_0v\alpha_1 \approx \rho_0^2. \tag{4.1b}$$

Since ultimately we need to know the distance x between the paraxial trajectory and the circular optic axis for any angle of deflection ϕ, we shall express the coordinates u and v of a point of the paraxial trajectory as

$$u = [\rho_0 + x(\phi)] \cos \phi, \tag{4.2a}$$

$$v = [\rho_0 + x(\phi)] \sin \phi. \tag{4.2b}$$

Thus, $u^2 + v^2$ equals $[\rho_0 + x(\phi)]^2$, or neglecting $x^2(\phi)$ compared to $2x(\phi)\rho_0\alpha_1$,

$$u^2 + v^2 - \rho_0^2 \approx 2x(\phi)\rho_0\alpha_1. \tag{4.3}$$

Substituting $u^2 + v^2 \approx \rho_0^2$ from Eq. (4.3) and $u \approx \rho_0 \cos \phi$, $v \approx \rho_0 \sin \phi$ from Eqs. (4.2) into Eq. (4.1b) we find

$$x(\phi) = x_1 \cos \phi + \alpha_1\rho_0 \sin \phi + \cdots, \tag{4.4a}$$

$$\tan \alpha(\phi) = -(x_1/\rho_0) \sin \phi + \alpha_1 \cos \phi + \cdots, \tag{4.4b}$$

with $\tan \alpha(\phi) = dx(\phi)/d(\rho_0\phi)$.

Equation (4.4) may also be written in the form of a transfer matrix as $X_2 = T_xX_1$, or

$$\begin{pmatrix} x_2 \\ a_2 \end{pmatrix} = \begin{pmatrix} \cos \phi_0 & \rho_0 \sin \phi_0 \\ -\rho_0^{-1} \sin \phi_0 & \cos \phi_0 \end{pmatrix} \begin{pmatrix} x_1 \\ a_1 \end{pmatrix}. \tag{4.5}$$

Similar to Eq. (1.14b), $\tan \alpha$ is replaced here by a, with [Eq. (2.7a)]

$$a = \sin \alpha/\sqrt{1 + \tan^2 \beta \cos^2 \alpha}.$$

The transfer matrix of Eq. (4.5) may be interpreted as relating the position and inclination of a paraxial trajectory in profile plane 1 [$X_1 = (x_1, a_1 \approx \tan \alpha_1)$ at $\phi = 0$] and in profile plane 2 [$X_2 = (x_2, a_2 \approx \tan \alpha_2)$ at $\phi = \phi_0$]. For this case, both profile planes are assumed to be positioned inside the magnetic field. Note that $\sin \phi_0$ vanishes for $\phi_0 = 180°$ (Claasen, 1908; Dempster, 1918), causing an object–image relation between the profile planes 1 and 2, as described in Section 1.3.2.1.

4.1.1.2 Trajectories in the Plane of Symmetry (y = 0) for Particles of Different Magnetic Rigidities

A beam normally contains particles of different magnetic rigidities $\chi_0 = p_0/(z_0 e)$ and $\chi = p/(ze) = \chi_0(1 + \Delta)$ [Eq. (2.20b)]. In a constant magnetic flux density B_0, these particles move along trajectories of different radii of curvature ρ_0 and ρ with

$$\rho = \rho_0(1 + \Delta).$$

Thus, Eq. (4.1a) reads $(u - x_1 - \rho_0 + \rho \cos \alpha_1)^2 + (v - \rho \sin \alpha_1)^2 = \rho^2$, or keeping only terms of first order in x_1, a_1, Δ,

$$u^2 - 2x_1 u + v^2 - 2a_1\rho_0(1 + \Delta)v \approx \rho_0^2(1 + 2\Delta). \tag{4.6}$$

Again using $u^2 + v^2 - \rho_0^2 \approx 2x\rho_0 a_1$ and $u \approx \rho_0 \cos \phi, v \approx \rho_0 \sin \phi$ [Eqs. (4.2) and (4.3)], Eq. (4.6) transforms to

$$x(\phi) = x_1 \cos \phi + \rho_0 a_1 \sin \phi + \rho_0 \Delta(1 - \cos \phi) + \cdots, \tag{4.7a}$$

$$a(\phi) = -(x_1/\rho_0) \sin \phi + a_1 \cos \phi + \Delta \sin \phi + \cdots, \tag{4.7b}$$

where again $a(\phi) \approx dx(\phi)/d(\rho_0\phi) \ll 1$. Since the magnetic field does not change the momentum or the charge of a particle, Δ stays constant while the particle moves from profile plane 1 to profile 2. Thus, Eqs. (4.7a) and (4.7b) may also be written as a 3×3 transfer matrix (Penner, 1961):

$$\begin{pmatrix} x_2 \\ a_2 \\ \Delta \end{pmatrix} = \begin{pmatrix} (x|x) & (x|a) & (x|\Delta) \\ (a|x) & (a|a) & (a|\Delta) \\ 0 & 0 & 1 \end{pmatrix} \begin{pmatrix} x_1 \\ a_1 \\ \Delta \end{pmatrix} = \begin{pmatrix} c_x & s_x & d_x \\ -s_x/\rho_0^2 & c_x & s_x/\rho_0 \\ 0 & 0 & 1 \end{pmatrix} \begin{pmatrix} x_1 \\ a_1 \\ \Delta \end{pmatrix},$$
$$\tag{4.8a}$$

$$c_x = \cos \phi_0, \qquad s_x = \rho_0 \sin \phi_0, \qquad d_x = \rho_0(1 - \cos \phi_0).$$

As in Eq. (4.5), the quantities x_1, $\alpha_1 \approx a_1$ and x_2, $\alpha_2 \approx a_2$ at $\phi = 0$ and $\phi = \phi_0$ denote the position and angle of inclination of a particle trajectory relative to the circular optic axis at profile planes 1 and 2 in the magnet. These profile planes could be positioned at the effective entrance and exit field boundaries of an idealized magnet with the flux densities rising at these effective boundaries like step functions (see also Section 7).

4.1.1.3 Particle Trajectories in the Cylinder Surface (x = 0)

Thus far, trajectories have been determined for particles that move perpendicularly to the magnetic flux density **B**. If particles move obliquely to **B**, their velocities **v** must be split into components $v_\parallel = |v| \cos \beta$ and $v_\perp = |v| \sin \beta$, respectively, parallel and perpendicular to the plane of symmetry $y = 0$ in the magnet air gap (Fig. 4.1). Here, b is defined as [Eq. (2.7b)],

$$b = \sin \beta / \sqrt{1 + \tan^2 \alpha \cos^2 \beta}.$$

From the Lorentz equation [Eqs. (2.9a) and (2.9b)] we see that the velocity component $v_\perp$ cannot be modified by **B** and that the velocity component $v_\parallel$ is changed in direction but not in magnitude. The particle under consideration thus moves along a helix, the projection of which on the said plane of symmetry coincides with the trajectory of a particle of equal mass and charge moving in this plane of symmetry with a velocity $\bar{v} = v_\parallel$. Altogether, the formulas of Sections 4.1.1.1 and 4.1.1.2 describe the projections of particle trajectories on the plane $y = 0$ if ρ_0 is obtained from $p_\parallel = m v_\parallel$ and not from $p = mv$. In the case of a homogeneous field, there are no forces acting perpendicular to the plane ($y = 0$) on the particles. Consequently, the particle motion in the cylinder surface ($x = 0$) is described by the drift-space transfer matrix of Eq. (1.3b). Exchanging that length l by the length of the optic axis in the magnetic sector $\rho_0 \phi_0$, we find $Y_2 = T_y Y_1$, or

$$\begin{pmatrix} y_2 \\ b_2 \end{pmatrix} = \begin{pmatrix} (y|y) & (y|b) \\ (b|y) & (b|b) \end{pmatrix} \begin{pmatrix} y_1 \\ b_1 \end{pmatrix} = \begin{pmatrix} 1 & \rho_0 \phi_0 \\ 0 & 1 \end{pmatrix} \begin{pmatrix} y_1 \\ b_1 \end{pmatrix}, \qquad (4.8b)$$

where $Y_1 = (y_1, b_1 \approx \tan \beta_1)$ and $Y_2 = (y_2, b_2 \approx \tan \beta_2)$ characterize an arbitrary trajectory at $\phi = 0$ and $\phi = \phi_0$.

4.1.2 Focusing and Dispersing Properties of Homogeneous Magnetic Sector Fields

Assume a magnetic sector field to be limited by entrance and exit boundaries at which the magnetic flux density rises discontinuously from zero to the full flux density B_0. Although such a distribution cannot be described by Maxwell's equations, and thus is an unrealistic assumption, such a discontinuous change from $B = 0$ to $B = B_0$ describes most first-order effects correctly if an effective field boundary is determined as outlined in Section 7.1.1. Assume further (as indicated in Fig. 4.1) that a source of particles is placed a distance l_1 upstream from the entrance-effective field boundary in a profile plane 1 from which particles of magnetic rigidity $\chi = \chi_0(1 + \Delta)$ emerge at distances x_1, y_1 from the optic axis at angles of inclination $\alpha_1 \approx \tan \alpha_1 \approx a_1, \beta_1 \approx \tan \beta_1 \approx b_1$. In a profile plane 4 a distance l_2 downstream from the exit-effective field boundary, the particle position (x_2, y_2) and inclination $(\tan \alpha_2 \approx a_2, \tan \beta_2 \approx b_2)$ can be found from the product of the transfer matrices of the object distance (l_1), the sector field (ρ_0, ϕ_0), and the image distance (l_2). Contrary to Eq. (1.3a), the action of a drift distance l here should be expressed by a 3×3 matrix connecting the variables x, a, Δ at the two corresponding profile planes 1 and 2 as

$$
\begin{pmatrix} x_2 \\ a_2 \\ \Delta \end{pmatrix} = \begin{pmatrix} 1 & l & 0 \\ 0 & 1 & 0 \\ 0 & 0 & 1 \end{pmatrix} \begin{pmatrix} x_1 \\ a_1 \\ \Delta \end{pmatrix}, \tag{4.9}
$$

since a_2 and Δ are independent of l in a drift space.

The transfer matrices that transform particle trajectories from profile plane 1 to profile plane 4 in Fig. 4.3 are thus obtained from Eqs. (4.8a), (4.8b), and (4.9) as

$$
\begin{pmatrix} (x|x) & (x|a) & (x|\Delta) \\ (a|x) & (a|a) & (a|\Delta) \\ 0 & 0 & 0 \end{pmatrix}
$$

$$
= \begin{pmatrix} 1 & l_2 & 0 \\ 0 & 1 & 0 \\ 0 & 0 & 1 \end{pmatrix} \begin{pmatrix} c_x & s_x & d_x \\ -s_x/\rho_0^2 & c_x & s_x/\rho_0 \\ 0 & 0 & 1 \end{pmatrix} \begin{pmatrix} 1 & l_1 & 0 \\ 0 & 1 & 0 \\ 0 & 0 & 1 \end{pmatrix}, \tag{4.10a}
$$

$$
\begin{pmatrix} (y|y) & (y|b) \\ (b|y) & (b|b) \end{pmatrix} = \begin{pmatrix} 1 & l_2 \\ 0 & 1 \end{pmatrix} \begin{pmatrix} 1 & \rho_0 \phi_0 \\ 0 & 1 \end{pmatrix} \begin{pmatrix} 1 & l_1 \\ 0 & 1 \end{pmatrix}, \tag{4.10b}
$$

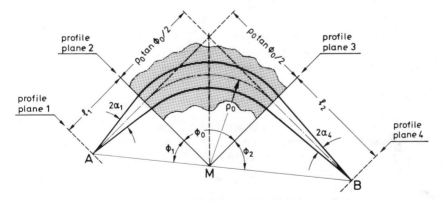

Fig. 4.3. The focusing of charged particles by a homogeneous magnetic sector field. Note that the points A, B, and M lie on a straight line (Barber, 1953).

again with $c_x = \cos \phi_0$, $s_x = \rho_0 \sin \phi_0$, $d_x = \rho_0(1 - \cos \phi_0)$. Equation (4.10a) describes the focusing and dispersing properties of a homogeneous magnetic sector field in the x direction, whereas Eq. (4.10b) describes the corresponding properties in y direction. Note that Eq. (4.10b) is nothing more than the transfer matrix of a drift distance of length $l_1 + \rho_0\phi_0 + l_2$, the path length of the optic axis between the profile planes 1 and 4 in Fig. 4.3.

4.1.2.1 Lens Properties of a Homogeneous Magnetic Sector Field

The focusing power $1/f_x$ of a homogeneous magnetic sector field in the x direction can be found from Eq. (4.10a) and the relation $1/f_x = -(a|x)$, as known from Eq. (1.24):

$$-\rho_0/f_x = \sin \phi_0. \tag{4.11}$$

The positions l_{f1} and l_{f2} of the focal planes 1 and 2 are both found relative to the entrance and exit boundaries of the sector field (Fig. 4.3) by setting both $(x|x)$ and $(a|a)$ in Eq. (4.10a) to zero (Sections 1.3.2.3 and 1.3.2.4):

$$l_f/\rho_0 = l_{f1}/\rho_0 = l_{f2}/\rho_0 = \rho_0(c_x/s_x) = \cotan \phi_0. \tag{4.12}$$

The principal planes of a lens both must be a distance f_x apart from the focal planes. Thus, their distances from the corresponding effective field boundaries are

$$l_p/\rho_0 = l_{p1}/\rho_0 = l_{p2}/\rho_0 = l_f - f_x/\rho_0 = (\cos \phi_0 - 1)/\sin \phi_0, \tag{4.13a}$$

or using simple trigonometric relations,

$$l_p/\rho_0 = l_{p1}/\rho_0 = l_{p2}/\rho_0 = -\tan (\phi_0/2). \tag{4.13b}$$

From Fig. 4.3 and Eqs. (4.11) and (4.13b), it is evident that the x-focusing action of a sector field can be approximated by a thin lens of focal length $f_x = \rho_0/\sin \phi_0$ placed at the center of the sector field, since the distance $\rho_0 \tan (\phi_0/2)$ between the entrance boundary and the principal plane is only slightly larger than the length $\rho_0\phi_0/2$ of the optic axis taken from the boundary of the sector field to its center line.

4.1.2.2 Relations between Object and Image Distances

If an object–image relation exists between the profile planes 1 and 4 in Fig. 4.3, the matrix element $(x|a)$ must vanish in Eq. (4.10a) (Section 1.3.2.1). This implies a lateral x magnification

$$(x|x) = M_x$$

and a corresponding angular magnification

$$(a|a) = 1/M_x.$$

Substituting $(x|x)$ and $(a|a)$ from Eq. (4.10a) into these relations we find the relative object and image distances for a given magnification M_x

$$\frac{l_2}{\rho_0} = \frac{\cos \phi_0 - M_x}{\sin \phi_0} \tag{4.14a}$$

$$\frac{l_1}{\rho_0} = \frac{\cos \phi_0 - (1/M_x)}{\sin \phi_0}. \tag{4.14b}$$

With $M_x = -1$, Eqs. (4.14) simplify to

$$l_1/\rho_0 = l_2/\rho_0 = \cotan(\phi_0/2). \tag{4.14c}$$

To give some idea of all the possible solutions for a focusing system with a lateral magnification M_x, the magnitudes of l_1/ρ_0 or l_2/ρ_0, as calculated from Eqs. (4.14a) and (4.14b) are plotted in Fig. 4.4.

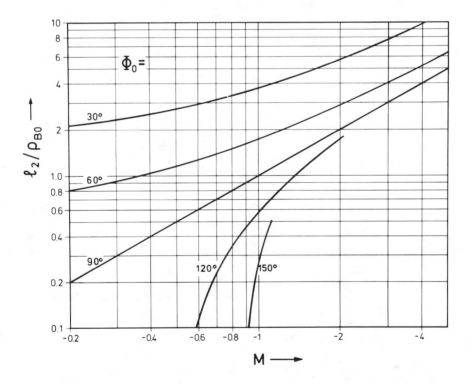

Fig. 4.4. The relative image distance l_2/ρ_{B0} as calculated from Eq. (4.14a) is plotted as a function of the x magnification $M = M_x$ and the angle of deflection ϕ_0. Because of the symmetry between Eqs. (4.14a) and (4.14b), this plot can also be interpreted as describing the relative object distance l_1/ρ_{B0} as calculated from Eq. (4.14b) where, however, M must be understood to describe $1/M_x$.

There is a very simple geometric construction for finding the image distance for a homogeneous sector field from a given object distance, as was demonstrated by Barber (1953). It states that the object A, the center of curvature M and the image B lie on a straight line, as indicated in Figs. 4.1 and 4.3. To prove the accuracy of this relation, we may look at the focusing condition $(x|a) = 0$ in Eq. (4.10a), which may be written as:

$$\frac{l_1}{\rho_0} \cot\phi_0 + 1 = \frac{l_2}{\rho_0}\left(\frac{l_1}{\rho_0} - \cot\phi_0\right).$$

On the other hand, Fig. 4.3 tells us that $\phi_0 + \phi_1 + \phi_2 = \pi$, which transforms to

$$\cot\phi_2 = \cot(\phi_0 + \phi_1) = \frac{1 - \cot\phi_0\cot\phi_1}{\cot\phi_0 + \cot\phi_1}$$

by simple trigonometry. Because of $\cot\phi_1 = \rho_0/l_1$ and $\cot\phi_2 = \rho_0/l_2$ (Fig. 4.3), the two relations are indeed identical.

4.1.2.3 The x-Transfer Matrix for a Homogeneous Magnetic Sector Field in Special Cases

The transfer matrix between the first and the second focal planes of a homogeneous magnetic sector field is found by replacing l_1 and l_2 in Eq. (4.10a) by l_{f_1} and l_{f_2} of Eq. (4.12):

$$\begin{pmatrix}(x|x) & (x|a) & (x|\Delta) \\ (a|x) & (a|a) & (a|\Delta) \\ 0 & 0 & 1\end{pmatrix} = \begin{pmatrix} 0 & \rho_0^2 s_x & \rho_0 \\ -s_x/\rho_0^2 & 0 & s_x/\rho_0 \\ 0 & 0 & 1\end{pmatrix}$$

$$= \begin{pmatrix} 0 & \rho_0\sin\phi_0 & \rho_0 \\ -\rho_0^{-1}\sin\phi_0 & 0 & \sin\phi_0 \\ 0 & 0 & 1\end{pmatrix}. \quad (4.15)$$

Note that $(x|a)$ equals $-1/(a|x) = f_x$ and that $(x|\Delta) = \rho_0$ is independent of ϕ_0.

The transfer matrix between the first and the second principal plane is found by replacing l_1 and l_2 in Eq. (4.10a) by l_{p1} and l_{p2} of Eq. (4.13a):

$$\begin{pmatrix}(x|x) & (x|a) & (x|\Delta) \\ (a|x) & (a|a) & (a|\Delta) \\ 0 & 0 & 1\end{pmatrix} = \begin{pmatrix} 1 & 0 & 0 \\ -s_x/\rho_0^2 & 1 & s_x/\rho_0 \\ 0 & 0 & 1\end{pmatrix}$$

$$= \begin{pmatrix} 1 & 0 & 0 \\ -\rho_0^{-1}\sin\phi_0 & 1 & \sin\phi_0 \\ 0 & 0 & 1\end{pmatrix}. \quad (4.16)$$

Note that both coefficients $(x|a)$ and $(x|\Delta)$ vanish in Eq. (4.16).

The transfer matrix which connects profile planes between which an object-image relation exists $[(x|a) = 0]$, is found by introducing the corresponding l_1 and l_2 from Eqs. (4.14a) and (4.14b) into the matrix elements of Eq. (4.10a):

$$\begin{pmatrix} (x|x) & (x|a) & (x|\Delta) \\ (a|x) & (a|a) & (a|\Delta) \\ 0 & 0 & 1 \end{pmatrix} = \begin{pmatrix} M_s & 0 & \rho_0(1 - M_x) \\ -s_x/\rho_0^2 & 1/M_x & s_x/\rho_0 \\ 0 & 0 & 1 \end{pmatrix}$$

$$= \begin{pmatrix} M_x & 0 & \rho_0(1 - M_x) \\ -\rho_0^{-1}\sin\phi_0 & 1/M_x & \sin\phi_0 \\ 0 & 0 & 1 \end{pmatrix}. \quad (4.17)$$

Note that $(x|x) = 1/(a|a)$, as well as $(x|\Delta)$, are independent of ϕ_0. Note also that all transfer matrices of Eqs. (4.15)-(4.17) are very simple. They are therefore preferable to the transfer matrix of Eq. (4.10a), whenever feasible.

4.1.2.4 Dispersion, Resolution, and Resolving Power of a Magnetic Sector Field

A particle of the magnetic rigidity $\chi = \chi_0(1 + \Delta)$ is separated from the optic axis in the image plane of a spectrometer by a distance $(x|\Delta)\Delta$ if we assume that this particle originated from the point $x_1 = y_1 = 0$, i.e., from the middle of the object of area $2x_{10}2y_{10}$. For $\Delta = 1$, this distance $(x|\Delta)\Delta$ is called the *rigidity dispersion*, defined by

$$D_\Delta = (x|\Delta). \quad (4.18a)$$

In order to separate particles of different magnetic rigidities χ and χ_0 from each other (Fig. 4.5), the distance $\Delta(x|\Delta)$ must be larger than the width of the image $2x_{10}(x|x)$. Thus, the minimum resolvable Δ is known as the *rigidity resolution*, defined by

$$\Delta_{min} = \frac{(x|x)}{(x|\Delta)} 2x_{10}. \quad (4.18b)$$

The reciprocal of this smallest resolvable Δ is also known as the *rigidity resolving power*, defined by

$$R_\Delta = -\frac{1}{\Delta_{min}} = -\frac{(x|\Delta)}{(x|x)2x_{10}}. \quad (4.18c)$$

Equations (4.18a)-(4.18c) apply to any imaging system that includes dispersive elements. In the case of a focusing single sector field separator, we find

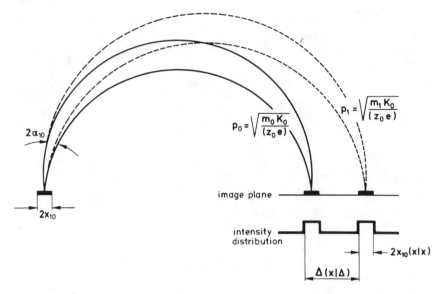

Fig. 4.5. The focusing and dispersion of a homogeneous 180° sector magnet is illustrated. Note especially the particle intensity recorded in the image plane across two ideal images of the entrance slit formed by particles of momenta p_0 and p_1 with $\Delta = (p_1 - p_0)/p_0$.

$(x|x) = M_x$ and $(x|\Delta) = \rho_0(1 - M_x)$, according to Eq. (4.17). As shown, for instance, by Wollnik (1980), Eqs. (4.18) yield

$$D_\Delta = \rho_0(1 - M_x), \tag{4.19a}$$

$$R_\Delta = (\rho_0/2x_{10})[1 - (1/M_x)]. \tag{4.19b}$$

Comparing Eqs. (4.19a) and (4.19b), we find very generally [for similar results, see also Ewald and Hintenberger (1953) or Chavet (1967)],

$$R_\Delta 2x_{10} = \bar{D}_\Delta, \tag{4.19c}$$

where $\bar{D}_\Delta = \rho_0[1 - (1/M_x)]$ is the matrix element $(x|\Delta)$ of the reversed system for which object and image are exchanged, so that M_x is replaced by $1/M_x$.

Note that none of Eqs. (4.19a)–(4.19c) depend explicitly on the angle of deflection ϕ_0 of the sector field. A sector field with a large angle of deflection thus yields the same resolving power as a sector field with a small angle of deflection as long as the lateral magnification M_x and the quotient x_{10}/ρ_0 both remain constant. As already pointed out by Brück (1960), the ϕ_0 independency of Δ_{min} or R_Δ for constant M_x relates to the fact that the area a beam of monoenergetic particles occupies in the magnetic sector field is independent of ϕ_0. As ϕ_0 decreases, the object and image distances

l_1 and l_2 in Eqs. (4.14a) and (4.14b) increase so that the width of the beam grows in x direction, whereas the path length in the sector field decreases (Fig. 4.3).

For a particle spectrometer, not only an entrance slit of width $2x_{10}$ is used but also an exit slit, the width of which is normally chosen equal to the width of the image $(x|x)2x_{10}$ of the entrance slit. Changing the magnetic flux density slightly sweeps the image of the entrance slit across the exit slit. The recordable intensity distribution behind the exit slit is then triangular, whereas since the full beam fits through the exit slit only at one instant, at all other times its jaws intersect a portion of the beam. One often requires that at least for a certain period the full beam intensity must be recorded behind the exit slit. Thus, the exit slit is normally chosen a little wider than $(x|x)2x_{10}$. For the case shown in Fig. 4.6, the width of the exit slit is 1.2 times $(x|x)2x_{10}$.

4.1.2.5 Momentum, Energy, and Mass Dispersion of a Magnetic Separator or Spectrometer

The magnetic rigidity χ_0 of a reference particle of rest mass m_{00}, charge $(z_0 e)$, and energy K_0 and the magnetic rigidity $\chi = \chi_0(1 + \Delta)$ of an arbitrary particle of rest mass m_0, charge (ze), and energy K were given earlier in Eqs. (2.17b) and (2.18b) as

$$\chi_0 = \frac{m_{00}}{(z_0 e)} \sqrt{\frac{K_0}{m_{00}} 2(1 + \eta_0)}, \qquad (4.20a)$$

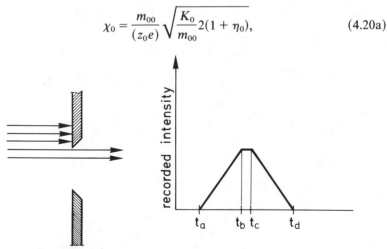

Fig. 4.6. The intensity recorded behind an exit slit during a sweep of a particle beam across this slit. Here it is assumed that the width of the exit slit is 1.2 times larger than the beam width $2x_{00}(x|x)$. At the time t_a the right edge of the beam reaches the left jaw of the exit slit. At the time t_d the left edge of the beam reaches the right jaw of the exit slit. The times t_b and t_c are the first and last instants at which the full beam passes through the exit slit.

$$\chi = \chi_0(1 + \Delta) = \chi_0 \sqrt{(1 + \delta_K) \left[1 + \frac{\eta_0 \delta_K + \delta_m}{1 + \eta_0} \right]}. \qquad (4.20b)$$

with [Eqs. (2.15) and (2.16b)];

$$\frac{K}{(ze)} = \frac{K_0}{(z_0 e)}(1 + \delta_K)_{m = \text{const}}, \qquad \frac{m_0}{(ze)} = \frac{m_{00}}{(z_0 e)}(1 + \delta_m)_{K = \text{const}},$$

and $\eta_0 = K_0 / 2m_{00}c^2$. The numerical value of η_0 is found in tesla meters (Tm) from Eq. (2.12) as $(0.144/z_0)\sqrt{\bar{m}_{00}\bar{K}_0(1 + \bar{\eta}_0)}$, with $\bar{\eta}_0 = \bar{K}_0/1863\bar{m}_{00}$, where $\bar{K}_0$ is K_0 in mega-electron-volts, and $\bar{m}_{00}$ is m_{00} in mass units (u).

In many cases, it is advantageous to use the relative energy over charge and mass over charge deviations δ_K and δ_m instead of the relative rigidity deviations Δ. Thus, a four-element position vector $(x_0, a_0, \delta_K, \delta_m)$ may be useful instead of (x_0, a_0, Δ). The x-transfer matrix that corresponds to Eq. (4.10a) then becomes

$$\begin{pmatrix} x_2 \\ a_2 \\ \delta_K \\ \delta_m \end{pmatrix} = \begin{pmatrix} (x|x) & (x|a) & (x|\delta_K) & (x|\delta_m) \\ (a|x) & (a|a) & (a|\delta_K) & (a|\delta_m) \\ 0 & 0 & 1 & 0 \\ 0 & 0 & 0 & 1 \end{pmatrix} \begin{pmatrix} x_1 \\ a_1 \\ \delta_K \\ \delta_m \end{pmatrix}. \qquad (4.21)$$

Since Eq. (4.20) yields $\Delta = [\delta_K(1 + 2\eta_0) + \delta_m]/2(1 + \eta_0)$ for δ_K, $\delta_m \ll 1$, the coefficients $(x|\delta_K)$, $(a|\delta_K)$, $(x|\delta_m)$, and $(a|\delta_m)$ are found as

$$\frac{(x|\delta_K)}{1 + 2\eta_0} = (x|\delta_m) = \frac{(x|\Delta)}{2(1 + \eta_0)}, \qquad \frac{(a|\delta_K)}{1 + 2\eta_0} = (a|\delta_m) = \frac{(a|\Delta)}{2(1 + \eta_0)}. \qquad (4.22)$$

Introducing these relations into Eqs. (4.18a) and (4.18c), we find

$$D_K/(1 + 2\eta_0) = D_m = D_\Delta/2(1 + \eta_0), \qquad (4.23a)$$

$$R_K/(1 + 2\eta_0) = R_m = R_\Delta/2(1 + \eta_0). \qquad (4.23b)$$

For relativistically slow particles $[x \ll c$ or $(K_0/m_{00}) \ll 931$ MeV/u$]$ or $\eta_0 \ll 1$, Eq. (4.20b) simplifies to

$$\chi = \chi_0(1 + \Delta) = \chi_0[1 + \tfrac{1}{2}(\delta_K + \delta_m) + \cdots,$$

as was derived before in Eq. (2.20bs). This yields

$$(x|\delta_K) = (x|\delta_m) = \tfrac{1}{2}(x|\Delta), \qquad (a|\delta_K) = (a|\delta_m) = \tfrac{1}{2}(a|\Delta).$$

At first glance, it may not appear useful to employ transfer matrices for which two columns and rows do not contain more information than the one before. The advantage of this method will become evident, however, when calculations must be performed for instruments that use both magnetic and electrostatic sector fields or in which flight times are of interest.

4.1.2.6 Example of a Homogeneous Magnetic Sector Field Spectrometer

Assume that a single homogeneous sector field separates singly charged molecular ions of kinetic energy $K_0 = 3000$ eV, with masses $m_{00} = 2000$ u and $m_{00}(1 + \delta_m) = 2001$ u. This would require a mass resolving power R_m of at least 2000. The factor $\bar{\eta}_0$ in this case equals $0.003/1.863 \times 2000 \; 10^{-9}$ characterizing relativistically slow particles. The magnetic rigidity of the specified molecular ions is found from Eq. (2.8s) as $\chi_0 \approx 0.144\sqrt{2000} \times 0.003 \approx 0.35$ Tm. For a maximum magnetic flux density of 1.4 T, we must thus choose a radius of deflection of the optic axis of $\rho_0 > 0.25 \; m$. With $\eta_0 = 0$, we calculate the achievable mass resolving power from Eqs. (4.19b) and (4.23b) as $R_m = (\rho_0/4x_{00})(1 - 1/M_x)$. For the case of a symmetric system with equal object and image distances ($l_1 = l_2$), Eqs. (4.14a) and (4.14b) yield $M_x = -1$ so that R_m and R_K both equal $\rho_0/2x_{00}$. For $\rho_0 = 0.25$ m, we thus obtain the required mass resolving power $R_m = 2000$ for an entrance slit of width $2x_{00} = 125 \; \mu$m. For ions that leave the source with an energy spread of $\Delta K = \pm 0.3$ V, the ion energy is $K_0(1 + \delta_K) = 3000(1 \pm 0.0001)$ eV. Due to the energy dispersion $D_K = \rho_0(1 - M_x)/2 \approx 0.25$ m of the magnetic sector field, the size of the image is 25 μm for an object of no width at all. In order to obtain the required mass resolving power, it is thus necessary to decrease the width of the entrance slit from $2x_{00} \approx 125 \; \mu$m to $2x_{00} \approx 100 \; \mu$m so that $|2x_{00}M_x| + |2\delta_K D_K|$ stays smaller than 125 μm.

The angle of deflection ϕ_0 is open to choice. For $\phi_0 = 180°$, the Eqs. (4.14a) and (4.14b) postulate $l_1 = l_2 = 0$. This was the design (Fig. 4.5) used by Dempster (1918) in his mass spectrometer and by Lawrence in the large-scale separators, the calutrons (Smythe, 1945). A 180° sector field, however, requires that both the particle source and the particle collector be placed in the deflecting field. Conversely, it is often required that the magnetic field of the ion source be adjusted independently of the deflecting magnet and that there be no strong magnetic field at the collector, so that all types of ion sources as well as electron multiplier detectors can be used. Choosing for such a system $\phi_0 = 60°$ and $l_1 = l_2$ so that M_x equals -1, the object and image distances are found from Eqs. (4.14) as $l_1 = l_2 \approx (0.5 + 1)\rho_0/0.866 \approx 0.43$ m for $\rho_0 = 0.25$ m.

4.1.3 Oblique Entrance and Exit of a Particle Beam into a Sector Magnet

For some applications it is advantageous if the particle beam obliquely crosses the entrance or exit field boundaries of a sector field. This modifies the focusing action of the sector field both in the x and y directions.

4.1.3.1 Particle Trajectories in the Plane of Symmetry (y = 0)

In the case of a bundle of parallel particle trajectories entering a deflecting magnet obliquely ($\varepsilon' > 0$), as shown in Fig. 4.7, the individual trajectories reach the deflecting field earlier or later, depending on their x coordinates. Consequently, all trajectories entering the sector field at positive x values are less deflected toward the optic axis. Trajectories characterized by negative x values enter the deflecting field somewhat earlier and thus experience a predeflection away from the optic axis. Both these effects also can be described by assuming that the beam enters and exits a sector field perpendicularly but that wedge-shaped field sections of a positive and a negative direction of deflection are placed at these boundaries, as shown in Fig. 4.7.

A particle trajectory that is initially parallel to the optic axis will.be deflected by the first wedge-field region by an angle $\Delta\alpha$. With $\dot{x} = dx/dt$, this $\Delta\alpha$ is found from $\dot{x}/v = \sin\Delta\alpha$, or explicitly:

$$\Delta\alpha \approx \frac{\dot{x}}{v} = \frac{1}{v}\int_{t_0}^{t_1} x\, dt \approx \frac{1}{\rho_0}\int_{-x\tan\varepsilon'}^{0} dz = \frac{x}{\rho_0}\tan\varepsilon',$$

where $dt = dz/v$. Here, t_0 and t_1 are the times at which the particle enters and exits the wedge-shaped field sections shown in Fig. 4.7, and v describes the magnitude of the particle velocity v. Integrating over the above $\Delta\alpha \approx \Delta a$, we obtain an additional parallel shift of the particle trajectory proportional

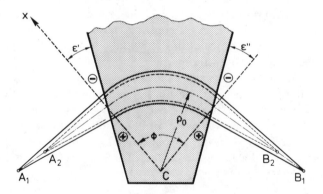

Fig. 4.7. The defocusing effect of oblique magnetic field boundaries is illustrated by showing particle trajectories which leave the point A_1 divergingly to be focused at B_1. As dotted curves orthogonal field boundaries are also indicated together with trajectories which leave the point A_2 divergingly to be focused at B_2. Note that the angles ε' and ε'' at the entrance and at the exit side are positive (as is the case here) if the normal to the oblique field boundary is further away from C, the center of curvature, as compared to the optic axis.

to the x coordinate. In a first-order approximation, however, this small shift can be neglected.

Since the above $\Delta\alpha$ is proportional to x, it can be interpreted as the deflection of a trajectory by a thin defocusing lens placed at the entrance field boundary where the magnetic flux density $B(z)$ is assumed to rise abruptly from 0 to B_0. According to Eq. (1.6), the relative refractive power of this lens equals

$$-\rho_0/f'_x = \tan \varepsilon'. \tag{4.24a}$$

For a particle beam leaving a sector field obliquely under an angle ε'', we can prove analogously that a thin defocusing lens placed at the exit field boundary describes all corresponding effects. The refractive power of this lens is

$$-\rho_0/f''_x = \tan \varepsilon''. \tag{4.24b}$$

The x transfer matrix of a sector field that has inclined entrance and exit field boundaries finally reads from in front of the entrance to behind the exit field boundary [Eqs. (1.6) and (4.8a)],

$$
\begin{pmatrix} (x|x) & (x|a) & (x|\Delta) \\ (a|x) & (a|a) & (a|\Delta) \\ 0 & 0 & 1 \end{pmatrix}
$$

$$
= \begin{pmatrix} 1 & 0 & 0 \\ -1/f''_x & 1 & 0 \\ 0 & 0 & 1 \end{pmatrix} \begin{pmatrix} c_x & s_x & d_x \\ -s_x/\rho_0^2 & c_x & s_x/\rho_0 \\ 1 & 0 & 1 \end{pmatrix} \begin{pmatrix} 1 & 0 & 0 \\ -1/f'_x & 1 & 0 \\ 0 & 0 & 1 \end{pmatrix}, \tag{4.25}
$$

with $c_x = \cos \phi_0$, $s_x = \rho_0 \sin \phi_0$, and $d_x = \rho_0(1 - \cos \phi_0)$. Explicitly, these matrix elements become for a single sector field separator with Eq. (4.8a),

$$(x|x) = \frac{\cos(\phi_0 - \varepsilon')}{\cos \varepsilon'}, \qquad (a|x) = -\frac{\sin(\phi_0 - \varepsilon' - \varepsilon'')}{\rho_m \cos \varepsilon' \cos \varepsilon''},$$

$$(x|a) = \rho_0 \sin \phi_0, \qquad (a|a) = \frac{\cos(\phi_0 - \varepsilon'')}{\cos \varepsilon''},$$

$$(x|\Delta) = \rho_0(1 - \cos \phi_0), \qquad (a|\Delta) = \tan \varepsilon'' + \frac{\sin(\phi_0 - \varepsilon'')}{\cos \varepsilon''},$$

[Ewald and Hintenberger (1953)]. In case a 4×4 transfer matrix is advantageous, as in Eq. (4.21) we find the corresponding coefficients $(x|\delta_K)$, $(a|\delta_K)$, $(x|\delta_m)$, $(a|\delta_m)$ from the derived $(x|\Delta)$ and $(a|\Delta)$ by using Eq. (4.22).

4.1.3.2   Particle Trajectories in the Cylinder Surface (x = 0)

Neither in the field-free region nor in the region of a homogeneous magnet does a charged particle experience forces in the y direction. However, as shown by Herzog (1950), such forces act on a particle while it traverses a fringing field under an angle ε, as indicated in Fig. 4.8. Such deflections are treated in Section 7.1.3. To first order they can be described by thin focusing lenses that exhibit focusing actions in the y direction for positive ε' and ε''. These lenses are placed at the entrance and exit effective field boundaries at which the magnetic flux density was assumed to change abruptly from 0 to B_0. According to Eq. (7.6), the focusing powers of these thin lenses are

$$\rho_0/f_y' = \tan \varepsilon', \tag{4.26a}$$

$$\rho_0/f_y'' = \tan \varepsilon''. \tag{4.26b}$$

Comparing Eqs. (4.26) and (4.24), we find

$$f_y' = -f_x' = \rho_0/\tan \varepsilon', \qquad f_y'' = -f_x'' = \rho_0/\tan \varepsilon''.$$

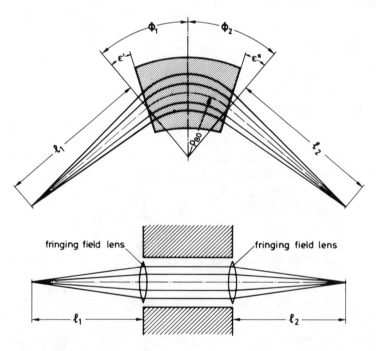

Fig. 4.8.   A stigmatic focusing homogeneous sector field using the fringing field y focusing of inclined field boundaries.

The y transfer matrix of a sector field that has inclined entrance and exit field boundaries thus reads [Eqs. (1.6) and (4.8b)],

$$\begin{pmatrix} (y|y) & (y|b) \\ (b|y) & (b|b) \end{pmatrix} = \begin{pmatrix} 1 & 0 \\ -1/f_y'' & 1 \end{pmatrix}\begin{pmatrix} 1 & \rho_0\phi_0 \\ 0 & 1 \end{pmatrix}\begin{pmatrix} 1 & 0 \\ -1/f_y' & 1 \end{pmatrix}, \quad (4.27)$$

with the explicit matrix elements being

$$(y|y) = 1 - \phi_0 \tan \varepsilon', \quad (b|y) = \rho_0^{-1}(\phi_0 \tan \varepsilon' \tan \varepsilon'' - \tan \varepsilon' - \tan \varepsilon''),$$

$$(y|b) = \rho_0\phi_0, \qquad\qquad (b|b) = 1 - \phi_0 \tan \varepsilon''.$$

4.1.3.3 A Stigmatic Focusing Homogeneous Sector Field with Inclined Field Boundaries

Inclined field boundaries in a sector field cause a defocusing action in the x direction and a focusing action in the y direction for positive ε' and ε'', as illustrated in Fig. 4.8. These focusing actions can reduce the beam height in the magnet air gap. For this reason, it is a good choice to place an object in the focal plane 1 of the entrance fringing field so that we find

$$l_1 = f_y' = \rho_0 \tan \varepsilon'. \quad (4.28a)$$

In this case all particles originating from a point move through the magnet along parallel trajectories, as indicated in Fig. 4.8. Later, these trajectories can be focused to a point at the focal plane 2 by an exit fringing field lens characterized by

$$l_2 = f_y'' = \rho_0 \tan \varepsilon''. \quad (4.28b)$$

Using these l_1, l_2 as object and image distances, the corresponding y-transfer matrix becomes with Eq. (4.27),

$$\begin{pmatrix} (y_f|y) & (y_f|b) \\ (b_f|y) & (b_f|b) \end{pmatrix}$$

$$= \begin{pmatrix} 1 & \rho_0 t'' \\ 0 & 1 \end{pmatrix}\begin{pmatrix} 1 & 0 \\ -t''/\rho_0 & 1 \end{pmatrix}\begin{pmatrix} 1 & \rho_0\phi_0 \\ 0 & 1 \end{pmatrix}\begin{pmatrix} 1 & 0 \\ -t'/\rho_0 & 1 \end{pmatrix}\begin{pmatrix} 1 & \rho_0 t' \\ 0 & 1 \end{pmatrix}, \quad (4.29)$$

for which the element $(y_f|b)$ vanishes, showing that an object–image relation exists between the profile planes 1 and f insofar as the y direction is concerned. The corresponding y magnification equals

$$M_y = (y_f|y) = -\tan \varepsilon'/\tan \varepsilon''.$$

Introducing the l_1, l_2 of Eqs. (4.28) as the object and image distances and splitting the sector field in two parts of deflection angles ϕ_1 and ϕ_2 as indicated in Fig. 4.8, the x transfer matrix of Eq. (4.25) reads

$$
\begin{pmatrix} (x_f|x) & (x_f|a) & (x_f|\Delta) \\ (a_f|x) & (a_f|a) & (a_f|\Delta) \\ 0 & 0 & 1 \end{pmatrix}
$$

$$
= \begin{pmatrix} (x_2|x) & (x_2|a) & (x_2|\Delta) \\ (a_2|x) & (a_2|a) & (a_2|\Delta) \\ 0 & 0 & 1 \end{pmatrix} \begin{pmatrix} (x_1|a) & (x_1|a) & (x_1|\Delta) \\ (a_1|x) & (a_1|a) & (a_1|\Delta) \\ 0 & 0 & 1 \end{pmatrix},
$$

Explicitly, the last two transfer matrices are written as

$$
\begin{pmatrix} (x_1|x) & (x_1|a) & (x_1|\Delta) \\ (a_1|x) & (a_1|a) & (a_1|\Delta) \\ 0 & 0 & 0 \end{pmatrix}
$$

$$
= \begin{pmatrix} c_{x1} & s_{x1} & \rho_0(1-c_{x1}) \\ -s_{x1}/\rho_0^2 & c_{x1} & s_{x1}/\rho_0 \\ 0 & 0 & 1 \end{pmatrix} \begin{pmatrix} 1 & 0 & 0 \\ t'/\rho_0 & 1 & 0 \\ 0 & 0 & 1 \end{pmatrix} \begin{pmatrix} 1 & \rho_0 t' & 0 \\ 0 & 1 & 0 \\ 0 & 0 & 1 \end{pmatrix},
$$

$$
\begin{pmatrix} (x_2|x) & (x_2|a) & (x_2|\Delta) \\ (a_2|x) & (a_2|a) & (a_2|\Delta) \\ 0 & 0 & 1 \end{pmatrix}
$$

$$
= \begin{pmatrix} 1 & \rho_0 t'' & 0 \\ 0 & 1 & 0 \\ 0 & 0 & 1 \end{pmatrix} \begin{pmatrix} 1 & 0 & 0 \\ t''/\rho_0 & 1 & 0 \\ 0 & 0 & 1 \end{pmatrix} \begin{pmatrix} c_{x2} & s_{x2} & \rho_0(1-c_{x2}) \\ -s_{x2}/\rho_0^2 & c_{x2} & s_{x2}/\rho_0 \\ 0 & 0 & 1 \end{pmatrix}.
$$

Here, the abbreviations $c_{x1} = \cos \phi_1$, $s_{x1} = \rho_0 \sin \phi_1$ and $c_{x2} = \cos \phi_2$, $s_{x2} = \rho_0 \sin \phi_2$ are used with $\phi_0 = \phi_1 + \phi_2$ (Fig. 4.8) and $t' = \tan \varepsilon'$, $t'' = \tan \varepsilon''$.

The angles ϕ_1 and ϕ_2 can be chosen such that particles originating from a point source are parallel in the x direction after a deflection of ϕ_1. This requires $(a_1|a) = 0$, or explicitly,

$$
\tan \phi_1 = 2 \tan \varepsilon'. \tag{4.30a}
$$

For the overall system to be point-to-point focusing, we must postulate $(x_f|a) = (x_2|x)(x_1|a) + (x_2|a)(a_1|a) = 0$. Because $(a_1|a) = 0$, this postulate requires $(x_2|x) = 0$ and thus analogously to Eq. (4.30a),

$$
\tan \phi_2 = 2 \tan \varepsilon''. \tag{4.30b}
$$

Writing $M_x = (x_f|x)$ for the x magnification and $M_y = (y_f|y)$ for the corresponding y magnification, we find with Eqs. (4.29) and (4.30),

$$M_x = (x_f|x) = -\sin \phi_1/\sin \phi_2, \tag{4.31a}$$

$$M_y = (y_f|y) = -\tan \phi_1/\tan \phi_2, \tag{4.31b}$$

as well as the dispersion $D_\Delta = (x_f|\Delta)$ and the resolving power $R_\Delta = (x_f|\Delta)/2x_{00}M_x$:

$$D_\Delta = 2\rho_0(1 - M_x), \tag{4.31c}$$

$$R_\Delta = \frac{2\rho_0}{2x_0[(1 - (1/M_x)]}. \tag{4.31d}$$

Comparing Eqs. (4.31b) and (4.31c) with Eqs. (4.18a) and (4.18c) yields the unexpected result:

For a stigmatic focusing sector field with inclined field boundaries (Fig. 4.8), the momentum dispersion and resolving power are simply twice as large as those for a homogeneous sector field with perpendicular field boundaries (see, for instance, Wollnik, 1980).

In both cases, however, the dispersions and resolving powers are independent of the angle of deflection $\phi_0 = \phi_1 + \phi_2$.

In addition to the increased resolving power, the y-focusing properties of inclined field boundaries normally allow a larger particle intensity to pass between given magnet poles. Consequently, a sector field that has inclined field boundaries is normally preferred to a sector field in which a particle beam crosses the field boundaries perpendicularly. However, for magnets with limited pole widths, very oblique entrance or exit angles ε' and ε'' should be avoided, since fringing field distortions could occur that cause the flux distribution to vary for different distances x from the optic axis. To describe such deviations and include them in a theory is difficult. Therefore, ε' and ε'' are normally limited to values below 30°. Angles up to 45°, and in some cases up to 60°, however, may be tolerable for carefully designed magnets. Should we want to employ larger angles of ε' and ε'', theory can only roughly predict the final performance of a magnet; one must therefore perform three-dimensional magnet calculations or experimental tests.

Altogether there are two solutions for symmetric sector magnet systems with $\varepsilon' = \varepsilon''$ and $M_x = -1$. First, we may choose a magnetic sector field with $\varepsilon' = \varepsilon'' = 0$. In this case, Eqs. (4.14c) and (4.19c) yield $l_1 = l_2 = \rho_0 \cotan(\phi_0/2)$ and $R_\Delta = 2\rho_0/2x_{10}$, and it therefore follows for $\phi_0 = 90°$ that both l_1 and l_2 must equal ρ_0. Second, we may choose a magnetic sector field with $\varepsilon' = \varepsilon'' \neq 0$. In this case, Eqs. (4.26), (4.28), (4.30), and (4.31) yield $2 \tan \varepsilon' = 2 \tan \varepsilon'' = \tan(\phi_0/2)$, as well as $l_1 = l_2 = \rho_0/\tan \varepsilon'$ and $R_\Delta = 4\rho_0/2x_{10}$, so that for $\phi_0 = 90°$ or $\varepsilon' = \varepsilon'' \approx 26.6°$, both l_1 and l_2 must equal $2\rho_0$. The height $2y_{10}$ of the object may be assumed to be $\pm 0.002\rho_0$ in both cases and the maximum angle of inclination $\beta_{10} = \pm 5$ mrad. The first system thus requires an exit slit that is $\pm[(l_1 + \rho_0\phi_0 + l_2)5 \times 10^{-3} + y_{10}] \approx \pm 0.019\rho_0$ high and requires a magnet air gap of at least $2G_0 = \pm[l_1 + \rho_0\phi_0 + 5 \times 10^{-3} + y_{10}] \approx \pm 0.014\rho_0$. The second system achieves an image formation not only in the x direction, but also in the y direction, so that the height of the image is $\pm y_{00}M_y = \pm 0.001\rho_0$ and a magnet air gap of only $2G_0 = \pm[l_1 \times 5 \times 10^{-3} + y_{10}] \approx 0.011\rho_0$. For $\rho_0 = 1$ m, the magnet air gaps thus must be ± 14 or ± 11 mm.

4.2   INHOMOGENEOUS MAGNETIC SECTOR FIELDS FORMED BY INCLINED PLANAR POLE FACES (WEDGE MAGNETS)

A magnet that has planar inclined pole faces, as shown in Fig. 4.9, creates a linearly inhomogeneous magnetic field. Such magnets—we shall call them wedge magnets—were investigated by Kofoed-Hansen *et al.* (1947), Richardson (1947), Jaffey *et al.* (1960), Ioanoviciu (1975), and others. In such magnets, the flux density **B** is inversely proportional to the distance r from the axis, where the two pole faces intersect virtually. Assuming a

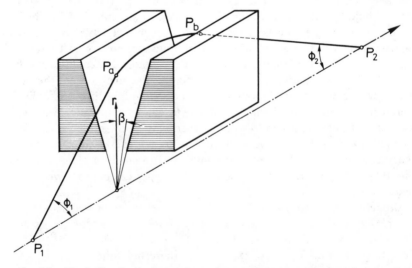

Fig. 4.9. The inclined pole faces of a wedge magnet. Note the cylindrical coordinate system r, β, z. The field entrance and exit points are marked P_a and P_b, respectively.

magnetic flux density B_0 at $r = r_0$, we find

$$B = (r_0/r)B_0. \tag{4.32}$$

A particle of the magnetic rigidity $\chi_0 = B_0\rho_0$ moves in such a flux density distribution along a trajectory of varying radius of curvature ρ. With $B_0\rho_0 = B\rho$ and Eq. (4.32), this radius of curvature is found to be

$$\rho = (B_0/B)\rho_0 = rN. \tag{4.33}$$

4.2.1 Particle Trajectories in Wedge Magnets

With the momentary radius of curvature $\rho = rN$ of a particle trajectory described by Eq. (4.33), we find from Fig. 4.10,

$$dr = -rN \sin \phi \, d\phi, \qquad dz = -rN \cos \phi \, d\phi.$$

Assuming that at the point $P_a(r_a, z_a)$ the trajectory under investigation formed an angle ϕ_a with the axis ($r = 0$), these differential equations determine the point $P_b(r_b, z_b)$ at which the trajectory forms the angle ϕ_b with the axis ($r = 0$),

$$r(\phi) = R_c \, e^{N \cos \phi}, \tag{4.34a}$$

$$z(\phi_b) = z(\phi_a) + NR_cI_{ab}. \tag{4.34b}$$

Here, R_c is some integration constant and I_{ab} equals

$$I_{ab} = \int_{\phi_a}^{\phi_b} -\cos \psi \, e^{N \cos \psi} \, d\psi,$$

or because of the symmetry of the cosine function,

$$I_{ab}(N, \phi_a, \phi_b) = I(N, \phi_a) - I(N, \phi_b), \tag{4.35a}$$

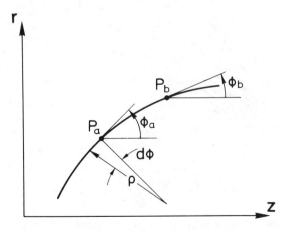

Fig. 4.10. The varying radius of curvature ρ of a particle in a wedge magnet.

$$I(N, \phi) = \int_0^\phi \cos \psi \, e^{N \cos \psi} \, d\psi. \tag{4.35b}$$

To get some idea of the magnitude of the integral $I(N, \phi)$, its numerical value is plotted as function of N and ϕ in Fig. 4.11. For a symmetric deflection, i.e., from ϕ_a to $\phi_b = -\phi_a$, we find with Eq. (4.34b), $z(\phi_b) = z(\phi_a) + 2NR_c I(N, \phi_a)$.

4.2.1.1 The Optic Axis

Assume that as shown in Fig. 4.12, a reference particle enters the deflecting field at the point $P_1(r_1, z_1)$ at an angle ϕ_1 with respect to the z axis. From Eq. (4.34a) we find for this point that $R_c = r_1 e^{-N \cos \phi_1}$, where N equals ρ_0 / r_1, with the magnetic rigidity of the particles under consideration being $\chi_0 = B_0 \rho_0$. Substituting this R_c back into Eqs. (4.34a) and (4.34b), we can determine the coordinates r_2 and z_2 of the point P_2 at which this trajectory

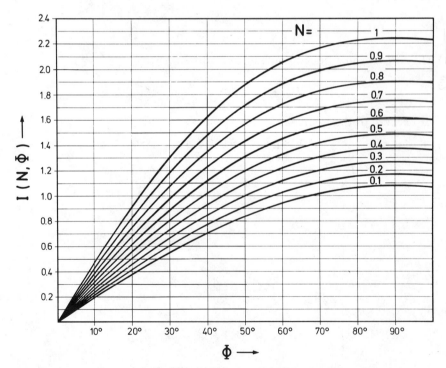

Fig. 4.11. The numerical values of $I(N, \phi)$ as calculated from Eq. (4.35b). For the definition of N and ϕ, see Eq. (4.33) and Fig. 4.10.

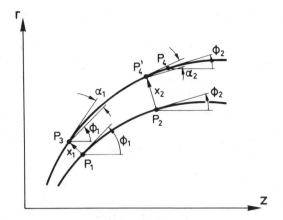

Fig. 4.12. Two particle trajectories are shown in a magnetic field $B = (B_1/r_1)r$, where r_1 is the r coordinate of point P_1 and B_1, the magnetic flux density at this point. The first trajectory is the optic axis along which a reference particle moves from P_1 to P_2. The second trajectory is the one along which an arbitrary particle moves from P_3 to P_4.

forms an angle ϕ_2 with the z axis as

$$r_2 = r_1 \exp[N(\cos \phi_2 - \cos \phi_1)], \tag{4.36a}$$

$$z_2 = z_1 + NI_{12}r_1 \exp(-N \cos \phi_1). \tag{4.36b}$$

4.2.1.2 The Transfer Matrix

To describe the motion of a particle of magnetic rigidity $\chi = \chi_0(1 + \Delta)$, Eq. (4.33) must be modified to read with $N_0 = \rho_0/r_1$,

$$\rho = rN = r_1N_0(1 + \Delta). \tag{4.37a}$$

As in the case of a homogeneous magnetic sector field, we now may describe a paraxial trajectory by its deviation x as well as by its inclination $\tan \alpha \approx a$ relative to the optic axis at any point r, z. As shown in Fig. 4.12, this particle under consideration enters the magnetic field at $P_3(r_3, z_3)$ a distance x_1 from and inclined under an angle $\alpha_1 \approx a_1$ with respect to the optic axis. Should the optic axis be perpendicular to the magnet boundary at P_1, the point P_3 is described in the r, z-coordinate system by

$$r_3 = r_1 + x_1 \cos \phi_1, \tag{4.37b}$$

$$z_3 = z_1 - x_1 \sin \phi_1, \tag{4.37c}$$

with the angle of inclination of the particle trajectory being

$$\phi_3 = \phi_1 + \alpha_1. \tag{4.37d}$$

After some straightforward calculations outlined in Section 4.5.1 of the Appendix to this chapter, we find the position x_2 and the inclination $\tan \alpha_2 \approx a_2$ of the trajectory under consideration relative to the optic axis at point P_2. These x_2 and a_2 terms may be expressed as functions of x_1, a_1, and Δ. Assuming that the magnet boundary is perpendicular to the optic axis at the entrance P_1 as well as at the exit P_2, the relation between (x_1, a_1) and (x_2, a_2) may be presented in the form of a transfer matrix similar to Eq. (4.8):

$$\begin{pmatrix} x_2 \\ a_2 \\ \Delta \end{pmatrix} = \begin{pmatrix} (x|x) & (x|a) & (x|\Delta) \\ (a|x) & (a|a) & (a|\Delta) \\ 0 & 0 & 1 \end{pmatrix} \begin{pmatrix} x_1 \\ a_1 \\ \Delta \end{pmatrix}. \tag{4.38}$$

The coefficients of this transfer matrix are here,

$$(x|x) = -N_0 s_2 c_1 \, e^{-(N_0 c_1)} I_{12} + c_2 c_1 \, e^{N_0(c_2 - c_1)} + s_2 s_1,$$

$$(x|a) = [-N_0 s_2 s_1 \, e^{-(N_0 c_1)} I_{12} + c_2 s_1 \, e^{N_0(c_2 - c_1)} - s_2 c_1] r_1 N_0,$$

$$(a|x) = [N_0 c_2 c_1 \, e^{-(N_0 c_2)} I_{12} - c_2 s_1 \, e^{N_0(c_1 - c_2)} + s_2 c_1]/(r_1 N_0),$$

$$(a|a) = N_0 c_2 s_1 \, e^{-(N_0 c_2)} I_{12} + c_2 c_1 \, e^{N_0(c_1 - c_2)} + s_2 s_1,$$

$$(x|\Delta) = [-s_2(1 - N_0 c_1) I_{12} - s_2 J_{12} + (c_2 - c_1) \, e^{(N_0 c_2)}] \, e^{-(N_0 c_1)} r_1 N_0,$$

$$(a|\Delta) = c_2(1 - N_0 c_1) I_{12} + c_2 J_{12} \, e^{-(N_0 c_2)} + s_2(c_2 - c_1),$$

with the abbreviations $c_1 = \cos \phi_1$, $s_1 = \sin \phi_1$, $c_2 = \cos \phi_2$, and $s_2 = \sin \phi_2$. The integral I_{12} was previously defined in Eq. (4.35a), whereas the new integral J_{12} must be taken from Eq. (4.83a) in the Appendix as being identical to I_{12}, except that the integrand is multiplied by $\cos \psi$. Analogous to Eqs. (4.35a), we find as a result of the symmetry of the cosine function, $I_{12} = I(N_0, \phi_1) - I(N_0, \phi_2)$ and $J_{12} = J(N_0, \phi_1) - J(N_0, \phi_2)$, with

$$I(N, \phi) = \int_0^\phi \cos \psi \, e^{N \cos \psi} \, d\psi, \tag{4.39a}$$

$$J(N, \phi) = \int_0^\phi \cos^2 \psi \, e^{N \cos \psi} \, d\psi. \tag{4.39b}$$

The numerical values of the integrals $I(N, \phi)$ and $J(N, \phi)$ are plotted in Figs. 4.11 and 4.13. For detailed, precise calculations, these values can be obtained from a numerical integration of Eqs. (4.39a) and (4.39b) or be taken from tables in the report by Jaffey *et al.* (1961) as $U(N, \pi - \phi)$ and $V(N, \pi - \phi)$.

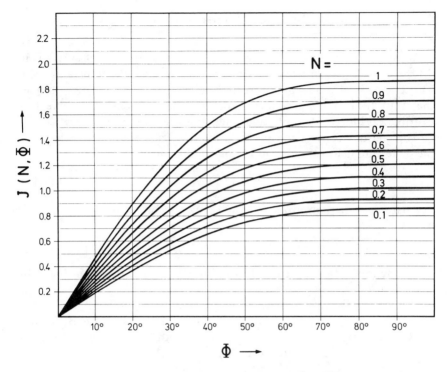

Fig. 4.13. The numerical values of $J(N, \phi)$ as calculated from Eq. (4.39b). For the definition of N and ϕ, see Eq. (4.33) and Fig. 4.11.

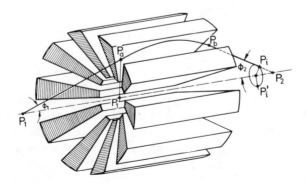

Fig. 4.14. A multigap spectrometer using several wedge magnets, a so-called orange spectrometer.

4.2.2 Focusing and Dispersing Properties of Wedge Magnets

Wedge magnets have often been used as multigap magnet arrays (Fig. 4.14) in which the particle source is placed at a point P_1 on the axis $r = 0$. This choice is also advantageous for one gap magnets, since all particles that start from a point source at the z axis experience the same field distribution independent of the azimuthal angle around the z axis.

The coordinate r_1 of the point P_a at which a reference particle enters the deflecting field is related to the object distance $l_1 = \overline{P_1 P_a}$ by

$$r_1 = l_1 \sin \phi_1. \tag{4.40a}$$

For particles emerging from P_1 at the z axis to be focused at P_i, it must hold that

$$r_2 = r_1 \exp[N_0(\cos \phi_2 - \cos \phi_1)] = -l_2 b \sin \phi_2, \tag{4.40b}$$

according to Eq. (4.36b). Here, $l_2 = \overline{P_b P_i}$ is the image distance in Fig. 4.14, and $b = \overline{P_b P_2} / \overline{P_b P_i} > 1$.

Particles originating from an object point are focused to a ring of radius $(b - 1)l_2 \sin \phi_2$. Here, each point of the ring corresponds to particles that are emitted from P_1 under angles $\phi_1 \pm \alpha_1$, with ϕ_1 describing the angle between the optic axis and the z axis. For $b = 1$, we find that the image occurs on the z axis, i.e., P_i and P_i' in Fig. 4.14 coincide.

Choosing the object and image distances l_1 and l_2 according to Eqs. (4.40a) and (4.40b), we find the transfer matrix of a particle spectrometer:

$$
\begin{pmatrix} (x_f|x) & (x_f|a) & (x_f|\Delta) \\ (a_f|x) & (a_f|a) & (a_f|\Delta) \\ 0 & 0 & 1 \end{pmatrix}
$$

$$
= \begin{pmatrix} 1 & l_2 & 0 \\ 0 & 1 & 0 \\ 0 & 0 & 1 \end{pmatrix} \begin{pmatrix} (x|x) & (x|a) & (x|\Delta) \\ (a|x) & (a|a) & (a|\Delta) \\ 0 & 0 & 1 \end{pmatrix} \begin{pmatrix} 1 & l_1 & 0 \\ 0 & 1 & 0 \\ 0 & 0 & 1 \end{pmatrix}. \tag{4.41}
$$

Here, $l_1 = r_1/s_1$ and $l_2 = -[r_1/(bs_2)] \, e^{N_0(c_2 - c_1)}$, with $s_1 = \sin \phi_1$, $c_1 = \cos \phi_1$, $s_2 = \sin \phi_2$, and $c_2 = \cos \phi_2$, are given by Eqs. (4.40a) and (4.40b), whereas the elements of the middle transfer matrix are given by Eq. (4.38).

In order to achieve an object–image relation between points P_1 and P_i of Fig. 4.14, we must postulate that the matrix element $(x_f|a)$ in Eq. (4.41) vanishes, yielding

$$(x|a) + (x|x)(r_1/s_1) - [(a|a) + (a|x)(r_1/s_1)](r_1/s_2 b) \, e^{N_0(c_2 - c_1)} = 0. \tag{4.42}$$

Introducing in Eq. (4.42) the coefficients of Eq. (4.38), we find a complex relation [see Eq. (4.85a) in Section 4.4.1 of the appendix to this chapter], which allows us to determine ϕ_2 for any choice of the quantities ϕ_1, N_0, and b. With these values, the lateral magnification $M_x = (x_f | x)$, the dispersion $(x_f | \Delta)$, and the distance $\overline{P_1 P_i} = l_1 c_1 + l_2 c_2 + z_B - z_A$ in Fig. 4.14 are all defined.

Wedge-magnet spectrometers often are restricted to symmetric systems

$$-\phi_2 = \phi_1 = 0, \qquad l_2 = l_1 = l, \qquad r_2 = r_1 = l \sin \phi,$$

so that b equals 1. In such cases, Eq. (4.42) becomes

$$N_0 e^{-(N_0 c)}(c + N_0 s^2) I(N_0, 0) = s(1 - N_0 c), \qquad (4.43a)$$

with $c = \cos \phi$ and $s = \sin \phi$. The lateral magnification equals $(x_f | x) = -1$, and for $-\phi_2 = \phi_1 < \pi/2$, the dispersion is

$$(x | \Delta) = 2(r_1 / N_0)[(1 - N_0 c)^2 + N_0 e^{-(N_0 c)}][(c/s) + N_0 s] J(N_0, \phi). \qquad (4.43b)$$

Finally, the distance between P_1 and $P_i = P_2$ is found to be

$$\overline{P_1 P_2} = 2l[N_0 s\, e^{-(N_0 c)} I(N_0, \phi) + c]. \qquad (4.43c)$$

Naturally, the Eqs. (4.43a)–(4.43c) are special cases ($l_1 = l_2$) of the general solution of Eq. (4.41). The corresponding solution for $l_1 \neq l_2$ is explicitly presented in Eqs. (4.85) in the appendix to this chapter.

Note also that in a symmetric system such as that shown in Fig. 4.14, all particles that have emerged from a point source move parallel to the z axis after half of the deflection. Omitting the second part of the system, the wedge magnet is thus point-to-parallel focusing. In the special symmetric system $\phi = \pi/2$ (Fig. 4.15), we find $l = r_1 = r_2$, as well as $c = 0$ and $s = 1$. Thus, the focusing condition of Eq. (4.43a) simplifies further to

$$I(N_0, \pi/2) = 1/N_0^2. \qquad (4.44a)$$

Equation (4.44a) postulates a fixed $N_0 \approx 0.7425$. Thus, the flux density $B(r_1)$ in the magnet must be chosen such that at the position $r_1 = l_1$, the charged particles under consideration move along a momentary radius of curvature of $\rho_1 = 0.7425 r_1$. The magnification here also equals $(x_f | x) = -1$, and the dispersion $(x_f | \Delta)$ follows from Eq. (4.43b) as

$$(x_f | \Delta) = 2(\rho_1 / N_0)[1 + N_0^2 J(N_0, \pi/2)] \approx 4.9 r_1, \qquad (4.44b)$$

because $J(0.742, \pi/2) \approx 1.49$. The distance between the points P_1 and P_2 in Fig. 4.16 is obtained with Eq. (4.44a) as

$$\overline{P_1 P_2} = 2l_1 / N_0 \approx 2.69 r_1. \qquad (4.44c)$$

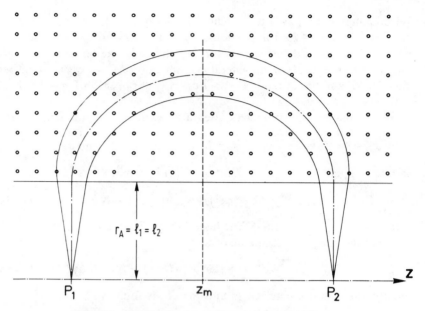

Fig. 4.15.   A 180° deflecting wedge-magnet spectrometer.

4.3   RADIALLY INHOMOGENEOUS SECTOR FIELDS FORMED BY CONICAL POLE FACES OR TOROIDAL ELECTRODES

In radially inhomogeneous sector fields (Figs. 4.16 and 4.17), a reference particle of rest mass m_{00}, charge $(z_0 e)$, and energy K_0 can move along a circular optic axis of radius ρ_0 in a constant magnetic flux density B_0 or a constant electrostatic field strength E_0. Such fields are created between conical pole faces or toroidal electrodes. These radii ρ_{B0} and ρ_{E0} of the optic axes are determined [Eqs. (2.11a) and (2.14)] as

$$\rho_{B0} = \frac{\chi_{B0}}{B_0} = \frac{1}{(z_0 e)B_0}\sqrt{m_{00}K_{00}2(1 + \eta_0)}, \qquad (4.45a)$$

$$\rho_{E0} = \frac{\chi_E}{E_0} = \frac{2K_0(1 + \eta_0)}{(z_0 e)E_0(1 + 2\eta_0)}, \qquad (4.45b)$$

in magnetic and electrostatic fields, respectively. Here, χ_{B0} and χ_{E0} are the magnetic and electrostatic rigidities of these reference particles, and we find from Eqs. (2.1c) and (2.3a), $\eta_0 = K_0/2m_{00}c^2$ or, numerically, $\bar{\eta}_0 = \bar{K}_0/1863\bar{m}_{00}$.

As is shown later, a radially inhomogeneous sector field has a reduced focusing power in the plane of deflection (x direction), causing longer

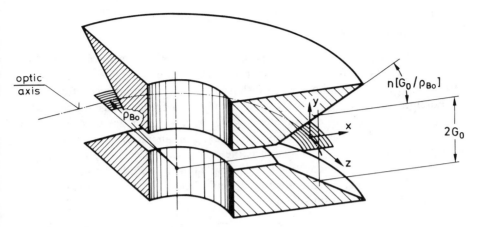

Fig. 4.16. A radially inhomogeneous magnetic sector field between conical pole faces. Note that the coefficient $n_1 = n_{B1}$ of Eqs. (4.48) and (4.58) determines the cone angle.

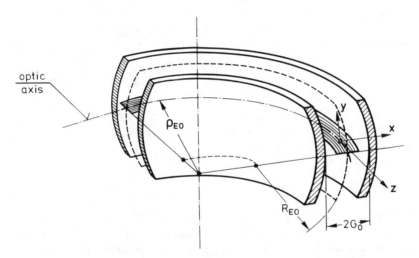

Fig. 4.17. A radially inhomogeneous electrostatic sector field between toroidal electrodes. Note that the coefficient $n_1 = n_{E1}$ of Eqs. (4.52), (4.53), and (4.58) equals ρ_{E0}/R_{E0}, as derived by Albrecht (1956) or Wollnik *et al.* (1972).

object and image distances in the xz plane. However, in the perpendicular y direction (Wallauschek, 1941; Siegbahn and Svartholm, 1946; Glaser, 1950, 1956; Gruemm, 1953; Ewald, 1957; Liebl and Ewald, 1959), it exhibits additional focusing that was not present in the case of a homogeneous magnet.

4.3.1 Equations of Motion in Radially Inhomogeneous Sector Fields

The equations of motion in radially inhomogeneous sector fields could be derived elegantly from the Euler–Lagrange equations [Eqs. (2.27a) and (2.27b)]. For clarity, however, we shall derive them here directly from the Lorentz equation [Eq. (2.6)]. In a radially inhomogeneous sector field optimally cylindrical coordinates $r = \rho_0 + x$, $z = \rho_0\phi$, y are used, with ρ_0 the radius of curvature of the optic axis. For an arbitrary particle, the resulting force in the x direction is the difference between the centrifugal force $mr\dot{\phi}^2$ and the electromagnetic centripetal force $-F_x$. In the y direction, there should only be the electromagnetic force F_y. Mathematically, the equations of motion thus read

$$m\ddot{x} = m(\rho_0 + x)\dot{\phi}^2 - F_x, \tag{4.46a}$$

$$m\ddot{y} = -F_y. \tag{4.46b}$$

We now assume that the main component of the velocity $v = (\rho_0 + x)\dot{\phi}$ is parallel to the optic axis and that the perpendicular component is small, of second order in x and y. In this case, Eqs. (4.46) can be rewritten to first order with $z = \rho_0\phi$ and $d/dt = (dz/dt)\, d/dz$,

$$\rho_0^2 x'' = (\rho_0 + x) - F_x(\rho_0 + x)^2/(mv^2) + \cdots, \tag{4.47a}$$

$$\rho_0^2 y'' = -F_y(\rho_0 + x)^2/(mv^2) + \cdots, \tag{4.47b}$$

where $z'' = d^2x/d(\rho_0\phi)^2 = d^2x/dz^2$ and $y'' = d^2y/d(\rho_0\phi)^2 = d^2y/dz^2$. The magnitude of the particle velocity v at the potential of the optic axis is found from Eq. (2.20a) to first order to be

$$v = v_0\left[1 + \frac{\delta_K + \hat{\delta}_K - \delta_m}{2(1 + \eta_0)(1 + 2\eta_0)} + \cdots\right]. \tag{4.47c}$$

Here, $v_0 = 2c\sqrt{\eta_0(1 + \eta_0)}/(1 + 2\eta_0)$ is the velocity [Eq. (2.2b)] of a reference particle moving along the optic axis, whereas δ_m and $\delta_K + \hat{\delta}_K$ denote the relative deviations in mass over charge and energy over charge [Eqs. (2.15) and (2.16c)] of the particle under consideration and of a reference particle. Since the electromagnetic forces F_x and F_y in radially inhomogeneous magnetic and electrostatic fields are different, the corresponding equations of motion must be derived separately.

4.3.1.1 Equations of Motion in a Magnetic Field between Conical Pole Faces

The field distribution in a magnet between conical pole faces is always characterized by $B_z(x, y, z) = 0$, i.e., a vanishing magnetic flux density in the direction of the optic axis. Thus, B_x and B_y are independent of z and $B_x(x, y)$ and $B_y(x, y)$ fully describe the field. Because of symmetry, $B_x(x, y = 0)$ vanishes, whereas $B_y(x, y = 0)$ is found to be inversely proportional to the width of the air gap since the width $2G_0$ of this gap (Fig. 4.16) increases linearly with x. To first order, we thus find

$$B_y(x, 0) = B_0[1 - n_{B1}(x/\rho_{B0}) + \cdots].$$

Because of curl $\mathbf{B} = 0$, it holds that

$$B_y(x, y) = B_0[1 - n_{B1}x/\rho_{B0} + \cdots], \tag{4.48a}$$

$$B_x(x, y) = B_0[n_{B1}y/\rho_{B0} + \cdots]. \tag{4.48b}$$

For a particle of charge (ze), the forces F_x and F_y in Eqs. (4.47a) and (4.47b) are thus $F_x = (ze)vB_y$ and $F_y = (ze)vB_x$, and we find

$$\rho_{B0}^2 x'' = (\rho_{B0} + x) - (\rho_{B0}^2 + 2x\rho_{B0} + \cdots)\frac{(ze)}{mv}B_0\left[1 - n_{B1}\left(\frac{x}{\rho_{B0}}\right)\right] + \cdots, \tag{4.49a}$$

$$\rho_{B0}^2 y'' = -\frac{(ze)}{mv}B_0 n_{B1}\left(\frac{y}{\rho_{B0}}\right) + \cdots. \tag{4.49b}$$

For a reference particle of charge (z_0e), rest mass m_{00}, and velocity v_0, the centrifugal and centripetal forces $m_{00}v_0^2/\rho_{B0}$ and $(z_0e)v_0B_0$ must balance along the optic axis. In this case, i.e. for $x = y = 0$, Eq. (4.49a) becomes the definition of the magnetic rigidity χ_{B0} of a reference particle [Eq. (2.10)]:

$$\chi_{B0} = B_0\rho_{B0} = m_0v_0/(z_0e).$$

For an arbitrary particle, the magnetic rigidity equals $\chi_B = mv/(ze)$, or

$$\chi_B = \chi_{B0}(1 + \Delta_B) = \chi_{B0}\left[1 + \frac{\delta_K(1 + 2\eta_0) + \delta_m}{2(1 + \eta_0)} + \cdots\right] \tag{4.50}$$

[Eq. (2.20b)], where the numerical value of η_0 is found from Eq. (2.3) to be $\bar{K}_0/1863\bar{m}_{00}$. With $\rho_{B0} = m_0v_0/[(z_0e)B_0]$, Eqs. (4.49a) and (4.49b) transform to

$$\rho_{B0}^2 x'' = -x(1 - n_{B1}) + \Delta_B\rho_{B0} + \cdots, \tag{4.51a}$$

$$\rho_{B0}^2 y'' = -yn_{B1} + \cdots, \tag{4.51b}$$

in a first-order approximation.

4.3.1.2 Equations of Motion in an Electrostatic Sector Field between Toroidal Electrodes

The potential distribution between toroidal electrodes (Fig. 4.17) is characterized by the ratio $n_{E1} = \rho_{E0}/R_E$ of the vertical R_E and the horizontal

ρ_{E0} radii of curvature of the midequipotential surface (Albrecht, 1956, Wollnik *et al.*, 1972). In detail, we find

$$V_E(x, y, z) = V_0 + E_0\rho_0\left[\frac{x}{\rho_{E0}} - \frac{1}{2}(1 + n_{E1})\left(\frac{x}{\rho_{E0}}\right)^2 + \frac{n_{E1}}{2}\left(\frac{y}{\rho_{E0}}\right)^2 + \cdots\right],$$

$$(4.52a)$$

where V_0 and E_0 describe the potential and the field strength at the optic axis of radius ρ_{E0}. In the case of a spherical condensor, the potential distribution is known to be $V_s = V_0\rho_{E0}/r$, which agrees with Eq. (4.52a) for $n_{E1} = \rho_{E0}/R_E = 1$. In the case of a cylindrical condensor (Fig. 4.18), the potential distribution is known to be $V_s = V_0 \ln(\rho_{E0}/r)$, which also agrees with Eq. (4.52a) for $R_E = \infty$ or $n_{E1} = \rho_{E0}/R_E = 0$.

Reference particles of charge (z_0e) and rest mass m_{00}, which are at rest at potential zero, have the kinetic energy $K_0 = -(z_0e)V_0$ at potential V_0, i.e., at the optic axis of a toroidal condensor. This results in an electrostatic rigidity $\chi_{E0} = m_{00}v_0^2/(z_0e)$ defined in Eq. (2.14),

$$\chi_{E0} = E_0\rho_{E0} = \frac{2K_0(1 + \eta_0)}{(z_0e)(1 + 2\eta_0)} = -2V_0\frac{1 + \eta_0}{1 + 2\eta_0},$$

with $2\eta_0 = K_0/m_{00}c^2 = 1/\sqrt{1 - (v_0/c)^2} - 1$, according to Eqs. (2.1b) and (2.1c). Using this relation, the potential V_0 can be replaced in Eq. (4.52a) by $E_0\rho_{E0}(1 + 2\eta_0)/2(1 + \eta_0)$, yielding

$$V_E(x, y, 0) = -E_0\rho_{E0}\left[\frac{1 + 2\eta_0}{2(1 + \eta_0)} - \left(\frac{x}{\rho_{E0}}\right) + \frac{1}{2}(1 + n_{E1})\left(\frac{x}{\rho_{E0}}\right)^2\right.$$

$$\left. - \frac{n_{E1}}{2}\left(\frac{y}{\rho_{E0}}\right)^2 + \cdots\right].$$

$$(4.52b)$$

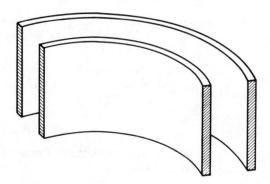

Fig. 4.18.   A cylinder condensor that has *x*- but no *y*-focusing properties.

Thus, the components of the electrostatic field strength read

$$E_x(x, y) = E_0[1 - (1 + n_{E1})(x/\rho_{E0}) + \cdots], \tag{4.53a}$$

$$E_y(x, y) = E_0[n_{E1}(y/\rho_{E0}) + \cdots], \tag{4.53b}$$

$$E_z(x, y) = 0. \tag{4.53c}$$

with the forces acting on a particle of charge (ze) being $F_x = (ze)E_x$ and $F_y = (ze)E_y$. This transforms Eqs. (4.47a) and (4.47b) to

$$\rho_{E0}x'' = (\rho_{E0} + x)\frac{(ze)}{mv^2}E_0\left[1 - (1 + \eta_{E1})\frac{x}{\rho_{E0}} + \cdots\right]$$

$$\times (\rho_{E0}^2 + 2x\rho_{E0} + \cdots) + \cdots, \tag{4.54a}$$

$$\rho_{E0}y'' = -\frac{(ze)}{mv^2}E_0\left[n_{E1}\frac{x}{\rho_{E0}} + \cdots\right](\rho_{E0}^2 + 2x\rho_{E0} + \cdots) + \cdots. \tag{4.54b}$$

Particles of rest mass m_0 and charge (ze), which were at rest at a point of potential ΔV, have the kinetic energy $K = -(ze)(V_e + \Delta V)$ at an arbitrary point $P(x, y)$ of potential V_E described in Eq. (4.52b). This results in an electrostatic rigidity [Eq. (2.14)],

$$\chi_E = \frac{mv^2}{(ze)} = \frac{2K(1 + \eta)}{(ze)(1 + 2\eta)}, \tag{4.55a}$$

with $2\eta = K/m_0c^2 = 1/\sqrt{1 - (v/c)^2} - 1$ and $m = m_0(1 + 2\eta)$. The quantities m_0 and K can also be written [Eqs. (2.15) and (2.16c)] as

$$\frac{K}{(ze)} = \frac{K_0}{(z_0e)}(1 + \delta_K + \hat{\delta}_K)_{m=\text{cont}}, \qquad \frac{m_0}{(ze)} = \frac{m_{00}}{(z_0e)}(1 + \delta_m)_{K=\text{const}},$$

where

$$\delta_K = -\frac{\Delta V}{V_0}, \qquad \hat{\delta}_K = -\frac{(\hat{V}_E - V_0)}{V_0} = -\frac{2x(1 + \eta_0)}{\rho_{E0}(1 + 2\eta_0)} + \cdots$$

according to Eq. (4.52b). With $\chi_{E0} = 2K_0(1 + \eta_0)/[(z_0e)(1 + 2\eta_0)]$ and $\eta = \eta_0(1 + \delta_K + \hat{\delta}_K)/(1 + \delta_m)$, Eq. (4.55a) thus becomes

$$\chi_E = \chi_{E0}\left[1 + \delta_K\frac{1 + 2\eta_0(1 + \eta_0)}{(1 + \eta_0)(1 + 2\eta_0)} + \delta_m\frac{\eta_0}{(1 + \eta_0)(1 + 2\eta_0)}\right.$$

$$\left. - \left(\frac{x}{\rho_{E0}}\right)\left(1 + \frac{1}{(1 + 2\eta_0)^2}\right) + \cdots\right] \tag{4.55b}$$

[see also Eq. (2.20c)], where the numerical value of η_0 is found from Eq. (2.3) as $\bar{K}_0/1863\bar{m}_{00}$. At that moment at which the particle under investigation crosses the optic axis ($x = y = 0$), the electrostatic rigidity is given as

$$\chi_{Er}(1 + \Delta_E) = \chi_{E0}\left[1 + \delta_K\frac{1 + 2\eta_0(1 + \eta_0)}{(1 + \eta_0)(1 + 2\eta_0)} + \delta_m\frac{\eta_0}{(1 + \eta_0)(1 + 2\eta_0)} + \cdots\right].$$

(4.55c)

Replacing $m_0v^2/(ze)$ in Eqs. (4.54a) and (4.56b) by the χ_E of Eq. (4.55b) and using $\chi_{E0} = E_0\rho_{E0}$ Wollnik (1967) obtained in a first-order approximation,

$$\rho_{E0}x'' = -x[1 - n_{E1} + (1 + 2\eta_0)^{-2}] + \Delta_E\rho_{E0}$$

$$= -x[2 - n_{E1} - (v_0/c)^2] + \Delta_E\rho_{E0},$$

(4.56a)

$$\rho_{E0}y'' = -yn_{E1},$$

(4.56b)

with $(1 + 2\eta_0)^{-2} = 1 - (v_0/c)^2$, according to Eqs. (2.1b) and (2.1c).

4.3.2 Particle Trajectories in Radially Inhomogeneous Magnetic or Electrostatic Sector Fields

The first-order equations of motion [Eqs. (4.51) and (4.56)] can be combined to read

$$x'' + xk_k^2 + \Delta/\rho_0 = 0,$$

(4.57a)

$$y'' + yk_x^2 = 0,$$

(4.57b)

with the side definitions [Eqs. (4.50), (4.51a), and (4.51b), (4.55c), (4.56a), and (4.56b)],

$$\rho_0^2k_x^2 = 1 - n_1 + h(1 + 2\eta_0)^2,$$

(4.58a)

$$\rho_0^2k_y^2 = n_1,$$

(4.58b)

$$\Delta = \frac{(1 + 2\eta_0)^2 + h}{2(1 + \eta_0)(1 + 2\eta_0)}\delta_k + \frac{(1 + 2\eta_0) - h}{2(1 + \eta_0)(1 + 2\eta_0)}\delta_m + \cdots.$$

(4.58c)

Here, h takes the values of 0 or 1 in the cases of magnetic or electrostatic sector fields, respectively; i.e.,

$$h_{\text{magnetic}} = 0,$$

(4.59a)

$$h_{\text{electrostatic}} = 1.$$

(4.59b)

For $k_k^2 > 0$ and $k_y^2 > 0$, the differential equations [Eqs. (4.57a) and (4.57b)] are satisfied by

$$x(z) = c_1 \cos(k_xz) + d_1 \sin(k_xz) + \Delta/k_x^2\rho_0,$$

(4.60a)

$$y(z) = c_2 \cos(k_yz) + d_2 \sin(k_yz),$$

(4.60b)

with $z = \rho_0\phi$ and undetermined coefficients c_1, c_2, d_1, d_2. From these $x(z)$ and $y(z)$, we find the angles of inclination $\alpha(z)$ and $\beta(z)$ (see Fig. 1.1) as $\tan \alpha(z) = dx/d(z) \approx a(z)$ and $\tan \beta(z) = dy/d(z) \approx b(z)$,

$$a(z) \approx \mp c_1 k_x \sin(k_x z) + d_1 k_x \cos(k_x z), \tag{4.60c}$$

$$b(z) \approx \mp c_2 k_y \sin(k_x z) + d_2 k_y \cos(k_y z). \tag{4.60d}$$

For $k_x^2 > 0$ or $k_y^2 > 0$, the upper signs are valid. For $k_x^2 < 0$ or $k_y^2 < 0$, the lower signs must be used, and it is understood that the cos and sin functions become cosh and sinh functions.

Knowing the positions x_1, y_1 and the angles of inclination $\alpha_1 \approx a_1$, $\beta_1 \approx b_1$, under which a particle entered a sector field at z_1

$$x(z_1) = x_1, \qquad y(z_1) = y_1, \qquad a(z_1) = \alpha_1, \qquad b(z_1) = \beta_1,$$

one can determine the coefficients c_1, c_2, d_1, d_2. For a sector field with an angle of deflection ϕ_0, finally the Eqs. (4.60) can be written for $a \ll 1$, $b \ll 1$, and $z_0 = w = \rho_0\phi_0$ as

$$\begin{pmatrix} x(w) \\ a(w) \\ \Delta(w) \end{pmatrix} = \begin{pmatrix} (x|x) & (x|a) & (x|\Delta) \\ (a|x) & (a|a) & (a|\Delta) \\ 0 & 0 & 1 \end{pmatrix} \begin{pmatrix} x_1 \\ a_1 \\ \Delta \end{pmatrix} = \begin{pmatrix} c_x & s_x & d_x \\ -s_x k_k^2 & c_x & s_x/\rho_0 \\ 0 & 0 & 1 \end{pmatrix} \begin{pmatrix} x_1 \\ a_1 \\ \Delta \end{pmatrix},$$
$$\tag{4.61a}$$

$$\begin{pmatrix} y(w) \\ b(w) \end{pmatrix} = \begin{pmatrix} (y|y) & (y|b) \\ (b|y) & (b|b) \end{pmatrix} \begin{pmatrix} y_1 \\ b_1 \end{pmatrix} = \begin{pmatrix} c_x & s_y \\ -s_y k_y^2 & c_y \end{pmatrix} \begin{pmatrix} y_1 \\ b_1 \end{pmatrix}, \tag{4.61b}$$

$$c_x = \cos(k_x w), \qquad s_x = \frac{\sin(k_x w)}{k_x}, \qquad d_x = \frac{1 - \cos(k_x w)}{\rho_0 k_x^2},$$

$$c_y = \cos(k_y w), \qquad s_y = \frac{\sin(k_y w)}{k_y},$$

$$k_y^2 \rho_0^2 = n_1, \qquad k_x^2 \rho_0^2 = 1 - n_1 + h/(1 + 2\eta_0)^2 = 1 - n_1 + h[1 - (v_0/c)^2],$$

and $k_x^2 \rho_0^2$ and $k_y^2 \rho_0^2$ are defined in Eqs. (4.58a) and (4.58b) and the rigidity deviation Δ in Eqs. (4.50) and (4.55c). Furthermore, it is again understood that for $k_x^2 < 0$ or $k_y^2 < 0$, the corresponding cos and sin functions become cosh and sinh functions.

Should we be interested in a mass and an energy dispersion rather than a rigidity dispersion, Eq. (4.61a) should be rewritten using Eq. (4.58c) as

$$\begin{pmatrix} x(w) \\ a(w) \\ \delta_K \\ \delta_m \end{pmatrix} = \begin{pmatrix} c_x & s_x & d_x N_K & d_x N_m \\ -s_x k_x^2 & c_x & (s_x/\rho_0)N_K & (s_x/\rho_0)N_m \\ 0 & 0 & 1 & 0 \\ 0 & 0 & 0 & 1 \end{pmatrix} \begin{pmatrix} x_1 \\ a_1 \\ \delta_K \\ \delta_m \end{pmatrix}, \tag{4.61c}$$

$$N_K = \frac{(1 + 2\eta_0)^2 + h}{2(1 + \eta_0)(1 + 2\eta_0)}, \qquad N_m = \frac{(1 + 2\eta_0) - h}{2(1 + \eta_0)(1 + 2\eta_0)},$$

where c_x, s_x, $d_x = (1 - c_x)/(\rho_0 k_x^2)$ are defined under Eq. (4.61a), with k_x^2 being given by Eq. (4.58a) and h being 0 or 1 for a magnetic or an electrostatic sector field, respectively. Note again that for relativistically slow particles, η_0 vanishes. Note also that for $\rho_0 = \infty$, Eqs. (4.61b) and (4.61c) describe the properties of a quadrupole lens of length w, where $k_x^2 = -k_y^2$ equals $(dB_y/dx)/\chi_{B0}$ or $(dE_x/dx)/\chi_{E0}$ [see also Eqs. (3.6) and (3.7)]. Note finally that Eqs. (4.61b) and (4.61c) also describe the properties of a field-free region for $\rho_0 = \infty$ and $k_x^2 = k_y^2 = 0$.

4.3.3   Focusing and Dispersing Properties of Radially Inhomogeneous Sector Fields

To determine the transfer matrices from a profile plane 1 a distance l_1 before a radially inhomogeneous sector field to a profile plane 4 a distance l_2 behind, we find with Eqs. (4.61a) and (4.61b),

$$
\begin{pmatrix}
(x_4|x) & (x_4|a) & (x_4|\Delta) \\
(a_4|x) & (a_4|a) & (a_4|\Delta) \\
0 & 0 & 1
\end{pmatrix}
$$

$$
= \begin{pmatrix}
1 & l_2 & 0 \\
0 & 1 & 0 \\
0 & 0 & 1
\end{pmatrix}
\begin{pmatrix}
c_x & s_x & d_x \\
-s_x k_x^2 & c_x & s_x/\rho_0 \\
0 & 0 & 1
\end{pmatrix}
\begin{pmatrix}
1 & l_1 & 0 \\
0 & 1 & 0 \\
0 & 0 & 1
\end{pmatrix}, \quad (4.62a)
$$

$$
\begin{pmatrix}
(y_4|y) & (y_4|b) \\
(b_4|y) & (b_4|b)
\end{pmatrix}
= \begin{pmatrix}
1 & l_2 \\
0 & 1
\end{pmatrix}
\begin{pmatrix}
c_y & s_y \\
-s_y k_y^2 & c_y
\end{pmatrix}
\begin{pmatrix}
1 & l_1 \\
0 & 1
\end{pmatrix}, \quad (4.62b)
$$

$$(x_4|x) = c_x - l_2 s_x k_x^2, \qquad\qquad (a_4|x) = -s_x k_0^2,$$

$$(x_4|a) = s_x + (l_1 + l_2)c_x - l_1 l_2 s_x k_x^2, \qquad (a_4|a) = c_x - l_1 s_x k_x^2,$$

$$(x_4|\Delta) = d_x + l_2 s_x/\rho_0, \qquad\qquad (a_4|\Delta) = s_x/\rho_0,$$

$$(y_4|y) = c_y - l_2 s_y k_y^2, \qquad\qquad (b_4|y) = -s_y k_y^2,$$

$$(y_4|b) = s_y + (l_1 + l_2)c_y - l_1 l_2 s_y k_y^2, \qquad (b_4|b) = c_y - l_1 s_y k_y^2,$$

where c_x, s_x, d_x, c_y and s_y are defined following Eqs. (4.61a) and (4.61b), as well as k_x and k_y. Except for $(x_4|\Delta)$ and $(a_4|\Delta)$, the coefficients of Eq. (4.62a) and (4.62b) are identical should x, a be exchanged for y, b. Consequently, it is sufficient to derive all optical properties for the x direction only.

4.3.3.1 Lens Properties of Radially Inhomogeneous Sector Fields

The focusing strength $1/f_x$ of a radially inhomogeneous sector field is found from Eq. (4.62a) as $(a_4|x) = -1f_x = -s_x k_x^2$. For $\rho_0^2 k_x^2 = [1 - n_1 + h/(1 + 2\eta_0)^2] > 0$, we obtain

$$\rho_0/f_x = \rho_0 k_x^2 s_x = k_x \sin(k_x w), \qquad (4.63a)$$

and for $\rho_0^2 k_x^2 = [1 - n_1 + h/(1 + 2n_0)^2] < 0$ with $k_x = \sqrt{-k_x^2}$,

$$\rho_0/f_x = -\rho_0 k_x^2 s_x = -k_x \sinh(k_x w). \qquad (4.63b)$$

The positions l_{fx1} and l_{fx2} of the focal planes relative to the entrance and exit field boundaries (see Sections 1.3.2.3 and 1.3.2.4) are found from $(a_4|a) = 0$ and $(x_4|x) = 0$, respectively, in Eq. (4.62a). For $k_x^2 > 0$, we obtain

$$l_{fx} = l_{fx1} = l_{fx2} = c_x/k_x^2 s_x = \cotan(k_x w)/k_x, \qquad (4.64a)$$

and for $k_x^2 < 0$ with $k_x = \sqrt{-k_x^2}$,

$$l_{fx} = l_{fx1} = l_{fx2} = c_x/k_x^2 s_x = \cotanh(k_x w)/k_x. \qquad (4.64b)$$

The positions of the first and second principal planes from the entrance and exit field boundaries, respectively, are then found as $l_{1px} = l_{2px} = l_{fx} - f_x$. For $k_x^2 > 0$, we obtain

$$l_{px} = l_{px1} = l_{px2} = \frac{c_x - 1}{k_x^2 s_x} = -\frac{\tan(k_x w/2)}{k_x}, \qquad (4.65a)$$

and for $k_x^2 < 0$ with $k_x = \sqrt{-k_x^2}$,

$$l_{px} = l_{px1} = l_{px2} = \frac{c_x - 1}{k_x^2 s_x} = \frac{\tanh(k_x w/2)}{k_x}. \qquad (4.65b)$$

Postulating the coefficient $(x_4|a)$ in Eq. (4.62a) to vanish, we can express the object and image distances l_1 and l_2 as functions of the lateral magnification $M_x = (x_4|x)$. These relations are only meaningful for $k_x^2 > 0$, yielding

$$l_2 = \frac{c_x - M_x}{k_x^2 s_x} = \frac{\cos(k_x w) - M_x}{k_x \sin(k_x w)}, \qquad (4.66a)$$

$$l_1 = \frac{c_x - 1/M_x}{k_x^2 s_x} = \frac{\cos(k_x w) - 1/M_x}{k_x \sin(k_x w)}. \qquad (4.66b)$$

4.3.3.2 The x Transfer Matrix in Special Cases

Analogous to Eqs. (4.15)–(4.17), quite simple characteristic transfer matrices also exist in the case of radially inhomogeneous sector fields. The transfer matrix between the first and second focal planes is found by replacing l_1 and l_2 in Eq. (4.62a) by the l_{fx} of Eq. (4.64a):

$$\begin{pmatrix} (x|x) & (x|a) & (x|\Delta) \\ (a|x) & (a|a) & (a|\Delta) \\ 0 & 0 & 1 \end{pmatrix} = \begin{pmatrix} 0 & 1/k_x^2 s_x & 1/k_x^2 \\ -k_x^2 s_x & 0 & s_x/\rho_0 \\ 0 & 0 & 1 \end{pmatrix}$$

$$= \begin{pmatrix} 0 & 1/[k_x \sin(k_x w)] & 1/k_x^2 \\ -k_x \sin(k_x w) & 0 & [\sin(k_x w)]/k_x \\ 0 & 0 & 1 \end{pmatrix}.$$

(4.67)

Note that in this case, the element $(x|\Delta)$ is independent of w.

The transfer matrix that relates the first and the second principal planes is found by replacing l_1 and l_2 in Eq. (4.62a) by the l_{px} of Eq. (4.65a):

$$\begin{pmatrix} (x|x) & (x|a) & (x|\Delta) \\ (a|x) & (a|a) & (a|\Delta) \\ 0 & 0 & 1 \end{pmatrix} = \begin{pmatrix} 1 & 0 & 0 \\ -k_x^2 s_x & 1 & s_x/\rho_0 \\ 0 & 0 & 1 \end{pmatrix}$$

$$= \begin{pmatrix} 1 & 0 & 0 \\ -k_x \sin(k_x w) & 1 & [\sin(k_x w)]/k_x \\ 0 & 0 & 1 \end{pmatrix},$$

(4.68)

Note that in this case, the element $(x|\Delta)$ vanishes.

The transfer matrix that relates an object and an image profile plane is found by replacing l_1 and l_2 in Eq. (4.62a) by the l_1 and l_2 of Eqs. (4.66):

$$\begin{pmatrix} (x|x) & (x|a) & (x|\Delta) \\ (a|x) & (a|a) & (a|\Delta) \\ 0 & 0 & 1 \end{pmatrix} = \begin{pmatrix} M_x & 0 & (1-M_x)/\rho_0 k_x^2 \\ -k_x^2 s_x & 1/M_x & s_x/\rho_0 \\ 0 & 0 & 1 \end{pmatrix}$$

$$= \begin{pmatrix} M_x & 1/[k_x \sin(k_x w)] & (1-M_x)/k_x^2 \\ -k_x \sin(k_x w) & 1/M_x & [\sin(k_x w)]/k_x \\ 0 & 0 & 1 \end{pmatrix}.$$

(4.69)

Radially inhomogeneous sector fields are mainly used in cases in which the beam of charged particles crosses the entrance and exit field boundaries perpendicularly. However, in principle it is also possible to use oblique field boundaries in such magnetic sector fieds. In those cases, the transfer matrices of Eqs. (4.25) and (4.27) must be used to describe the actions of the entrance and exit field boundaries. Although these equations were derived for homogeneous magnetic sector fields, they are also correct for radially inhomogeneous sector fields if the middle transfer matrices in both

Eqs. (4.25) and (4.27) are replaced by those of Eqs. (4.61a) and (4.61b), respectively.

4.3.3.3 Dispersion, Resolution, and Resolving Power of an x-focusing Radially Inhomogeneous Sector Field

From the definitions of Eqs. (4.10a) and (4.10b) and from Eq. (4.69), we find the dispersion D_Δ, the resolution Δ_{min}, and the resolving power $R_\Delta = -\Delta_{min}^{-1}$ for a single sector field transfer matrix for which $(x|a)$ vanishes as

$$D_\Delta = \frac{1 - M_x}{\rho_0 k_x^2} = \frac{\rho_0(1 - M_x)}{1 - n_1 + h/(1 + 2\eta_0)^2}, \tag{4.70a}$$

$$R_\Delta = \frac{1 - 1/M_x}{2x_{00}\rho_0 k_x^2} = \frac{\rho_0(1 - 1/M_x)}{2x_{00}[1 - n_1 + h/(1 + 2\eta_0)^2]}. \tag{4.70b}$$

Inspecting Eqs. (4.70a) and (4.70b), we also find that for a radially inhomogeneous sector field, neither the dispersion nor the resolution nor the resolving power do explicitly depend on the angle of deflection ϕ_0 of the sector field. Note, however, that D_Δ and R_Δ are $(\rho_0^2 k_x^2)$ times larger than the corresponding quantities for a homogeneous sector magnet as given in Eqs. (4.19a) and (4.19b). From Eqs. (4.70a) and (4.70b), we also find here

$$2x_{00}R_\Delta = \tilde{D}_\Delta, \tag{4.70c}$$

where $\tilde{D}_\Delta = (1 - 1/M_x)/[\rho_0 k_0^2]$ is the D_Δ of Eq. (4.70a) for the reversed system for which object and image are exchanged, so that M_x is replaced by $1/M_x$.

If we are interested in an energy (δ_K) or a mass (δ_m) dispersion rather than a dispersion in rigidity (Δ), we must multiply Eqs. (4.70a) and (4.70b) by the expression of Eq. (4.58c). In detail, we find

$$D_K = \frac{\rho_0(1 - M_x)}{1 - n_1 + h/(1 + 2\eta_0)^2} N_K, \qquad D_m = \frac{\rho_0(1 - M_x)}{1 - n_1 + h/(1 + 2\eta_0)^2} N_m, \tag{4.71a}$$

$$R_K = \frac{\rho_0(1 - 1M_x)}{2x_{00}[1 - n_1 + h/(1 + 2\eta_0)]^2} N_K,$$

$$R_m = \frac{\rho_0(1 - 1M_x)}{2x_{00}[1 - n_1 + h/(1 + 2\eta_0)]^2} N_m, \tag{4.71b}$$

with $N_k = [(1 + 2\eta_0)^2 + h]/[2(1 + \eta_0)(1 + 2\eta_0)]$ and $N_m = [(1 + 2\eta_0) - h]/[2(1 + \eta_0)(1 + 2\eta_0)]$, as defined in Eq. (4.61c) and h being

0 or 1 for a magnetic or an electrostatic sector field. Note that the mass and energy dispersions are independent of the deflecting angle ϕ_0 of the sector field under consideration as was the case in Eqs. (4.19a) and (4.19b). For relativistically slow particles for which η_0 vanishes, Eqs. (4.71a) and (4.71b) simplify to

$$D_K = \frac{\rho_0(1 - M_x)(1 + h)}{2(1 - n_1 + h)}, \qquad D_m = \frac{\rho_0(1 - M_x)(1 - h)}{2(1 - n_1 + h)}, \qquad (4.72a)$$

$$R_K = \frac{\rho_0(1 - 1/M_x)(1 + h)}{4x_{00}(1 - n_1 + h)}, \qquad R_m = \frac{\rho_0(1 - 1/M_x)(1 - h)}{4x_{00}(1 - n_1 + h)}. \qquad (4.72b)$$

Note that an electrostatic sector field exhibits only energy over charge ($D_K \neq 0$) but no mass over charge ($D_m = 0$) dispersion for low-energy particles. Consequently, an electrostatic sector field deflects slow particles of different mass to charge ratios equally as long as their energy to charge ratios are the same.

4.3.4 Examples of Radially Inhomogeneous Sector Fields

4.3.4.1 A Stigmatic Focusing Magnetic Sector Field

In many cases, it is of interest to refocus all monoenergetic particles emerging from one point of the particle source to one point of the image independent of the angles of inclination α_0, β_0 under which these particles left the source. In a radially inhomogeneous sector field, this can be achieved if k_x^2 equals k_y^2. For a magnetic system, this requires

$$n_1 = 1 - n_1 = 0.5 = \rho_0^2 k_x^2 = \rho_0^2 k_y^2.$$

Postulating the object and image distances l_1 and l_2 to be equal and both positive, the Eqs. (4.66a) and (4.66b) yield with $w = \rho_0\phi_B$,

$$M_x = -1, \qquad l_1/\rho_{B0} = l_2/\rho_{B0} = 2\cot(\phi_B/2\sqrt{2}). \qquad (4.73)$$

With $M_x = -1$, Eq. (4.72a) states that D_K and D_m both equal $2\rho_{B0}$ for $h = 0$. These $D_\Delta = 2D_K = 2D_m$ are twice as large as those of a symmetric homogeneous magnetic sector field [Eq. (4.19a)] and equally large as those of a symmetric stigmatic focusing homogeneous sector field system that has inclined field boundaries [Eq. (4.31c)].

For $\phi_0 = \pi\sqrt{2} \approx 254.5°$, we find from Eq. (4.71) the simple solution $l_1 = l_2 = 0$. For most applications, however, it is desirable to have both particle source and collector outside the sector field. A special case is the deflection angle $\phi_0 \approx 155°$ (Camplan et al., 1970) for which we find from Eq. (4.73), $l_1 = l_2 \approx \rho_{B0}$.

One shortcoming of a radially inhomogeneous magnet is that for increasing flux densities, the pole shoes do not saturate equally everywhere. At positions where the gap is narrower (Fig. 4.16), it saturates faster than at positions where the gap is wider (von Egidy, 1962). Such saturations cause field distortions and decrease the optical performance of the instrument with a consequent reduction of the resolving power. Thus, radially inhomogeneous magnets should be designed to operate at low flux densities to keep saturation effects to a minimum. As an alternative, we can also compensate for the produced field distortions by special devices such as correction coils (see Section 9.2).

As an example, we may investigate a radially inhomogeneous magnet that must deflect singly charged ions of mass 200 and energy 50 keV. The magnetic rigidity of such ions is $\chi_{B0} = B_0\rho_{B0} \approx 0.144\sqrt{200 \times 05 \times 0.455}$ Tm. Choosing the radius of deflection ρ_{B0} of such a system to be 1 m, we need a flux density of 0.455 T, which is sufficiently small to keep saturation effects within limits.

4.3.4.2 An Electrostatic Sector Field Energy Analyzer

As may be seen from Eq. (4.72a) an electrostatic sector field acts as an energy-over-charge analyzer for slow particles ($\eta_0 = 0$) with no mass separation. The simplest such electrostatic energy analyzer consists of a cylinder condensor (Fig. 4.18). In this case, $R_E = \infty$ or $n_{E1} = 0$ is postulated so that one finds $\rho_0 k_x = \sqrt{2}$ and $k_y = 0$. For a symmetric system ($l_1 = l_2$ and $M_x = -1$), Eqs. (4.66a) and (4.66b) yield with $w = \rho_0\phi_E$,

$$l_1 = l_2 = (\rho_0/2)\cotan(\phi_E/2), \tag{4.74a}$$

whereas Eq. (4.72a) reads $D_K = \rho_{E0}$ and $D_m = 0$. For $\phi_E = 1/\sqrt{2} \approx 127.3°$, both l_1 and l_2 vanish according to Eq. (4.74a), stating that (Hughes and Rojansky, 1929) particles originating from a point P_1 of the optic axis are focused to a point P_2 of the optic axis after a deflection of 127.3°.

One shortcoming of a cylinder condensor is that it lacks a focusing action in the y direction. For a stigmatic focusing energy analyzer, which refocuses all monoenergetic particles diverging from one point of the particle source to one corresponding point of the image, we must postulate $k_x = k_y$. For $\eta_0 = 0$ this requires $n_{E1} = 2 - n_{E1}$ [Eq. (4.58a)]. This corresponds to a spherical condensor for which R_E/ρ_{E0} equals unity (Fig. 4.17). For a symmetric system ($l_1 = l_2$ and $M_x = -1$), the Eqs. (4.66a) and (4.66b) yield

$$l_1 = l_2 = \rho_0\cotan(\phi_E/2), \tag{4.74b}$$

whereas Eq. (4.72a) reads $D_K = 2\rho_{E0}$ and $D_m = 0$. For $\phi_E = 180°$, Eq. (4.74b) yields $l_1 = l_2 = 0$, so that similar to a homogeneous magnetic field, the image of a point at the optic axis must be expected after a deflection of 180°. Note, however, that in the case of a spherical condensor, this image is stigmatic, whereas in the case of a homogeneous magnetic field, there was an image only in the x direction.

As an example, we may consider recording the energy spectrum of electrons generated by bombarding some specimen by X rays. The energy K_0 of these electrons reaches from a few electron volts to several thousand electron volts. For a spherical condensor of $\rho_{E0} = 0.15$ m radius, we need thus a field strength of $E_0 = K_0/\rho_{E0}(ze)$ of about 10–1000 V/m at the optic axis. Using electrodes separated by a distance $2G_0 = 0.020$ m, the potential difference $V_2 - V_1$ at these electrodes must be about 0.2–20 V. Because of these rather small voltages, any contaminating layer on the electrodes should be carefully avoided since such layers normally are nonconductive and can be charged by scattered electrons or ions, so that the electrostatic field is distorted.

4.4 PARTICLE FLIGHT TIMES IN RADIALLY INHOMOGENEOUS SECTOR FIELDS, QUADRUPOLES, AND FIELD-FREE REGIONS

The flight time of a particle moving with a velocity $v(x, y)$ along its trajectory s is found from Eq. (2.20a) as

$$T = \int \frac{ds}{v(s)} = \int \frac{\sqrt{(1 + x/\rho_0)^2 + x'^2 + y'^2}}{v(x, y)} \, dz$$

$$\approx \int \frac{1 + x/\rho_0 + \cdots}{1 + (\delta_K + \hat{\delta}_K - \delta_m)/[2(1 + \eta_0)(1 + 2\eta_0)] + \cdots} \left(\frac{dz}{v_0}\right), \quad (4.75)$$

where the approximate sign corresponds to a first-order approximation. Here, z is the coordinate along the circular optic axis of radius ρ_0, x and y are coordinates, as shown in Fig. 2.6, whereas, as before, $x' = dx/dz = \tan \alpha$ and $y' = dy/dz = \tan \beta$ denote inclinations of the trajectories under consideration.

For field-free drift distances as well as for quadrupole lenses of length l, Eq. (4.75) simplifies further because of $\rho_0 = \infty$. For magnetic fields or field-free regions, $\hat{\delta}_K$ vanishes. In the case of electrostatic quadrupoles, $\hat{\delta}_K$ varies with the electrostatic potential $\hat{V}$. Since this potential is proportional to x^2 and y^2 [Eq. (3.39a)], its effect can be neglected in a first-order approximation. Thus, Eq. (4.75) can be integrated for field-free regions or quadrupole lenses of lengths l, yielding

$$T = \frac{l}{v_0}\left[1 + \frac{\delta_m - \delta_K}{2(1 + \eta_0)(1 + 2\eta_0)} + \cdots\right].$$

The flight time T_0 of a reference particle ($\delta_K = \delta_m = 0$) moving along the optic axis here in both cases would have been l/v_0 so that we find

$$t = T - T_0 = T_0 \frac{\delta_m - \delta_K}{2(1 + \eta_0)(1 + 2\eta_0)}. \tag{4.76}$$

Very generally, this flight-time difference can be expressed by the elements of an additional row of a transfer matrix:

$$\begin{pmatrix} x_{i+1} \\ a_{i+1} \\ t_{i+i} \\ \delta_K \\ \delta_m \end{pmatrix} = \begin{pmatrix} (x|x) & (x|a) & 0 & (x|\delta_K) & (x|\delta_m) \\ (a|x) & (a|a) & 0 & (a|\delta_K) & (a|\delta_m) \\ (t|x) & (t|a) & 1 & (t|\delta_K) & (t|\delta_m) \\ 0 & 0 & 0 & 1 & 0 \\ 0 & 0 & 0 & 0 & 1 \end{pmatrix} \begin{pmatrix} x_i \\ a_i \\ t_i \\ \delta_K \\ \delta_m \end{pmatrix}. \tag{4.77}$$

For a field-free region or for a quadrupole of length l the coefficients of the third row of Eq. (4.77) are found from Eq. (4.76) to be

$$(t|x) = 0, \qquad v_0(t|\delta_K) = -l/[2(1 + \eta_0)(1 + 2\eta_0)],$$

$$(t|a) = 0, \qquad v_0(t|\delta_m) = l/[2(1 + \eta_0)(1 + 2\eta_0)].$$

For a quadrupole lens, the other coefficients [Eq. (3.6a)] read

$$(x|x) = (a|a) = c_x, \qquad (a|x) = -k_x^2(x|a),$$

$$(x|a) = s_x, \qquad (x|\delta_K) = (x|\delta_m) = (a|\delta_K) = (a|\delta_m) = 0,$$

with $c_x = \cos(k_x w)$, $s_x = k_x^{-1} \sin(k_x w)$, and $k_x = \sqrt{|k_x^2|}$, where it is understood that sin and cos become sinh and cosh for $k_x^2 < 0$. For $k_x^2 = 0$, Eq. (4.77) describes the action of a field-free region of length l.

For sector fields, ρ_0 is finite, and $\hat{\delta}_K$ vanishes for $\rho_0 = \rho_{B0}$, whereas it yields

$$\hat{\delta}_K = -\frac{\hat{V}_E - V_0}{V_0} = \frac{-2x(1 + \eta_0)}{\rho_{E0}(1 + 2\eta_0)} + \cdots$$

[Eq. (4.52b)] for $\rho_0 = \rho_{E0}$. To evaluate Eq. (4.75) in a first-order approximation, it is necessary to know the x deviation of the particle trajectory under consideration from the optic axis,

$$x(w) = x_0 c_x + a_0 s_x + d_x \left[\frac{(1 + 2\eta_0)^2 + h}{2(1 + \eta_0)(1 + 2\eta_0)} \delta_K + \frac{(1 + 2\eta_0) - h}{2(1 + \eta_0)(1 + 2\eta_0)} \delta_m \right]$$

$$+ \cdots,$$

with $c_x = \cos(k_x w)$, $s_x = k_x^{-1} \sin(k_x w)$, and $d_x = [1 - \cos(k_x w)]/k_x^2 \rho_0$, according to Eq. (4.61c) and $w = \rho_0 \phi_0$. Introducing this $x(w)$ into δ_K and then $x(w)$ as well as δ_K and δ_m into Eq. (4.75), we find $t = T - T_0$, with

$T_0 = w/v_0 = \rho_0\phi_0/v_0$ as

$$t = \int_0^{\rho_0\phi_0} \left\{ 1 + \frac{x[1 + h/(1 + 2\eta_0)^2]}{\rho_0} + \frac{\delta_m - \delta_K}{2(1 + \eta_0)(1 + 2\eta_0)} + \cdots \right\} \frac{dw}{v_0} - T_0$$

$$= x_0(t|x) + a_0(t|a) + \delta_K(t|\delta_K) + \delta_m(t|\delta_m). \tag{4.78}$$

Explicitly, these coefficients read (see also Wollnik and Matsuo, 1981, and Wollnik and Matsuda, 1981),

$$(t|x)v_0 = (s_x/\rho_0)N_t, \tag{4.79a}$$

$$(t|a)v_0 = [(1 - c_x)/k_x^2\rho_0]N_t = d_x N_t, \tag{4.79b}$$

$$(t|\delta_K)v_0 = -w/[2(1 + \eta_0)(1 + 2\eta_0)] + [(w - s_x)/k_x^2\rho_0^2]N_K N_t, \tag{4.79c}$$

$$(t|\delta_m)v_0 = w/[2(1 + \eta_0)(1 + 2\eta_0)] + [(w - s_x)/(k_x^2\rho_0^2)]N_m N_t, \tag{4.79d}$$

with

$$N_t = 1 + [h/(1 + 2\eta_0)^2] = [2(1 + \eta_0)/(1 + 2\eta_0)]N_K,$$

and N_K, as well as N_m defined in Eq. (4.61c). Note that with $\rho_0 = \infty$, the Eqs. (4.79a)–(4.79d) transform to the time coefficients of Eq. (4.77) describing the properties of a quadrupole lens of length w where $k_x^2 = -k_y^2$ equal $(dB_y/dx)/\chi_{B0}$ or $(dE_x/dx)/\chi_{E0}$. Consequently, the Eqs. (4.79a)–(4.79d) also describe the properties of a field-free region of length l with $\rho_0 = \infty$ and $k_x^2 = k_y^2 = 0$.

Comparing the coefficients of Eqs. (4.79a)–(4.79b) with those of Eq. (4.61c), Wollnik and Matsuo (1981) found for magnetic as well as for electrostatic sector fields,

$$-(t|x)v_0 = \frac{2(1 + \eta_0)}{1 + 2\eta_0}[(x|x)(a|\delta_K) - (a|x)(x|\delta_K)], \tag{4.80a}$$

$$-(t|a)v_0 = \frac{2(1 + \eta_0)}{1 + 2\eta_0}[(x|a)(a|\delta_K) - (a|a)(x|\delta_K)]. \tag{4.80b}$$

As is shown in Section 8.2, Eqs. (4.80a) and (4.80b) are a consequence of the fact that an optical transformation is a transformation between two sets of canonically conjugate variables (see also Wollnik and Berz, 1985). Consequently, Eq. (4.80) are also valid for any product transfer matrix involving field-free regions, quadrupole lenses, and sector fields.

APPENDIX

4.5.1 Paraxial Trajectories in Wedge Magnets

In Section 4.2.1.2, transfer matrices were cited that relate x_2, α_2, Δ with x_1, α_1, Δ for a wedge magnet. To derive these transfer matrices explicitly, first we must determine the coordinates r_4, z_4 of a point P_4 at which the trajectory under consideration forms the same angle ϕ_2 with the z axis in Fig. 4.12, as does the optic axis at P_2. These coordinates are found by replacing the indices 1, 2 in Eqs. (4.36a) and (4.36b) by 3, 4, with r_3, z_3, ϕ_3 and $N = N_0(1 + \Delta)$ defined in Eqs. (4.37a)-(4.37c)

$$r_4 = (r_1 + x_1 \cos \phi_1) \exp\{N_0(1 + \Delta)[\cos \phi_2 - \cos(\phi_1 + \alpha_1)]\},$$

$$z_4 = (z_1 - x_1 \sin \phi_1) + (r_1 + x_1 \cos \phi_1) N_0(1 + \Delta) I_{34}$$
$$\times \exp[-N_0(1 + \Delta) \cos(\phi_1 + \alpha_1)]$$

where I_{34} equals $I_{12}[N_0(1 + \Delta), \phi_1 + \alpha_1, \phi_2]$ [see also Eq. (4.35a)].

As in all optical systems, our main interest is to describe how much point P_4 deviates from point P_2,

$$r_4 - r_2 = (r_1 + x_1 c_1) \exp\{N_0(1 + \Delta)[c_2 - \cos(\phi_1 + \alpha_1)]\}$$
$$- r_1 \exp N_0(c_2 - c_1), \tag{4.81a}$$

$$z_4 - z_2 = -x_1 s_1 + (r_1 + x_1 c_1) N_0(1 + \Delta) I_{34} \exp[-N_0(1 + \Delta) \cos(\phi_1 + \alpha_1)]$$
$$- r_1 N_0 I_{12} \exp -(N_0 c_1) \tag{4.81b}$$

with $c_1 = \cos \phi_1$, $s_1 = \sin \phi_1$, and $c_2 = \cos \phi_2$, $s_2 = \sin \phi_2$. For $\alpha_1 \ll 1$ and $\Delta \ll 1$, these relations can be expanded in power series in α_1 and Δ around $\alpha_1 = \Delta = 0$. Retaining only linear terms, we find

$$r_4 - r_2 = e^{N_0(c_2 - c_1)}[x_1 c_1 + \alpha_1 r_1 N_0 s_1 + \Delta r_1 N_0(c_2 - c_1)] + \cdots, \tag{4.82a}$$

$$z_4 - z_2 = x_1[c_1 N_0 I_{12} e^{-N_0 c_1} - s_1] + \alpha_1 r_1 N_0[c_1 + N_0 s_1 I_{12} e^{-N_0 c_1}]$$
$$+ \Delta r_1 N_0 e^{-N_0 c_1}[I_{12}(1 - N_0 c_1) + N_0 J_{12}] + \cdots, \tag{4.82b}$$

where $(\partial I_{12}/\partial \alpha_1) = c_1 \exp N_0 c_1$ [Eq. (3.35b)] and $\partial I_{12}/\partial \Delta = N_0 J_{12}$ with

$$J_{12}(N_0, \phi_1, \phi_2) = \int_{\phi_1}^{\phi_2} -\cos^2 \psi \, e^{N_0 \cos \psi} \, d\psi.$$

Because of the symmetry of the cosine function, here one can write analogously to Eqs. (4.35a) and (4.35b):

$$J_{12}(N, \phi_1, \phi_2) = J(N, \phi_1) - J(N, \phi_2), \tag{4.83a}$$

$$J(N, \phi) = \int_0^{\phi} \cos^2 \psi \, e^{N \cos \psi} \, d\psi. \tag{4.83b}$$

Similar to a homogeneous magnetic sector field, we may describe a paraxial trajectory by its deviation x as well as by its inclination $\tan \alpha \approx \sin \alpha \approx a$ relative to the optic axis at any point r, z. The paraxial trajectory that enters the field perpendicularly at $P_3(r_3, z_3)$ is thus described by x_1 and a_1 relative to the optic axis at $P_1(r_1, z_1)$. Relative to the optic axis at point $P_2(r_2, z_2)$, we can then describe the paraxial trajectory by x_2 and a_2 (Fig. 4.12). These values must be determined from the coordinates of point P_4 at which the trajectory under investigation forms the same angle ϕ_2 with the z axis as the optic axis at point P_2. For this purpose, one may assume that in the neighborhood of point P_4, the paraxial trajectory can be approximated by a circle of constant radius of curvature $\rho_4 = N_0 r_4$, which in a zeroth-order approximation equals $\rho_2 = N_0 r_2$. Since the distance $\overline{P_4 P_4'}$ is certainly small compared to r_1 (Fig. 4.12), we find to first order in $\overline{P_4 P_4'}/\rho_1$ the values of x_2 and $a_2 \approx \tan \alpha_2$ as,

$$x_2 = \overline{P_2 P_4'} + \cdots, \qquad a_2 \approx \tan \alpha_2 = \overline{P_4 P_4'}/\rho_4 + \cdots, \qquad (4.84)$$

with $\overline{P_2 P_4'} = (r_4 - r_2)c_2 - (z_4 - z_2)s_2$ and $\overline{P_4 P_4'} = (z_4 - z_2)c_2 + (r_4 - r_2)s_2$ as well as $\rho_4 - \rho_2 = N_0 r_2$ [Eq. (4.81a)] $\rho_4 = r_1 N_0 \exp N_0(c_2 - c_1) + \cdots$. Here c_1, c_2, s_1, s_2 denote the same abbreviations as in Eq. (4.81a). Substituting Eqs. (4.82a) and (4.82b) into Eq. (4.84), we can express x_2 and α_2 as functions of x_1 and $\tan \alpha_1 \approx a_1$. These relations were already presented in the form of a transfer matrix in Eq. (4.38).

In order to achieve an object–image relation between the points P_1 and P_i of Fig. 4.14, we must postulate that the matrix element $(x_f|a)$ in Eq. (4.41) vanishes which explicitly yields

$$N_0 e^{-N_0 e_1} I_{12}[N_0^2 bs_2^2 s_1^2 + N_0(bs_2^2 c_1 + s_1^2 c_2) + c_1 c_2]$$
$$= s_2 e^{N_0(c_2 - c_1)} [N_0^2 bc_2 s_1^2 + N_0(bc_1 c_2 - s_1^2) - c_1]$$
$$+ s_1[N_0(b^2 s_2 - c_1 c_2) - N_0^2 bc_1 s_2^2 - c_2]. \qquad (4.85a)$$

When the angle ϕ_1 and the parameter N_0 are known for some choice of b the angle ϕ_2 can be determined from Eq. (4.85a). The lateral magnification of such a system $M_x = (x_f|x)$, the dispersion $(x_f|\Delta)$ and the distance $\overline{P_1 P_i} = l_1 c_1 + l_2 c_2 + z_1 - z_2$ in Fig. 4.14 are found to be (see also Ruedenauer, 1970):

$$(x_f|x) = s_1 - N_0 c_1 e^{-N_0 c_1} I_{12}\left(s_2 + \frac{c_2}{N_0 bs_2}\right)$$
$$+ c_1 e^{N_0(c_2 - c_1)}\left(c_2 - \frac{1}{bN_0}\right), \qquad (4.85b)$$

$$(x_f|\Delta) = -r_1 N_0 e^{-N_0 c_1}[(1 - N_0 c_1)I_{12} + J_{12}]\left(s_2 + \frac{c_2}{N_0 bs_2}\right)$$
$$+ (c_1 - c_2) e^{N_0(c_1 - c_2)}\left(c_2 - \frac{1}{bN_0}\right), \qquad (4.85c)$$

$$\overline{P_0 P_i} = r_1\left[N_0 e^{N_0 c_1} I_{12} + \frac{c_1}{s_1} - \frac{c_2}{bs_2}\exp N_0(c_2 - c_1)\right], \qquad (4.85d)$$

where the value of $z_2 - z_1$ is taken from Eq. (4.36b) and where l_1 and l_2 have been replaced by the expressions given in Eqs. (4.40a) and (4.40b).

REFERENCES

Albrecht, R. (1956). *Z. Naturforsch.* **11a**, 156.

Barber, N. F. (1953). *Proc. Leeds Philos. Soc. Sci. Sect.* **2**, 427.

Brück, H. (1960). "Optique Corpusculaire; Theorie et Technique des Accelerateurs de Particule," Fascicule IV, C.D.K., Paris.

Camplan, J., Meuniet, R., and Sarrouy, J. L. (1970). *Nucl. Instrum. Methods* **84**, 37.

Claasen, J. (1908). *Phys. Z.* **9**, 762.

Dempster, A. (1918). *Phys. Rev.* **11**, 316.

Ewald, H. (1957). *Z. Naturforsch.* **12a**, 28.

Ewald, H., and Hintenberger, H. (1953). "Methoden und Anwendungen der Massenspektroskopie." Verlag Chemie, Weinheim.

Glaser, W. (1950). *Oesterr. Inq.-Arch.* **4**, 354.

Glaser, W. (1956). *In* "Handbuch der Physik," Vol. XXXIII, S. Fluegge (ed.). Springer Verlag, Berlin.

Gruemm, H. (1953). *Acta Phys. Austriaca* **8**, 119.

Herzog, R. (1950). *Acta Phys. Austriaca* **4**, 431.

Hughes, A., and Rojansky, V. (1929). *Phys. Rev.* **34**, 284.

Ioanoviciu, D. (1975). *Int. J. Mass Spectrom. Ion Phys.* **18**, 289.

Jaffey, A. H., Mallmann, C. A., Suarez-Etchepare, J. and Suter, T. (1960). Argonne Natl. Lab. Rep. ANL-6222.

Kofoed-Hansen, O., Lindhard, J., and Nielsen, O. B. (1950). *Mat. Fys. Medd. Dan. Vid. Selsk.* **25**, 1.

Liebl, H., and Ewald, H. (1959). *Z. Naturforsch.* **14a**, 842.

Penner, S. (1961). *Rev. Sci. Instrum.* **32**, 150.

Richardson, H. O. W. (1947). *Proc. Phys. Soc.* **59**, 792.

Ruedenauer, F. (1970). *Int. J. Mass Spectrom. Ion Phys.* **4**, 181, 195.

Siegbahn, K., and Svartholm, N. (1946). *Nature* **157**, 872L.

Smythe, H. D. (1945). "Atomic Energy for Military Purposes," Princeton Univ. Press, Princeton.

von Egidy, T. (1962). *Ann. Phys. (München)* **9**, 221.

Wallauschek, R. (1941). *J. Phys.* **117**, 565.

Wollnik, H. (1967). *In* "Focusing of Charged Particles," A. Septier (ed.). Academic, New York.

Wollnik, H. (1980). *In* "Applied Charged Particle Optics, A. Septier (ed.). Academic, New York.

Wollnik, H., and Berz, M. (1985). *Nucl. Instrum. Methods* **238**, 127.

Wollnik, H., and Matsuda, H. (1981). *Nucl. Instrum. Methods* **189**, 361.

Wollnik, H., and Matsuo, T. (1981). *Int. J. Mass Spectrom. Ion Phys.* **37**, 209.

Wollnik, H., Matsuo, T., and Matsuda, H. (1972). *Nucl. Instrum. Methods* **102**, 13.

5

Charged Particle Beams and Phase Space

In Chapters 1-4, individual particle trajectories were determined in optical systems by solving the Lorentz equation [Eq. (2.6)]. In order to obtain solutions of this differential equation of second order, it was necessary to know the particle position (x, y, z) as well as the particle momentum (p_x, p_y, p_z) at a starting time t_1. Knowing this initial position of a particle in the six-dimensional so-called "phase space," we can thus determine all future positions by Eq. (2.6).

In Chapter 5, not the trajectory for a single particle but the motion of an ensemble of particles shall be investigated. The corresponding bundle of trajectories is usually referred to as a beam.* The most interesting property of the motion of such a beam in phase space is described by Liouville's

---

* In contrast to this common use for the word "beam," some authors speak of a beam only if each trajectory of a bundle interacts with the others. Should we here want to characterize such a situation, we shall refer to the special type of interaction and speak of a plasma beam or a beam with space charge.

theorem. This theorem is known from thermodynamics and reads

For a cloud of moving particles, the particle density $\rho(x, p_x, y, p_y, z, p_z)$ in phase space is invariable.

According to Liouville's theorem, a cloud of particles that at a time t_1 fills a certain volume in phase space may change its shape at later times t but not the magnitude of its volume (Fig. 5.1). Any attempt to reduce the size of this volume by electromagnetic fields is in vain, although it may often be desirable.

5.1 LIOUVILLE'S THEOREM AND FIRST-ORDER TRANSFER MATRICES

In a first-order approximation used up to now in Chapters 1, 3, and 4 and also here, the particle motions in the three space coordinates are independent of each other. This splits Liouville's theorem into the conservation of three two-dimensional phase-space areas at any time t_i:

$$\iint dx_i \, dp_{xi} = \text{const}_x, \qquad \iint dy_i \, dp_{yi} = \text{const}_y, \qquad \iint dz_i \, dp_{zi} = \text{const}_z.$$

At positions with vanishing vector potential, we read from Eqs. (2.22a) and (2.22b) $p_{xi} = a_i n_i p_r$ and $p_{yi} = b_i n_i p_r$ so that the first two of these statements become

$$p_r \iint dx_i n_i \, da_i = p_r \iint dx_i n_i \, d(\sin \alpha_i) = \text{const}_x, \qquad (5.1a)$$

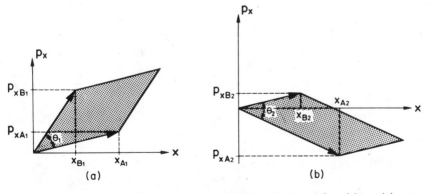

Fig. 5.1. The two-dimensional x, p_x phase-space area of a beam of particles at (a) $z = z_1$ and (b) $z = z_2$. Note the varied shapes and equal sizes of the indicated phase-space areas.

$$p_r \int\int dy_i n_i \, db_i = p_r \int\int dy_i n_i \, d(\sin\beta_i) = \text{const}_y. \tag{5.1b}$$

Here, n_i is the locally varying refractive index, and a_i as well as b_i are taken from Eqs. (2.22a) and (2.22b) for the cases of planar motions, i.e., $\beta_i = 0$ for Eq. (5.1a) and $\alpha_i = 0$ for Eq. (5.1b).

To accept Eq. (5.1a) as correct, remember that $\mathbf{X}_i = (x_i, a_i)$ of a particle trajectory at z_i can be determined from $\mathbf{X}_1 = (x_1, a_1)$ at z_1 by $\mathbf{X}_i = T_x \mathbf{X}_1$, which reads explicitly,

$$\begin{pmatrix} x_i \\ a_i \end{pmatrix} = \begin{pmatrix} (x|x) & (x|a) \\ (a|x) & (a|a) \end{pmatrix} \begin{pmatrix} x_1 \\ a_1 \end{pmatrix}.$$

Remembering now the integration rules in a multidimensional space (see, for instance, Strómberg, 1981), we can rewrite Eq. (5.1a) as

$$\iint_{G(z_i)} dx_i \, da_i = \iint_{G(z_1)} |T_x| \, dx_1 \, da_1,$$

where G describes regions in the x, a space at z_i and at z_1. Because of $|T_x| = n_1/n_2$ as illustrated in Eq. (2.23a) and proved in Eq. (8.25a), Eq. (5.1a) is thus correct, as is Eq. (5.1b) analogously. For optical systems in which no accelerations occur, i.e., in which n is constant throughout, we find $|T_x| = 1$ [Eq. (1.19a)] and thus a conservation of the x, p_x or x, a phase-space area, as is illustrated in Fig. 5.1 (see also Banford, 1966).

5.2 PHASE-SPACE AREAS OF PARTICLE BEAMS PASSING THROUGH OPTICAL SYSTEMS

Although the magnitude of the phase-space area of a beam does not vary by going from one standard profile plane to the next, the shape of this area may vary considerably. To investigate these shape variations, only the projections of particle trajectories on the xz plane or the yz plane are treated for simplicity; i.e., x, p_x and y, p_y as functions of z. Although this splitting of the full four-dimensional phase-space volume into two two-dimensional phase-space areas is quite common, it raises some problems (see Section 5.6).

Neighboring trajectories remain neighbors in all smooth mapping transformations, as are tansformations by transfer matrices. In particular, trajectories, which at some profile plane form points along a line in phase space, will again form points along a line in phase space at any other profile plane. If this line encompasses some phase-space area at the first profile plane, it will, according to Eq. (5.1a), encompass a phase-space area of equal size at the second profile plane.

Assume a particle source, extending, as illustrated in Fig. 5.2, over $\pm x_{10}$ in the x direction, where each point x_1 of this source emits charged particles under angles of inclination $\pm \alpha_1$. Of all these particles we shall look only at those which pass through an aperture, extending over $\pm x_{20} = \pm d_x$, placed a distance $l_x = z_2 - z_1$ downstream from the object. The size of the object as well as the size and position of this aperture, which we shall call an *entrance pupil*, determine the size and the shape of the phase-space area occupied by the particle beam at $z = z_1$. Checking the phase-space positions of the trajectories A, B, C, D at $(z = z_1)$, we find this phase-space area to be a parallelogram of size

$$A_x = 4\bar{\varepsilon}_x = 4(x_{10}a_{10}). \tag{5.2a}$$

Here, $\pm a_{10} \approx \pm \tan \alpha_{10} = \pm d_x/l_x$ corresponds to the maximum angle under which a particle that started at the center of the object can still pass through the entrance pupil.

For certain applications, we do not postulate that the optical system under consideration must transmit the full parallelogram-like phase-space area, but we are satisfied if it transmits that area which is enclosed by the ellipse inscribed within the parallelogram of Fig. 5.2. The area of this ellipse is

$$A_x = \pi \varepsilon_x = \pi(x_{10}a_{10}), \tag{5.2b}$$

and thus about 20% smaller than the area of the corresponding parallelogram. The beam diameter in this case, however, may be up to 40% smaller at critical positions in the optical system (see Section 5.3.3.3).

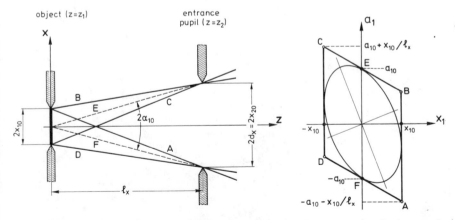

Fig. 5.2. Particle trajectories limited by a source of diameter $2x_{10}$ and a diaphragm of diameter $2x_{20} = 2d_x$ placed a distance l_x downstream from the source. Also shown is the corresponding parallelogram-like x, a phase-space area (including the inscribed phase-space ellipse) of the ensemble of trajectories at $z = z_1$.

5.2.1 Phase-Space Area of a Particle Beam in a Drift Space

In field-free regions, all particles move along straight trajectories. Thus, for a specific trajectory the angle of inclination α stays constant, whereas the deviation x from the optic axis increases linearly with the flight distance z and with the inclination $\tan \alpha$:

$$x(z) = x_1 + (z - z_1)\tan\alpha_1 \approx x_1 + (z - z_1)a_1, \qquad a(z) = a_1,$$

where as in Eq. (1.1a), we have $x_1 = x(z_1)$ and $a_1 = a(z_1)$. For a particle source of diameter $2x_{10}$ and an entrance pupil of diameter $2x_{20}$ as before in Fig. 5.2, the corresponding phase-space area is a parallelogram limited by the trajectories A, B, C, D. The change in shape (Fig. 5.3) of this phase-space area with increasing z can be interpreted as being caused by a shearing force parallel to the x axis in the x, a phase space. Since $x(z)$ increases

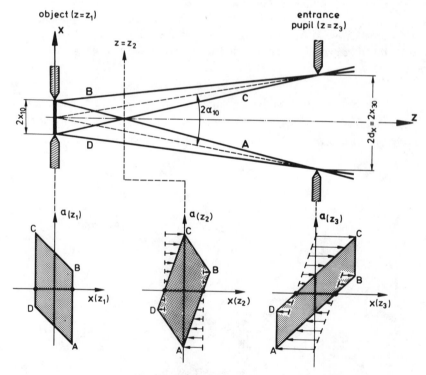

Fig. 5.3. Particle trajectories limited by a source ($z = z_1$) of diameter $2x_{10}$ and a diaphragm ($z = z_2$) of diameter $2x_{20} = 2d_x$ at $z = z_2$. Also shown is the x, a phase-space area occupied by the ensemble of trajectories at $z = z_1$, $z = z_i$, and $z = z_2$, where z_i is that value of z at which the trajectories A and C meet. Note the horizontal shearing action in the field-free region.

linearly with a_1, this force increases linearly with a_1, as indicated in Fig. 5.3. Note that the areas of the three phase-space parallelograms in Fig. 5.3 are all of equal size and that lines AB and DC are upright at the object ($z = z_1$), whereas lines BC and DA are upright at the entrance pupil ($z = z_2$).

5.2.2 Phase Space Area of a Particle Beam in an Image-Forming System

In Fig. 5.4, the formation of an image by a thin lens is shown both in real space and in phase space. Particles leaving the source but failing to pass through the indicated lens aperture at z_3 are not shown. The action of a thin lens, i.e., the transfer from a profile plane at z_3 just in front of the thin lens to a profile plane at z_4 just behind the thin lens of focal length f, is described by Eq. (1.6),

$$x_4 = x_3,$$

and $\tan \alpha_4 = \tan \alpha_3 - x_3/f$ or with $x_3 = x_4$, approximately

$$a_4 = a_3 - (x_3/f).$$

Here, x_3, a_3 and x_4, a_4 characterize the trajectory under consideration at z_3 and z_4, respectively. Since $\Delta a_4 = -x_3/f$ is proportional to x_4, the thin-lens action can be interpreted as a shearing force acting parallel to the a axis in the x, a phase-space diagram of Fig. 5.4. This shearing action must be combined with the shearing actions parallel to the x axis in the x, a phase space, caused by the field free regions $z_3 - z_1$ and $z_6 - z_4$ (see Section 5.2.1). By proper choices of the lengths of the drift distances $z_3 - z_1$ and $z_6 - z_4$, an object–image relation can be achieved between the profile planes z_1 and z_6 in Fig. 5.4. This object–image relation implies that all particles that originated from a point of the object are focused to a point of the image independent of the angle which the corresponding trajectory formed with the optic axis at z_1. In phase space, this requires that the lines AB and DC in Fig. 5.4, which were upright at the object, become upright again at the image.

5.2.3 Virtual Object Lens

In Fig. 5.5 a thin-lens imaging system is again indicated. Also, in Fig. 5.5 each point of an object of size $2x_{10}$ emits particles at different angles; however, only those particles which pass through an entrance pupil are allowed to enter the optical system. In phase space the sides AB, CD of the parallelogram are upright at the object ($z = z_1$), and the sides BC and

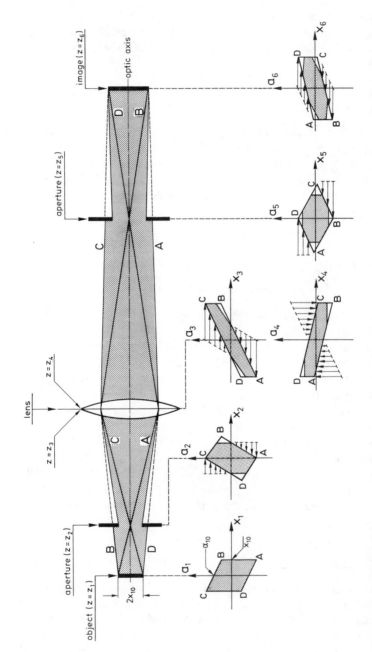

Fig. 5.4. Particle trajectories in an image-forming optical system in real space as well as in phase space. Note the horizontal shearing actions in the field-free regions and the vertical shearing action at the position of the thin lens. Note also that diaphragms limit the particle beam at z_2 and z_5, i.e., at positions at which the trajectories AC or BD cross the optic axis. If the widths of these diaphragms are chosen to be $1/\sqrt{2}$ times or about 70% of the total beam widths at these locations, the parallelogram-like phase-space area becomes a symmetric octogon-like area, as indicated.

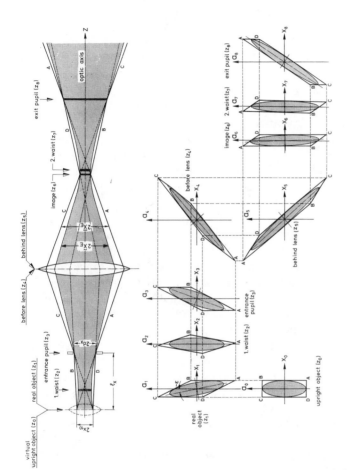

Fig. 5.5. A beam moving through an image-forming system that includes a virtual object lens. Note the horizontal shearing actions on the phase-space area in the field-free regions (from z_1 to z_4 and from z_5 to z_8), as well as the vertical shearing actions caused by the virtual thin-object lens (from z_0 to z_1) and by the main thin lens (from z_4 to z_5). Note that the parallelogram-like phase-space areas are transformed such that their sides AB and CD are upright at z_0, z_1 and z_6, i.e., at object and image, and that its sides AD and CB are upright at z_3 and z_8, i.e., at the entrance and exit pupils. Note further that the elliptical phase-space areas are transformed such that they are upright at z_0, z_2 and z_7, i.e., at the beam waists. In the case of a thick lens, there would be another upright ellipse between z_4 and z_5, characterizing a beam width maximum.

145

DA are upright at the pupil ($z = z_3$). All trajectories that started from a point of the object at z_1 are finally focused to a corresponding point of the image so that at z_6 in the x, a phase space, the parallelogram sides *AB* and *CD* are upright again. All trajectories that had passed through one point of the entrance pupil at z_3 are analogously focused to some point of a so-called exit pupil at z_8 where in phase space the parallelogram sides *BC* and *DA* are upright again. Note that an image as well as a pupil are characterized by discontinuities of the beam envelope and by upright sides of the phase-space parallelogram. Thus, it is not possible to distinguish between an image and a pupil if only the beam behind the optical system is accessible.

One could now change the position $l_x = z_3 - z_1$ and the diameter $2d_x$ of the entrance pupil so that $\pm d_x/l_x = \pm\tan \alpha_{10} \approx \pm a_{10}$ stays constant. Decreasing l_x, that is, moving the entrance pupil toward the object, causes the exit pupil to move toward the image. For negative values of l_x, the entrance pupil occurs before the image, i.e., the particle source is placed a distance l_x upstream from z_1 and thus illuminates the object. For $l_x = \infty$, the phase-space area at the object is an upright rectangle, and the exit pupil occurs at the second focal plane of the focusing system.

To sketch the phase-space areas of a given particle beam at different profile planes within an optical system, all particle trajectories must be determined that started at the object ($z = z_1$) within a parallelogram-like phase-space area as in Fig. 5.5. The resulting oblique phase-space areas are relatively difficult to describe mathematically. This description would be much simpler if one could postulate the original phase-space area at the object to have the shape of a rectangle instead of a parallelogram. On the other hand, it is quite simple to transform an assumed initial rectangular phase-space area into the really existing oblique parallelogram by the shearing action of an also assumed thin lens indicated at the object in Fig. 5.6. However, this simple, though rigorously correct, method of calculation requires the optical system under consideration to be virtually implemented by a thin lens (Wollnik, 1976). This virtual thin lens must be positioned at an infinitesimally short distance behind the real object while its focal length must be determined such that the assumed rectangular x, a phase-space area is sheared in the a direction by an amount such that the resulting shape is identical to the observed shape of the phase-space area at the real object. This same procedure also applies for an elliptical phase-space area, in which case an initially upright ellipse is transformed into the really existing phase-space ellipse by an appropriate shearing action of a virtual thin lens.

To determine the position vector $\mathbf{X}_n(z_n) = (x_n, a_n)$ of an arbitrary trajectory at the nth profile plane of an optical system, there are thus two

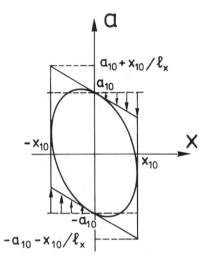

Fig. 5.6. The action of a virtual thin-object lens is shown which transforms an assumed upright rectangular phase-space area into the really existing parallelogram of Fig. 5.2.

procedures. Both require knowledge of the transfer matrix

$$T_x(z_n, z_1) = \begin{pmatrix} (x_n|x_1) & (x_n|a_1) \\ (a_n|x_1) & (a_n|a_1) \end{pmatrix}$$

between the profile plane n and the profile plane 1 at which a real object is located. In the first case, we find $X_n(z_n) = (x_n, a_n)$ simply from

$$\begin{pmatrix} x_n \\ a_n \end{pmatrix} = \begin{pmatrix} (x_n|x_1) & (x_n|a_1) \\ (a_n|x_1) & (a_n|a_1) \end{pmatrix} \begin{pmatrix} x_1 \\ a_1 \end{pmatrix}, \tag{5.3a}$$

where $X_1 = (x_1, a_1)$ describes a trajectory within the really existing phase space parallelogram at $z = z_1$ of area $4x_{10}a_{10}$ in Fig. 5.6 or within the inscribed ellipse of area $\pi x_{10}a_{10}$. Here, it holds that

$$|x_1| < x_{10}, \qquad |a_1| < a_{10} + x_1/l_x,$$

where l_x is the distance between the object and the entrance pupil of diameter $\pm dx = \pm l_x a_{10}$, as indicated in Fig. 5.5. In the second case in which a virtual thin object lens is used, we find $X_n(z_n) = (x_n, a_n)$ from

$$\begin{pmatrix} x_n \\ a_n \end{pmatrix} = \begin{pmatrix} (x_n|x_0) & (x_n|a_0) \\ (a_n|x_0) & (a_n|a_0) \end{pmatrix} \begin{pmatrix} x_0 \\ a_0 \end{pmatrix}$$

$$= \begin{pmatrix} (x_n|x_1) & (x_n|a_1) \\ (a_n|x_1) & (a_n|a_1) \end{pmatrix} \begin{pmatrix} 1 & 0 \\ -1/f_x & 1 \end{pmatrix} \begin{pmatrix} x_0 \\ a_0 \end{pmatrix}, \tag{5.3b}$$

where $\mathbf{X} = (x_0, a_0)$ describes an arbitrary trajectory at z_0 within the assumed virtual phase-space rectangle (or inscribed ellipse) of area $4x_{00}a_{00}$ in Fig. 5.6. Here, the simple inequality relations hold

$$|x_0| < x_{00}, \qquad |a_0| < a_{00}.$$

For $x_{00} = x_{10}$ and $a_{00} = a_{10}$, both methods yield the same phase-space area at z_n if f_x, the focal length of the virtual object lens, is chosen to be the distance l_x between the object and the entrance pupil of diameter $\pm d_x = \pm l_x a_{10}$ as in Fig. 5.2. This statement is easily proven if one determines the positions of trajectories characterizing the four corners of the phase-space parallelogram once by Eq. (5.3a) and once by Eq. (5.3b). Note that at z_6, i.e., at the position of the image in Fig. 5.5, it holds that

$$(x_6|a_0) = (x_6|a_1) = 0,$$

and that at z_8, i.e., at the position of the exit pupil in Fig. 5.5, we find

$$(x_8|x_0) = (x_8|x_1) - (x_8|a_1)/l_x = 0.$$

5.3   BEAM ENVELOPES

A particle beam is defined by the magnitude of its cross section and the range of angles under which individual particle trajectories pass through arbitrary points of this cross section. In the x, a phase space, such beams normally have the following shapes:

(1)   *Parallelograms,*
(2)   *Ellipses,* inscribed within these parallelograms with the midpoints of the parallelogram sides touching the ellipse (see Figs. 5.2, 5.5, and 5.6) as tangents, and
(3)   *Octogons,* as in Fig. 5.4 with all sides tangential to the inscribed ellipse. The midpoints of the octogon sides here touch the ellipse and the lengths of the octogon sides are $\sqrt{2} - 1$ times the length of the parallelogram sides with which they overlap (Fig. 5.7).

As shown in Fig. 5.2, x, a phase-space parallelograms are formed by the size $2x_{10}$ of the object and the size $2d_x$ of an angle-limiting aperture a distance l_x downstream in Fig. 5.4, while x, a phase-space octogons are formed by two beam-limiting diaphragms positioned at z_2 and z_5, i.e., at those z values at which the trajectories B and D or the trajectories A and C intersect. The apertures of these diaphragms must be chosen such that their widths are $1/\sqrt{2} \approx 0.7$ times the widths of the original particle beam at z_2 and z_5, respectively. This ratio is found by inspecting in Fig. 5.7 the length of the diagonal of a square and the diameter of an inscribed circle

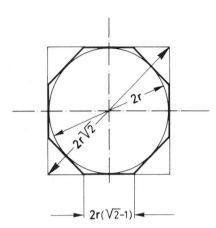

Fig. 5.7. A square is shown with an inscribed circle and octogon.

and noting that all discussed phase-space transformations cause only linear distortions of Fig. 5.7. Note also that x, a phase-space ellipses are formed by many diaphragms that scrape off particles in the corners of a parallelogram, an octogon, a hexadekagon, etc.

When designing optical systems with beams of elliptical phase-space areas, we should remember that the calculation asumes this ellipse to exist from the very beginning, whereas in reality this is true only behind many beam-limiting diaphragms. Unfortunately, the intercepted particles may be partially scattered on these obstacles so that they are not eliminated and can still contribute to the background in the system. When designing systems that are not infinitely long, it is thus advisable to use the full parallelogram-like phase-space area to determine the performance of all optical elements or to introduce diaphragms, as in Fig. 5.4, to approximate an elliptical phase-space area by an octogon. Since the scattering cross section is especially large for small-angle scattering events (Moliere, 1948), it is always advisable to limit a beam by razor-blade-like diaphragms that are perpendicular to the optic axis and avoid surfaces that are more or less parallel to it. In case of low-energy beams, such razor-blade-like diaphragms also show reduced charge-up effects caused by almost unavoidable layers of insulating deposits since the diaphragm surfaces close to the beam are then extremely short (Fig. 5.8).

5.3.1 Envelopes of Beams with Parallelogram-Like Phase-Space Areas

The envelope of the particle beam in Fig. 5.5 is either formed by the trajectories B, D or by the trajectories A, C. At an arbitrary z_n, these four

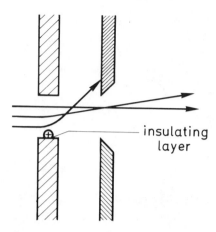

Fig. 5.8. Normal and razor-blade-like diaphragms. The number of small-angle scattering events for ions impinging on the inner wall of the diaphragm is proportional to the thickness of the diaphragm so that a sharp-edged diaphragm should always be preferred. This is even more so if one considers the build-up of an insulating layer on a diaphragm, for instance, some polymerized organic deposit. Such a layer can easily be charged by spurious ions and as illustrated deflect the following ions from their usual trajectories.

trajectories are separated from the optic axis by x_{An}, x_{Bn}, x_{Cn}, x_{Dn}, with,

$$x_{An} = -x_{Cn}, \qquad x_{Bn} = -x_{Dn}.$$

In Section 5.2.3, we had postulated each optical system to start with a virtual thin object lens that transforms an assumed upright rectangular phase-space area into the really existing one of parallelogram-like shape. Thus, all calculations are started from an upright rectangular phase-space area with corners at the coordinates $\pm x_{00}$, $\pm a_{00}$. For the trajectories A and B of Fig. 5.5, we find from Eq. (5.3a),

$$x_{An} = (x_n|x_0)x_{00} - (x_n|a_0)a_{00},$$

$$x_{Bn} = (x_n|x_0)x_{00} + (x_n|a_0)a_{00}.$$

The absolute value of the larger of these two values is directly the magnitude of the envelope $\bar{R}_{xn} = x_{max}(z_n)$ of the particle beam at z_n:

$$\bar{R}_{xn} = |(x_n|x_0)|x_{00} + |(x_n|a_0)|a_{00} \tag{5.4a}$$

(Wollnik, 1976). Analogously, we find the magnitude $S_{xn} = a_{max}(z_n) = \sin \alpha_{max}(z_n)$, where α_{max} is the maximum angle of inclination of a trajectory of the particle beam at z_n:

$$\bar{S}_{xn} = |(a_n|x_0)|x_{00} + |(a_n|a_0)|a_{00}. \tag{5.4b}$$

5.3.2 Envelopes of Beams with Octogon-Like Phase-Space Areas

Analogous to Eq. (5.4a), we find the beam envelope for an octogon-like phase-space area at z_n. For an initial phase-space rectangle with corners at the coordinates $\pm x_{00}$, $\pm a_{00}$, the corresponding phase-space octogon is characterized by the points $P_1[\pm x_{00}, \pm(\sqrt{2} - 1)a_{00}]$ and $P_2[\pm(\sqrt{2} - 1)x_{00}, \pm a_{00}]$, according to Fig. 5.7. Thus, the beam envelope is given by the larger of the expressions

$$\bar{R}_{xn1} = |(x_n|x_0)|x_{00} + |(x_n|a_0)|a_{00}(\sqrt{2} - 1), \tag{5.5a}$$

$$\bar{R}_{xn2} = |(x_n|x_0)|x_{00}(\sqrt{2} - 1) + |(x_n|a_0)|a_{00}. \tag{5.5b}$$

The relative decrease in beam widths as compared to Eq. (5.4a) is thus given by either one of the relations

$$\frac{\bar{R}_{xn} - \bar{R}_{xn1}}{\bar{R}_{xn}} = \frac{(2 - \sqrt{2})|(x_n|x_0)|x_0}{|(x_n|x)|x_0 + |(x_n|a_0)|a_0}, \tag{5.6a}$$

$$\frac{\bar{R}_{xn} - \bar{R}_{xn2}}{\bar{R}_{xn}} = \frac{(2 - \sqrt{2})|(x_n|a_0)|a_0}{|(x_n|x)|x_0 + |(x_n|a_0)|a_0}. \tag{5.6b}$$

For most beam positions z with large beam diameters, the quantity $|(x_n|a_0)|a_0$ is larger than $|(x_n|x_0)|x_0$. Thus, $\bar{R}_{xn1}$ is normally larger than $\bar{R}_{xn2}$, and the decrease in beam width must be determined from Eq. (5.6a).

5.3.3 Envelopes of Beams with Elliptical Phase-Space Areas

In the case of an elliptical phase-space area, the beam envelope is formed by different trajectories for each position z_n. Although at first sight this situation seems more, difficult than the one of Section 5.3.1 or 5.3.2, the mathematical treatment eventually becomes simpler. The reason is that for a beam-filling parallelogram- or octogon-like phase-space areas, the envelope is a discontinuous function, whereas for a beam-filling elliptical phase-space area, the envelope varies smoothly with z.

5.3.3.1 Twiss Parameters A_T, B_T, C_T

Consider an ellipse as in Fig. 5.9 with major and minor axes of lengths l_a and l_b. In a $\bar{x}$, $\bar{a}$-coordinate system that coincides with these axes the ellipse is described by

$$(\bar{x}/l_a)^2 + (\bar{a}/l_b)^2 = 1.$$

In an x, a-coordinate system rotated (as in Fig. 5.9) by an angle θ with respect to the $\bar{x}, \bar{a}$ coordinates, this ellipse of area $\pi \varepsilon_x = \pi l_a l_b$ is described by

$$x^2 C_T + 2xa A_T + a^2 B_T = \varepsilon_x, \tag{5.7a}$$

where A_T, B_T, C_T are the so-called Twiss parameters (often referred to as α, β, γ; see Courant and Snyder, 1958), which read explicitly,

$$A_T = (l_a/l_b - l_b/l_a) \sin\theta \cos\theta,$$

$$B_T = (l_b/l_a) \sin^2\theta + (l_a/l_b) \cos^2\theta,$$

$$C_T = (l_a/l_b) \sin^2\theta + (l_b/l_a) \cos^2\theta.$$

Combining these relations, we find for any value of θ,

$$B_T C_T - A_T^2 = 1. \tag{5.8}$$

Replacing C_T by $(1 + A_T^2)/B_T$, according to Eq. (5.8), we can also rewrite Eq. (5.7a) as either

$$x^2 + (xA_T + aB_T)^2 = B_T \varepsilon_x, \tag{5.7b}$$

or

$$(xC_T + aA_T)^2 + a^2 = C_T \varepsilon_x. \tag{5.7c}$$

Also note that for known Twiss parameters A_T, B_T, C_T, the inclination of the phase-space ellipse is found from simple trigonometry as

$$\tan 2\theta = \frac{2A_T}{C_T - B_T} = \frac{2A_T B_T}{1 + A_T^2 B_T^2}. \tag{5.9}$$

The trajectory with the largest x value at some z_n, i.e., $x_{max}(z_n) = R_{xn}$ in Fig. 5.9, is found at that point $(x = R, a = R')$ of the ellipse of Eq. (5.7a) for which $dx/da = (\partial\varepsilon_x/\partial a)/(\partial\varepsilon_x/\partial x)$ vanishes, i.e., for which $RA_{Tn} + R'B_{Tn} = 0$, or with Eq. (5.7b),

$$R_{xn} = \pm\sqrt{B_{Tn}\varepsilon_x}, \qquad R'_{xn} = \mp A_{Tn}\sqrt{\varepsilon_x/B_{Tn}}. \tag{5.10a}$$

and $\tan\theta_x = R'_{xn}/R_{xn}$. Analogous to Eq. (5.10a), we find the trajectory with the maximum a value at z_n, i.e., $a_{max}(z_n) = S_n$ (Fig. 5.9), at that point $(x = S, a = S')$ of the ellipse of Eqs. (5.7) for which $da/dx = (\partial\varepsilon_x/\partial x)/(\partial\varepsilon_x/\partial a)$ vanishes, i.e.,

$$S_{xn} = \mp A_{Tn}\sqrt{\varepsilon_x/C_{Tn}}, \qquad S'_{xn} = \pm\sqrt{C_{Tn}\varepsilon_x}, \tag{5.10b}$$

and $\tan\theta_a = S'_{xn}/S_{xn}$. Minima of the beam envelope R are called beam waists (see also Fig. 5.10). At such beam waists $(z = z_w)$, the phase-space

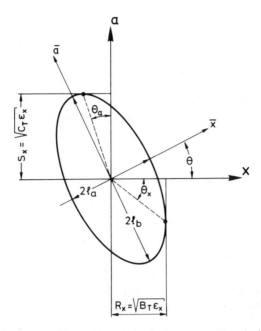

Fig. 5.9. A phase-space ellipse with major and minor axes of lengths l_a and l_b. Note the x, a- and $\bar{x}$, $\bar{a}$-coordinate systems rotated relative to each other by an angle θ. Note also R_x, the maximal x value as well as S_x, the maximal a value of any trajectory of the particle beam.

ellipse is upright, since clearly R_n in Fig. 5.9 is minimal for $\theta = 0$, π, 2π, $3\pi, \ldots$. For such cases, the coefficients of Eq. (5.7a) become

$$A_{Tw} = 0, \tag{5.11a}$$

$$B_{Tw} = 1/C_{Tw} = x_w/a_{\bar{w}}, \tag{5.11b}$$

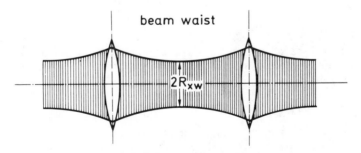

Fig. 5.10. The beam envelope is shown in a lens channel exhibiting waists in the middle between lenses.

where x_w and a_w denote the magnitudes of the major l_a and minor l_b axes, respectively, of the upright phase-space ellipse at $z = z_w$. Note that because A_T is proportional to $\sin 2\theta$ [see below Eq. (5.7a)], the condition $A_T = 0$ characterizes not only beam waists at $\theta = 0, \pi, 2\pi, \ldots$, but also beam maxima at $\theta - \pi/2 = 0, \pi, 2\pi, 3\pi, \ldots$. Note also that the products $x_w a_w$ at beam waists and at beam maxima both equal ε_x, i.e., the magnitude of the elliptical phase-space area $A_x = \pi x_w a_w$ of the particle beam [Eq. (5.2b)] divided by π.

5.3.3.2 Twiss Parameters and Transfer Matrices

Assume that the relation between the coordinates x_n, a_n and x_1, a_1 of an arbitrary trajectory at z_n and z_1, i.e., $\mathbf{X}_n = T_x \mathbf{X}_1$, is given as

$$\begin{pmatrix} x_n \\ a_n \end{pmatrix} = \begin{pmatrix} (x|x) & (x|a) \\ (a|x) & (a|a) \end{pmatrix} \begin{pmatrix} x_1 \\ a_1 \end{pmatrix}.$$

Assume further that the Twiss parameters A_{T1}, B_{T1}, C_{T1} are known, characterizing a phase-space ellipse of area $\pi\varepsilon_x$ at z_1 as

$$x_1^2 C_{T1} + 2x_1 a_1 A_{T1} + a_1^2 B_{T1} = \varepsilon_x. \tag{5.12a}$$

The Twiss parameters A_{Tn}, B_{Tn}, C_{Tn} of the phase-space ellipse of the same particle beam at z_n is then determined as

$$x_n^2 C_{Tn} + 2x_n a_n A_{Tn} + a_n^2 B_{Tn} = \varepsilon_x. \tag{5.12b}$$

By the above relation $\mathbf{X}_n = T_x \mathbf{X}_1$, the position vector $\mathbf{X}_n = (x_n, a_n)$ is given as function of the position vector $\mathbf{X}_1 = (x_1, a_1)$. Alternatively, $\mathbf{X}_1$ can be determined from $\mathbf{X}_n$ by the inverse relation $\mathbf{X}_1 = T_x^{-1} \mathbf{X}_n$, which reads explicitly,

$$\begin{pmatrix} x_1 \\ a_1 \end{pmatrix} = \begin{pmatrix} (a|a) & -(x|a) \\ -(a|x) & (x|x) \end{pmatrix} \begin{pmatrix} x_n \\ a_n \end{pmatrix}.$$

Introducing these x_1, a_1 into Eq. (5.12a) and comparing the coefficients of x_n^2, $x_n a_n$, a_n^2 with those in Eq. (5.12b), we find (Bovet et al., 1970; Brown, 1970)

$$\begin{pmatrix} B_{Tn} \\ A_{Tn} \\ C_{Tn} \end{pmatrix} = \begin{pmatrix} (x|x)^2 & -2(x|x)(x|a) & (x|a)^2 \\ -(x|x)(a|x) & (x|x)(a|a) + (x|a)(a|x) & -(x|a)(a|a) \\ (a|x)^2 & -2(a|x)(a|a) & (a|a)^2 \end{pmatrix} \begin{pmatrix} B_{T1} \\ A_{T1} \\ C_{T1} \end{pmatrix}. \tag{5.13}$$

Starting all calculations from z_1 with an upright elliptical phase-space area of size $\pi\varepsilon_x = \pi x_{00}a_{00}$, we find from Eqs. (5.11a) and (5.11b): $A_{T1} = 0$, $B_{T1} = 1/C_{T1} = x_{00}/a_{00}$. In this case, Eq. (5.13) simplifies to

$$\varepsilon_x B_{Tn} = (x|x)^2 x_{00}^2 + (x|a)^2 a_{00}^2, \tag{5.14a}$$

$$-\varepsilon_x A_{Tn} = (x|x)(a|x)x_{00}^2 + (x|a)(a|a)a_{00}^2, \tag{5.14b}$$

$$\varepsilon_x C_{Tn} = (a|x)^2 x_{00}^2 + (a|a)^2 a_{00}^2. \tag{5.14c}$$

The magnitude of the envelope R_{xn} of the particle beam at $z = z_n$ is determined from Eqs. (5.10a) and (5.14a) as

$$R_{xn} = \sqrt{(x|x)^2 x_{00}^2 + (x|a)^2 a_{00}^2}. \tag{5.15a}$$

Analogously, the maximum angle of inclination of any trajectory of the beam is found from Eqs. (5.10b) and (5.14c) as

$$S_{xn} = \sqrt{(a|x)^2 x_{00}^2 + (a|a)^2 a_{00}^2}. \tag{5.15b}$$

If the initial phase-space ellipse is oblique, Eqs. (5.15a) and (5.15b) apply as well, if the optical system under investigation is implemented by a virtual thin-object lens as in Eq. (5.3b). This object lens transforms the assumed upright ellipse [characterized by $A_{T0} = 0$, $B_{T0} = 1/C_{T0} \neq 0$] into the really existing phase-space ellipse at z_1 [characterized by $A_{T1} \neq 0$, $B_{T1} = (1 + A_{T1}^2)/C_{T1} \neq 0$]. The relations between these two sets of Twiss parameters are found by introducing the matrix elements of a virtual thin lens of focal length f_x, i.e. the elements of Eq. (1.6), into Eq. (5.13):

$$\begin{pmatrix} B_{T1} \\ A_{T1} \\ C_{T1} \end{pmatrix} = \begin{pmatrix} 1 & 0 & 0 \\ f_x^{-1} & 1 & 0 \\ f_x^{-2} & 2f_x^{-1} & 1 \end{pmatrix} \begin{pmatrix} B_{T0} \\ 0 \\ B_{T0}^{-1} \end{pmatrix}. \tag{5.16}$$

For a given phase-space ellipse at z_1, i.e., for known Twiss parameters A_{T1}, B_{T1}, C_{T1}, Eq. (5.16) yields

$$B_{T1} = B_{T0}, \tag{5.17a}$$

$$f_x = B_{T0}/A_{T1}. \tag{5.17b}$$

This states that the optical system under consideration must be implemented by a virtual thin-object lens of focal length B_{T1}/A_{T1}. The transfer matrix of this virtual thin-object lens thus reads

$$\begin{pmatrix} 1 & 0 \\ -A_{T1}/B_{T1} & 1 \end{pmatrix}.$$

Because a thin lens exerts only a shearing action parallel to the a axis in Fig. 5.6, the maximum x values of the upright ($A_{T0} = 0$) and of the oblique ($A_{T0} \neq 0$) ellipses are identical. With $B_{T0} = B_{T1}$, this x_1 value is found from Eq. (5.12a) for $a_1 = 0$ as

$$x_{00} = R_{x1} = \pm\sqrt{\varepsilon_x B_{T1}}. \tag{5.17c}$$

Furthermore, a_{00} equls that a_1 value of the oblique ellipse for which x_1 vanishes. This a_1 value is found from Eq. (5.12a) for $x_1 = 0$ as

$$a_{00} = S_{x1} = \pm\sqrt{\varepsilon_x / B_{T1}}. \tag{5.17d}$$

Using Eqs. (5.17a) and (5.17b), the Twiss parameters $A_{T1}, B_{T1}, C_{T1} = (1 + A_{T1}^2/ C_{T1})$ of the initial oblique ellipse are determined as

$$B_{T1} = x_{00}^2/ \varepsilon_x = x_{00}/ a_{00}, \tag{5.18a}$$

$$A_{T1} = B_{T1}/f_x = x_{00}/f_x a_{00}. \tag{5.18b}$$

5.3.3.3 The Envelope Equation

The Twiss parameters A_{Tn} and B_{Tn} in Eq. (5.7b) can also be expressed as functions of the magnitude of the beam envelope $R_x = \pm\sqrt{\varepsilon_x B_T}$ and of its inclination $dR_x/ dz = R_x' = \mp A_T\sqrt{\varepsilon_x/ B_T}$, as given in Eq. (5.10a). Thus, we find $B_T = R_x^2/ \varepsilon_x$, $A_T = - R_x R_x'/ \varepsilon_x$, and describe the phase-space ellipse with Eq. (5.8) and $x' \approx a$ as

$$[(\varepsilon_x^2/ R_x^2) + R_x'^2]x^2 - 2R_x R_x' xx' + R_x^2 x'^2 = \varepsilon_x^2. \tag{5.19}$$

This relation can be rewritten as $(R_x' x - R_x x')^2 + \varepsilon_x^2 x^2/ R_x^2 = \varepsilon_x^2$, which after a differentiation with respect to z, yields

$$R_x'' x - R_x x'' - (\varepsilon_x^2/ R_x^3)x = 0. \tag{5.20}$$

This is a differential equation for the beam envelope $R_x(z)$ where x and x'' denote the position $x(z)$ and its second derivative $d^2x(z)/ dz^2$ for some arbitrary particle trajectory.

If the arbitrary particle trajectory $x(z)$ is described by Eq. (1.16), the very general Hill's equation, $x'' + xk_x^2(z) = 0$, Eq. (5.20) becomes (see, for instance, Steffen, 1965) the so-called *envelope equation*:

$$R_x'' + k_x^2(z)R_x - (\varepsilon_x^2/ R_x^3) = 0. \tag{5.21}$$

For most problems of interest, the quantity $k_x(z)$ in this equation can be approximated by a piecewise constant function. Note that Eq. (5.21) differs from the trajectory equations [Eqs. (1.16), (3.4) and (4.57)] for a constant k_x. The difference is rather small as long as the beam diameter is large. In the neighborhood of beam waists, however, where R_x is small, the term ε_x^2/ R_x^3 becomes large so that R_x can never vanish.

5.4 POSITIONS AND SIZES OF ENVELOPE MINIMA

Assume a particle beam which at z_0 fills an upright phase-space rectangle of area $4\varepsilon_x = 4x_{00}a_{00}$ or an upright phase-space ellipse of area $\pi\varepsilon_x = \pi x_{00}a_{00}$. After passing this beam through an arbitrary optical system, the sizes as well as the positions of minima of the beam envelope can be determined from Eqs. (5.4a) and (5.15a). If such an arbitrary system starts with a virtual object lens as in Eq. (5.3b) and is followed by a drift distance $l = z_k - z_n$, the total transfer matrix reads from z_0 to z_k,

$$\begin{pmatrix} (x_k|x) & (x_k|a) \\ (x_k|x) & (a_k|a) \end{pmatrix} = \begin{pmatrix} 1 & l \\ 0 & 1 \end{pmatrix} \begin{pmatrix} (x|x) & (x|a) \\ (a|x) & (a|a) \end{pmatrix}. \tag{5.22}$$

5.4.1 Images and Pupils

In the case of a beam filling a parallelogram-like phase-space area, the matrix elements of Eq. (5.22) define the beam envelope at z_k according to Eq. (5.4a) as

$$\bar{R}_{xk} = |(x|x) + l(a|x)|x_{00} + |(x|a) + l(a|a)|a_{00}.$$

With increasing z_k, this envelope R_{xk} varies discontinuously (see also Fig. 5.5) at an image where $(x_k|a)$ vanishes, i.e., at $z_k = z_i$ with $l_i = z_i - z_n$,

$$l_i = z_i - z_n = -(x|a)/(a|a), \tag{5.23a}$$

or at a pupil at which $(x_k|x)$ vanishes, i.e., at $z_k = z_p$ with $l_p = z_p - z_n$,

$$l_p = z_p - z_n = -(x|x)/(a|x). \tag{5.23b}$$

Introducing l_i from Eq. (5.23a) or l_p from Eq. (5.23b) into the above $\bar{R}_{xk}$ yields the half-diameters of the corresponding image and pupil as

$$\bar{R}_{xi} = x_{00}/(a|a), \tag{5.24a}$$

$$\bar{R}_{xp} = a_{00}/(a|x), \tag{5.24b}$$

where use is made of $(x|x)(a|a) - (x|a)(a|x) = 1$, taken from Eq. (1.19b).

5.4.2 Beam Waists

In the case of a beam filling an elliptical phase-space area, the matrix elements of Eq. (5.22) define the beam envelope at the position z_k according to Eq. (5.15a) as

$$R_{xk} = \sqrt{[(x|x) + l(a|x)]^2 x_{00}^2 + [(x|a) + l(a|a)]^2 a_{00}^2}.$$

This envelope R_{xk} varies continuously with z_k through beam minima (waists) and maxima (see also Fig. 5.5). At any beam waist, i.e., at $z_k = z_w$ with $l = l_w = z_w - z_n$, it must be postulated that $(dR_{xk}/dl)_w$ vanishes so that

$$l_w = -\frac{(x|x)(a|x)x_{00}^2 + (x|a)(a|a)a_{00}^2}{(a|x)^2 x_{00}^2 + (a|a)^2 a_{00}^2}. \tag{5.25a}$$

Comparing the numerator and denominator of Eqs. (5.25a), (5.14b) and (5.14c), we find

$$l_w = A_{Tn}/C_{Tn}. \tag{5.25b}$$

Here, A_{Tn} and $C_{Tn} = (1 + A_{Tn}^2)/B_{Tn}$ characterize the phase-space ellipse at z_n, i.e., at the end of the optical system under consideration but before the additional drift distance l_w. Note that for negative values of l_w, a beam waist must be expected upstream from z_n.

The beam half-diameter R_{xw} at a beam waist is found by introducing the l_w of Eq. (5.25a) into the above R_{xk}, yielding with $\varepsilon_x = x_{00} a_{00}$,

$$R_{xw} = \frac{\varepsilon_x}{\sqrt{(a|x)^2 x_{00}^2 + (a|a)^2 a_{00}^2}}. \tag{5.26}$$

Comparing the denominator of this expression with Eq. (5.14c), we find

$$R_{xw} = \sqrt{\frac{\varepsilon_x}{C_{Tn}}} = \sqrt{\frac{\varepsilon_x B_{Tn}}{(1 + A_{Tn}^2)}}.$$

Knowing the transfer matrix from z_0 to z_n as in Eq. (5.22), we can thus determine the position $z_w = z_n + l_w$ of a beam waist as well as the waist diameter $2R_{xw}$.

If an image is formed in the neighborhood of a beam waist, we find the distance between this image at $l_i = z_i - z_n$ and the beam waist at $l_w = z_w - z_n$ from Eqs. (5.23a) and (5.25a) as

$$l_i - l_w = \frac{x_{00}^2(a|x)/(a|a)}{(a|x)^2 x_{00}^2 + (a|a)^2 a_{00}^2}. \tag{5.27}$$

5.4.3   Pupils and Waists for Point-to-Parallel Focusing Systems

For a vanishing coefficient $(a|a)$, the optical system under consideration is point-to-parallel focusing (see Section 1.3.2.4). Particle trajectories which originated from some object point with different angles are thus forced to be parallel behind the optical system. Introducing this $(a|a) = 0$ into Eqs. (5.23b) and (5.25a), we find $l_i = \infty$, and

$$l_p = l_w = -(x|x)/(a|x). \tag{5.28a}$$

Using these l_p and l_w, Eqs. (5.24b) and (5.26) yield equally large beam diameters at the positions of the pupil and the beam waist:

$$\bar{R}_{xp} = R_{xw} = a_{00}/(a|x). \tag{5.28b}$$

Introducing the length $l_p = l_w$ of Eq. (5.28a) into Eq. (5.22) further yields

$$\begin{pmatrix} (x_k|x) & (x_k|a) \\ (a_k|x) & (a_k|a) \end{pmatrix} = \begin{pmatrix} 0 & (x|a) \\ (a|x) & 0 \end{pmatrix}. \tag{5.29}$$

Since both $(x_k|x)$ and $(a_k|a)$ vanish, one can state the following:

A transfer matrix which expresses a waist-to-waist (or an upright object-to-upright-pupil relation) also connects the two focal planes [see Eq. (1.9)] of the optical system under consideration.

5.5 A MINIMAL SIZE BEAM ENVELOPE AT A POSTULATED LOCATION

In Section 5.4 it was shown how the minimum beam envelope can be found downstream from a given optical system. In Section 5.5 it shall now be shown how a minimum beam envelope can be achieved at a given location, for instance, a target position. Assume that at z_0, i.e., before some virtual object lens, a particle beam fills an upright phase-space rectangle of area $4\varepsilon_x = 4x_{00}a_{00}$ or an upright phase-space ellipse of area $\pi\varepsilon_x = \pi x_{00}a_{00}$. Assume further that some arbitrary optical system transports the beam to an arbitrary z_m. According to Eqs. (5.4a) and (5.15a) then, the beam half-diameters at z_m are found to be

$$\bar{R}_{xm} = |(x|x)|x_{00} + |(x|a)|\varepsilon_x/x_{00}, \tag{5.30a}$$

$$R_{xm} = \sqrt{(x|x)^2 x_{00}^2 + (x|a)^2 \varepsilon_x^2/x_{00}^2}. \tag{5.30b}$$

Evidently both $\bar{R}_{xm}$ and R_{xm} become large for both very small and very large values of x_{00} and are minimal for intermediate values. These optimal values are found from $d\bar{R}_{xm}/dx_{00} = 0$ and $dR_{xm}/dx_{00} = 0$, which both yield

$$(x_{00})_{opt} = \pm\sqrt{\varepsilon_x[(x|a)/(x|x)]}. \tag{5.31}$$

Equation (5.31) can be interpreted as either one or the other of the two following statements.

(1) For a given optical system and a beam with $\varepsilon_x = x_{00}a_{00}$, we should increase x_{00} and decrease a_{00} such that x_{00}/a_{00} equals $(x|a)/(x|x)$.

(2) For given values of x_{00} and a_{00}, we should modify the optical system such that $(x|a)/(x|x)$ equals x_{00}/a_{00}.

Introducing the $(x_{00})_{opt}$ of Eq. (5.31) into Eqs. (5.30a) and (5.30b), we find for the cases of a parallelogram-like and of an elliptical phase-space area,

$$(\bar{R}_{xm})_{opt} = \pm 2\sqrt{\varepsilon_x(x|x)(x|a)}, \qquad (5.32a)$$

$$(R_{xm})_{opt} = \pm\sqrt{2\varepsilon_x(x|x)(x|a)}, \qquad (5.32b)$$

showing that $(\bar{R}_{xm})_{opt}$ is $\sqrt{2}$ times larger than $(R_{xm})_{opt}$.

As an example, Eqs. (5.31), (5.32a) and (5.32b) may be used to design a magnet with a minimal air gap (Brown, 1970; Wollnik, 1970; Martin et al., 1984), whereas in the plane of deflection, i.e., in the xz plane, a wide particle beam is desired (see Section 9.2). In the heretofore perpendicular y direction, we clearly want a particle beam that is as narrow as possible so that the magnet air gap can be small and the magnet price low. Analogous to Eq. (5.22), the particle motion in the y direction can be described for a system as shown in Fig. 5.11 by

$$\begin{pmatrix} (y_m|y) & (y_m|b) \\ (b_m|y) & (b_m|b) \end{pmatrix} = \begin{pmatrix} 1 & l \\ 0 & 1 \end{pmatrix} \begin{pmatrix} (y|y) & (y|b) \\ (b|y) & (b|b) \end{pmatrix}. \qquad (5.33)$$

The elements of the right-hand matrix describe an optical system starting with a virtual object lens and ending at the entrance to the deflection magnet of radius ρ_0 and deflection angle ϕ_0. In this special case, l characterizes a position $\rho_0\phi$ at the optic axis in the magnet with $0 < \phi < \phi_0$. For $l = 0$ or $\phi = 0$, i.e., at the entrance of the magnet, and for $l = \rho_0\phi_0$, i.e., at the exit of the magnet, one may postulate the y envelopes of the beam to be equal and at the same time to be as small as possible.

To simplify this discussion, we assume that $(b|b)$ vanishes in Eq. (5.33), which is—at least approximately—correct in most cases of interest. The postulate of equal beam cross sections at the entrance and at the exit of

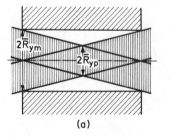
(a)

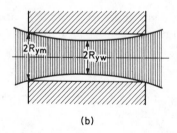
(b)

Fig. 5.11. The y envelopes are shown for two beams that are formed by some lens to pass optimally through the poles of a narrow gap magnet. Left: (a) a beam is assumed that occupies a parallelogram-like y, b phase-space area and Right: (b) a beam is assumed that occupies an elliptical y, b phase-space area.

the magnet then requires a pupil or a waist at the center of the magnet, that is, after exchanging x for y in Eq. (5.28a),

$$l_p = l_w = \rho_0 \phi_0 / 2 = -(y|y)/(b|y) = (y|y)(y|b). \qquad (5.34a)$$

Here we used the fact that Eq. (5.1b) postulates $(y|b) = 1/(b|y)$ for $(b|b) = 0$. Exchanging x for y also in Eq. (5.28b), we find the beam half-diameter at the position $l = \rho_0 \phi_0 / 2$ as

$$\bar{R}_{yp} = R_{yw} = b_{00}/(b|y). \qquad (5.34b)$$

The postulate of a minimal beam cross section at the entrance to the magnet requires an object of half-height,

$$(y_{00})_{opt} = \pm\sqrt{\varepsilon_y[(y|b)/(y|y)]}, \qquad (5.35)$$

as is found from Eq. (5.31) after exchanging x for y. Exchanging x for y also in Eqs. (5.32a) and (5.32b) yields the beam half heights at the entrance to the magnet as

$$(\bar{R}_{ym})_{opt} = \pm 2\sqrt{\varepsilon_y(y|b)(y|y)}, \qquad (5.36a)$$

$$(R_{ym})_{opt} = \pm\sqrt{2\varepsilon_y(y|b)(y|y)}, \qquad (5.36b)$$

in the cases of parallelogram-like and elliptical phase-space areas.

Replacing $(y|y)(y|b)$ in Eqs. (5.36a) and (5.36b) by $\rho_0 \phi_0 / 2$ according to Eq. (5.34a) yields the beam heights at the entrance and the exit of the sector magnet as

$$(\bar{R}_{ym})_{opt}/\sqrt{2} = (R_{ym})_{opt} = \pm\sqrt{\varepsilon_y \rho_0 \phi_0}. \qquad (5.37a)$$

Thus, the magnet air gap $+G_0$ must have at least the size of $\pm(\bar{R}_{ym})_{opt}$ or $\pm(R_{ym})_{opt}$, which are both sole functions of the length $\rho_0 \phi_0$ of the magnet and the phase-space area ε_y of the particle beam. Because of $b_{00} y_{00} = \varepsilon_y$ at the upright object, we can introduce into Eq. (5.34b) $(b_{00})_{opt} = \varepsilon_y/(y_{00})_{opt}$ with y_{00} determined in Eq. (5.35) and find the beam cross section at the pupil or waist as

$$\bar{R}_{yp} = R_{yw} = \pm\sqrt{\varepsilon_y(y|b)(y|y)} = \pm\sqrt{\varepsilon_y \rho_0 \phi_0 / 2}, \qquad (5.37b)$$

a value which is a factor of 2 or $\sqrt{2}$, respectively, smaller than the beam half-diameters of Eq. (5.37a), i.e., at the entrance or the exit to the magnet.

An optical system chosen to fulfill Eq. (5.37b) matches a given particle beam optimally to a magnet independent of the fact whether the occupied phase-space is parallelogram-like or elliptical. Only the necessary magnet air gap deviates by a factor $\sqrt{2}$ for the two cases. Since such an optical system must fulfill two conditions, at least two variables of the optical system must be open to choice. These two variables, for instance, can be the field strengths in two quadrupoles of a doublet preceding the sector magnet under consideration.

5.6 LIOUVILLE'S THEOREM AND ITS APPLICATION TO WIDE-ANGLE BEAMS

In Eqs. (5.1a) and (5.1b) we had assumed that the particle motions in the three space coordinates are independent of each other. A more general case is that only the motion in the z direction is independent of the x, y motion. In this case, we find $\iiint dx_i\, dp_{xi}\, dy_i\, dpy_i$ = constant, or with Eqs. (2.21d), (2.22a), and (2.22b) in the case of vanishing vector potentials,

$$p_0^2 \iiint n_i^2\, dx_i\, dy_i\, da_i\, db_i = \text{const}, \tag{5.38a}$$

where n_i is the refractive index at z_i. If both the magnetic vector potential and the electrostatic field vanish in the profile plane z_i, we speak of a standard profile plane. Since then n_i has the same value for all x_i, y_i, the quantity n_i may be taken out of the integral so that Eq. (5.38a) can be evaluated at $z_i = z_1$ and $z_i = z_2$, yielding

$$n_1^2 A_{xy}(z_1) = n_2^2 A_{xy}(z_2), \tag{5.38b}$$

$$A_{xy}(z_i) = \iiint dx_i\, d\left(\frac{\sin\alpha_i}{\sqrt{1+\tan^2\beta_i\cos^2\alpha_i}}\right) dy_i\, d\left(\frac{\sin\beta_i}{\sqrt{1+\tan^2\alpha_i\cos^2\beta_i}}\right),$$

where a_i and b_i are replaced by the expressions of Eqs. (2.7a) and (2.7b) and only monoenergetic particles of equal mass, i.e., $\delta_m = \delta_K = 0$, are taken into account.

5.6.1 Abbe's "Sine-Law" for a Two-Dimensional Phase Space

For two-dimensional wide-angle beams (i.e., for $y = \beta = 0$), Eq. (5.38b) describes the total phase-space area of a particle beam in the coordinates x and $a = \sin\alpha$ as $n \iint dx\, da = \text{const}_x$. Now assume that each point x_0 of an object with $|x_0| < x_{00}$ emits particles under angles α_0 with $|\alpha_0| < \alpha_{00}$ and that these particles are focused to an image of size $2x_{i0}$ each point x_i of which is illuminated under angles α_i with $|\alpha_i| < \alpha_{i0}$. In this case, Eq. (5.48b) becomes $n_0 \iint dx_0\, da_0 = n_i \iint dx_i\, da_i$, or

$$n_0 x_{00} \sin\alpha_{00} = n_i x_{i0} \sin\alpha_{i0}. \tag{5.39}$$

This is Abbe's sine law for charged particle optics, which may be expressed in words as follows:

The products of the beam half-diameters x_{i0} multiplied by the sines of the corresponding maximal angles of inclination α_{i0} are equal for all profile planes at the same electrostatic potential in which images or pupils exist. For profile *planes at different electrostatic potentials, this product varies as the quotient of the corresponding refractive indices.*

As an example, consider a particle source of dimaeter $2x_{00}$, which emits particles into all directions ($\alpha_{00} = \pm\pi/2$ or $\sin\alpha_{00} = \pm 1$) and assume that all these particles have energies eV_s. After an acceleration by a potential difference V_0, these particles may in some other standard profile plane form an image of diameter $2x_{i0} = M_x 2x_{00}$ the center of which is illuminated with maximum angles $\pm\alpha_{i0}$. With n proportional to $\sqrt{V_s}$ and $\sqrt{V_0 + V_s}$ according to Eq. (2.21d) for the two profile planes under consideration, Eq. (5.39) transforms to

$$M_x \sin\alpha_{i0} = \sqrt{V_s/(V_0 + V_s)} \approx \sqrt{V_s/V_0}. \tag{5.40}$$

For $V_s = 1$ V, $V_0 = 10,000$ V, and $M_x = -0.1$, we thus find $\sin\alpha_{i0} \approx \pm 0.1$, stating that the center point of this image is illuminated under a maximal angle $\alpha_{i0} \approx \pm 5.7°$.

5.6.2 Current Density

For realistic wide-angle beams in a four-dimensional phase space, the areas occupied by a particle beam at different z are characterized by Eq. (5.38b). At the positions of the object z_0 and of a stigmatic image z_i, the corresponding integrals can be solved, since in these profile planes the boundaries of the phase-space volumes occupied by the particle beam are perpendicular to the x and y axes. If the particles are emitted from each point of an object of area $\pm x_{00}y_{00}$ under solid angles $\pm\alpha_{00}\beta_{00}$ and focused to an image of area $\pm x_{i0}y_{i0}$ each point of which is illuminated under solid angles $\pm\alpha_{i0}\beta_{i0}$, Eq. (5.38b) yields $n_0^2 2x_{00}2y_{00}2a_{00}2b_{00} = n_i^2 2x_{i0}2y_{i0}2a_{i0}2b_{i0}$, or explicitly with Eqs. (2.7a) and (2.7b),

$$\frac{n_0^2 2x_{00}2y_{00}2 \tan\alpha_{00}2 \tan\beta_{00}}{1 + \tan^2\alpha_{00} + \tan^2\beta_{00}} = \frac{n_i^2 2x_{i0}2y_{i0}2 \tan\alpha_{i0}2 \tan\beta_{i0}}{1 + \tan^2\alpha_{i0} + \tan^2\beta_{i0}}, \tag{5.41}$$

where n_0 and n_i describe the refractive indices at object and image.

If there are no particle losses, as one normally assumes, the total particle current J_0 leaving the object must be identical to the current J_i arriving at the image. For a beam in which the current density j is equal at every point of a given cross section, we find $j_0 = J_0/(2x_{00}2y_{00})$ and $j_i = J_i/(2x_{i0}2y_{i0})$. Thus, Eq. (5.41) yields $j_i = j_0(x_{00}y_{00})/(x_{i0}y_{i0}) = j_0(n_i/n_0)^2(a_ib_i)/(a_{00}b_{00})$, or explicitly,

$$j_i = j_0 \left(\frac{n_i}{n_0}\right)^2 \left(\frac{\tan\alpha_{i0} \tan\beta_{i0}}{\tan\alpha_{00} \tan\beta_{00}}\right) \left(\frac{1 + \tan^2\alpha_{00} + \tan^2\beta_{00}}{1 + \tan^2\alpha_{i0} + \tan^2\beta_{i0}}\right). \tag{5.42}$$

For a system with equal magnifications $M_x = M_y$ in the x and y directions, we find $x_{00}/x_{i0} = y_{00}/y_{i0}$ and consequently, $\alpha_{00}/\alpha_{i0} = \beta_{00}/\beta_{i0}$. If, furthermore, z_0 is a position before the particles are accelerated, i.e., $\alpha_{00} = \pm \beta_{00} = \pm \pi/2$ or $a_{00} = b_{00} = 1/\sqrt{2}$, we find $\alpha_{i0} = \beta_{i0}$, so that Eq. (5.42) yields $j_i = j_0(n_i/n_0)^2 2a_{i0}^2$, or explicitly,

$$j_i = j_0 \left(\frac{n_i}{n_0}\right)^2 \frac{2\tan^2 \alpha_{i0}}{1 + 2\tan^2 \alpha_{i0}} = j_0 \left(\frac{n_i}{n_0}\right)^2 \sin^2 u_{i0}, \qquad (5.43)$$

where u_{i0} is the angle formed by one of the bundle-limiting trajectories and the optic axis at $z = z_i$, so that $\tan^2 u_{i0} = 2 \tan^2 \alpha_{i0}$.

As an example, we may try to find the current density limit for an oscilloscope or television tube with equal magnifications in the directions of x and y so that Eq. (5.43) is valid. At the cathode, the electrons have an energy kT so that n_0 is proportional to $\sqrt{kT}$. At the scintillation screen, the electrons are accelerated by an anode–cathode voltage V_i to an energy $kT + eV_i$ so that n_i is proportional to $\sqrt{kT + eV_i}$. With this condition Eq. (5.43) yields with $kT + eV_i \approx eV_i$,

$$j_i = j_0(1 + eV_i/kT) \sin^2 u_{i0}. \qquad (5.44)$$

Equation (5.43) is identical to the "Langmuir limit"—the maximum current density that Langmuir (1937) obtained—under the assumption that the distribution of electron energies in a beam is Maxwellian.

In case of a system that has different lateral magnifications M_x and M_y, we find

$$x_{i0} = M_x x_{00}, \qquad y_{i0} = M_y y_{00}, \qquad a_{i0} = a_{00}/M_x, \qquad b_{i0} = b_{00}/M_y.$$

With $\alpha_{00} = \beta_{00} = \pi/2$ or $a_{00} = b_{00} = 1/\sqrt{2}$, Eq. (5.43) yields $j_i = j_0(n_i/n_0)^2 2a_{i0}b_{i0}$. For $M_y \gg M_x$ or $\beta_{i0} \ll \alpha_{i0}$, we find from Eqs. (2.7a) and (2.7b),

$$a_{i0} = \sin \alpha_{i0}, \qquad b_{i0} = \tan \beta_{i0} \cos \alpha_{i0}.$$

Thus, Eq. (5.42) becomes

$$\begin{aligned} j_i &= j_0(n_i/n_0)^2 2 \sin \alpha_{i0} \cos \alpha_{i0} \tan \beta_{i0} \\ &= j_0(n_i/n_0)^2 (M_x/M_y) 2 \sin^2 \alpha_{i0}, \end{aligned} \qquad (5.45)$$

since $M_x/M_y = b_{i0}/a_{i0} = \tan \beta_{i0}/\tan \alpha_{i0}$.

5.7 BEAMS WITH SPACE CHARGE

If two charged particles move parallelly to another, they both experience a repelling force so that the resulting particle trajectories become divergent

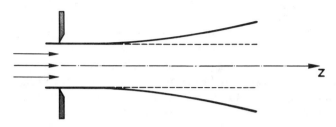

Fig. 5.12. Particle trajectories are shown that are originally parallel and become increasingly divergent due to space-charge forces. Dashed lines mark the same trajectories in the absence of space charge.

in Fig. 5.12. As long as the separation between the trajectories is large, this effect is negligible. However, if the distances between neighboring trajectories become small, these so-called space-charge forces are large and in extreme cases dominant over the forces of external fields.

Assume for the moment a beam with a constant charge density q and a circular cross section that varies only slightly along the optic axis. This space charge causes a radial field strength E_r in the beam with

$$\int_0^A E_r \, dA = \int_0^V (q/\varepsilon_0) \, dV.$$

Here, $A = 2\pi r l$ is the area of the inner cylinder surface of Fig. 5.13 in square meters, $V = \pi r^2 l$ is the corresponding cylinder volume in cubic meters, and $\varepsilon_0 = 8.85 \times 10^{-12}$ Cb/Nm^2. Carrying out the integrations, we find

$$E_r = qr/2\varepsilon_0 = Ir/2\varepsilon_0\pi R^2 v, \tag{5.46}$$

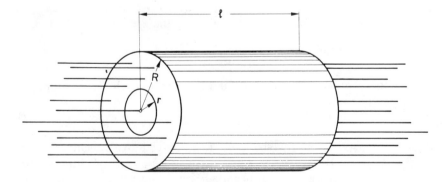

Fig. 5.13. A cylinder-like portion of an ion beam is indicated. The beam axis coincides with the axis of the cylinder of length l and radius R.

with the total current being $I = qR^2\pi v$, where $v = l/\Delta t$ is the particle velocity. Note here that the space-charge force increases linearly with r as the force in a lens [Eq. (1.4)].

In most realistic cases, the cross section of a particle beam is not circular but at the best elliptical, with the major and minor axes R_x and R_y orientated in the x and y directions. In case the beam envelopes do not vary appreciably along the optic axis, the space charge produces an electric field with the components (Lawson, 1977)

$$E_x = \frac{qR_y x}{\varepsilon_0(R_x + R_y)} = \frac{Ix}{\varepsilon_0 \pi v R_x(R_x + R_y)}, \tag{5.47a}$$

$$E_y = \frac{qR_x y}{\varepsilon_0(R_x + R_y)} = \frac{Iy}{\varepsilon_0 \pi v R_y(R_x + R_y)}, \tag{5.47b}$$

$$E_z = 0, \tag{5.47c}$$

where the total beam current I equals $q(\pi R_x R_y)v$.

Using Eqs. (5.47a)–(5.47c), the equations of motion [Eqs. (2.6) or (4.57a) and (4.57b)] become with $v = \sqrt{\eta(1 + \eta)}[2c/(1 + 2\eta)]$, according to Eq. (2.2b),

$$x'' = \left[k_x^2 - \frac{I}{v\varepsilon_0 \pi R_x(R_x + R_y)} \right] x = 0, \tag{5.48a}$$

$$y'' = \left[k_y^2 - \frac{I}{v\varepsilon_0 \pi R_y(R_x + R_y)} \right] y = 0. \tag{5.48b}$$

This shows that the space-charge forces reduce or overcompensate any focusing action ($k_x^2 > 0$ and/or $k_y^2 > 0$) caused by external fields. These relations differ from Hill's equation in that the force on the particle is not only proportional to k_x, which may vary with z, but that this force also depends on the beam envelopes R_x and R_y. To determine R_x, we might introduce Eq. (5.48a) into Eq. (5.20), yielding the x-envelope equation,

$$R_x'' + k_x^2 R_x - \frac{I}{v\varepsilon_0 \pi(R_x + R_y)} - \frac{\varepsilon_x^2}{R_x^3} = 0. \tag{5.49a}$$

Analogously, we may determine R_y and find the y-envelope equation,

$$R_y'' + k_y^2 R_y - \frac{I}{v\varepsilon_0 \pi(R_x + R_y)} - \frac{\varepsilon_y^2}{R_y^3} = 0. \tag{5.49b}$$

Equations (5.49a) and (5.49b) were first derived by Kapchinskij and Vladimirskij (1959). As we see from Eqs. (5.49a) and (5.49b) the space-charge effects increase with the beam current I and with $1/v$ the particle velocity. Thus we can state that these effects are proportional to $I\sqrt{m/K}$ and consequently are most important for low-energy particles of high mass.

In the case of nonnegligible space-charge forces, we are normally satisfied if we can determine the envelopes of a particle beam by Eqs. (5.49a) and (5.49b). However, single-particle trajectories can be determined from Eqs. (5.48a) and (5.48b) if necessary, but even such calculations do not describe the real world, since for both cases the current density was assumed to be constant over the beam cross section. For more realistic calculations, we therefore rely on numerical calculations of many individual particle trajectories. One corresponding computer code is BEAM-TRACE (see Berz, 1984 or Wollnik *et al.*, 1987). From an assumed initial current density distribution across the particle beam at the entrance to the optical system under consideration, this code determines the space-charge distribution and calculates individual particle trajectories under the assumption that this space-charge distribution is constant for a short section dz of the optical system. From the resulting trajectory distribution it then determines a new space-charge distribution and particle trajectories as before, however, now in a linearly varying space-charge distribution. This procedure is followed iteratively until the resulting trajectory distribution at dz no longer changes. As a next step, this code assumes the now-obtained linearly varying space-charge distribution to exist not only from 0 to dz but also from dz to $2\,dz$. Then it (1) determines particle trajectories from dz to $2\,dz$; (2) determines a new space-charge distribution at $2\,dz$ from the obtained trajectory distribution; and (3) calculates a different but again linearly varying space-charge distribution between dz and $2\,dz$. This procedure is followed iteratively until the resulting space-charge distribution at $2\,dz$ no longer varies. Analogously, the beam is followed from $2\,dz$ to $3\,dz$ to $4\,dz$, and so on, until eventually the trajectory distribution is known throughout the entire optical system. Current density distributions of a particle beam calculated in this fashion are indicated in Fig. 5.14 for various positions of z in a region free of external fields.

As outlined above, the space-charge effects can be taken into account when the effects of external electromagnetic fields are determined. However, it is also possible to compensate for the positive charges of ions by negative charges of low-energy electrons whereby the electrons are held in the ion beam by the positive space charge. Such electrons can be taken from some electron source, or can be produced from collisions of energetic ions with rest gas atoms resulting in a plasma throughout the beam. This situation more or less prevails when the beam passes through a magnetic field, since the electrons then move along small circles, which stay completely inside the ion beam. In an electrostatic field, however, the low-energy electrons are accelerated toward the positive electrode, which destroys the space-charge compensation completely. For this reason, electrostatic fields are normally shielded by a negatively charged diaphragm that is placed between

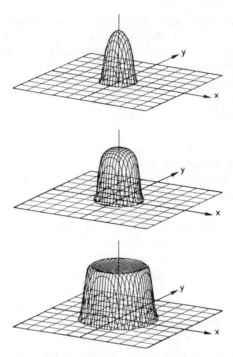

Fig. 5.14.   For an intense charged-particle beam propagating through a field-free region from $z = 0$ to $z = z_0$, the current density distribution is shown at several intermediate positions. Although the beam cross section increases rapidly with z, the lateral distribution of the beam is always scaled here to equal areas in order to demonstrate the effects of space charge. For the same reason the current density distributions are also always shown relative to the maximally occurring value at each z. Note that the originally parabolic distribution converts more and more to a homogeneous one, whereby, however, the current density is slightly higher in the outer sections of the beam.

the ground electrode and the actual electrostatic field, as indicated in Fig. 7.7, since this potential keeps all low-energy electrons from leaving the beam. The time Δt [in microseconds] necessary to produce equally many electrons as there are ions of energy K [in mega-electron-volts] in the beam depends (see Bernas *et al.*, 1954, or Wollnik, 1980) on the residual-gas pressure p [in millibars] with typically $\Delta t\, p\sqrt{K} \approx 10^{-5}$ [in μsec mbar$\sqrt{\text{MeV}}$]. An ion beam that has high-frequency intensity oscillations thus cannot easily be space-charge compensated.

Space-charge forces taken into account in Eqs. (5.49a) and (5.49b) or in computer codes like BEAM-TRACE assume a smooth variation of current density across the particle beam as does a space-charge compensation. However, in reality, the distance between two individual charged particles can become very small. In such a case, both particles experience very strong repelling forces. As Boersch (1954) observed in electron beams, such effects cause the current density distribution across the beam to acquire long tails (Rose and Spehr, 1983). Most probably the intense tails of space-charge compensated finely focused ion beams (see Camplan, 1981) are also due to the Boersch effect. This is even more so because of the finding of Osher (1977) that along with the low-energy electrons, positively charged low-energy ions are also formed that must be compensated for by even more electrons so that the final number of particles in a space-charge-compensated ion beam can greatly exceed the initial number of high-energy ions.

REFERENCES

Banford, A. P. (1966). "The Transport of Charged Particle Beams." E. & F. N. Spoon Ltd, London.
Bernas, R., Kaluszyner, L., and Druaux, J. (1954). *J. Phys. Radium* **15**, 273.
Berz, M. (1984). Thesis, Univ. Giessen.
Börsch, H. (1954). *Z. Phys.* **139**, 115.
Bovet, C., Gouivan, R., Gumowski, I., and Reich, K. H. (1970). CERN/MPS-SI/Int. DL/70/4.
Brown, K. L. (1970). *Private communication.*
Brown, K. L., and Howry, S. K. (1970). SLAC Rept. No. 91.
Camplan, J. (1981). *Nucl. Instr. Meth.* **187**, 157.
Courant, E. D., and Snyder, H. S. (1958). *Ann. Phys.* **3**, 48.
Helmer, J. C. (1966). *Am. J. Phys.* **34**, 222.
Kapchinskij, I. M., and Vladimirskij, V. V. (1959). *Proc. Conf. High Energy Acc.*, CERN, Geneva, 274.
Langmuir, D. (1937). *Proc. IRE* **25**, 977.
Lawson, J. D. (1977). "The Physics of Charged-Particle Beams." Clarendon Press, Oxford.
Martin, S., Hardt, A., Meissburger, F., Berg, G., Hacker, K., Hürlimann, W., Römer, J., Sagetka, T., Retz, A., Schnult, O., Brown, K. L., and Halbach, K. (1983). *Nucl. Instrum. Methods* **214**, 281.
Molière, G. (1948). *Z. Naturforsch.* **3a**, 78.
Osher, J. E. (1977). *Inst. Phys. Conf. Ser.* **38**, 201.
Rose, H., and Spehr, R. (1983). *In* "Applied Charged Particle Optics" (A. Septier, ed.), Vol. C, p. 475. Academic Press, New York.
Steffen, K. G. (1965). "High Energy Beam Optics." Wiley (Interscience), New York.
Stromberg, K. (1981). "An Introduction to Classical Real Analysis." Wadsworth, New York.
Wollnik, H. (1970). *Proc. Int. Conf. Elec. Isotope Separators, Marburg*; BMBW-Forschungsbericht K70-28, 282.
Wollnik, H. (1976). *Nucl. Instrum. Methods* **137**, 169.
Wollnik, H., Brezina, J., and Berz, M. (1987). *Nucl. Instrum. Method* (in print).

6

Particle Beams in
Periodic Structures

Particle trajectories in optical systems are described by Hill's equation [Eq. (1.16)], which reads for independent motions in the x and y directions,

$$x'' + k_x^2(z)x = 0, \tag{6.1a}$$

$$y'' + k_y^2(z)y = 0. \tag{6.1b}$$

The coefficients k_x and k_y describe restoring forces that drive a charged particle back to the optic axis with the magnitudes of these forces varying along the z coordinate. For $k_x = k_y = 0$, this restoring force vanishes so that Eqs. (6.1) describe particle trajectories in a field-free region. For non-vanishing k_x and k_y, Eqs. (6.1a) and (6.1b) describe particle trajectories in quadrupole lenses ($k_x^2 + k_y^2 = 0$) according to Eqs. (3.4a) and (3.4b), as well as in sector fields with rotationally symmetric pole faces ($k_x^2 = 1 - k_y^2$) or with toroidal electrodes [$k_x^2 = 1 - k_y^2 + (1 + 2\eta)^2$], according to Eqs. (4.57a) and (4.57b).

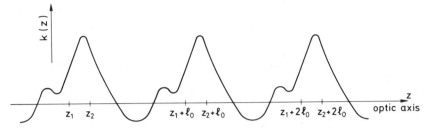

Fig. 6.1.   A periodically varying parameter k with period length l_0.

In this chapter we shall discuss Hill's general equation for optical systems that consist of many individual cells. In this case, the k_x and k_y of Eqs. (6.1a) and (6.1b) are periodic functions of z, as indicated in Fig. 6.1 and given by

$$k_x(z) = k_x(z + l_0), \qquad (6.2a)$$

$$k_y(z) = k_y(y + l_0). \qquad (6.2b)$$

6.1   SINGLE-PARTICLE TRAJECTORIES AND BEAM ENVELOPES

Assume that one cell of a periodic optical structure is described by the x-transfer matrix,

$$T_x = \begin{pmatrix} (x|x) & (x|a) \\ (a|x) & (a|a) \end{pmatrix}.$$

Assume further that the phase-space area of the particle beam under consideration is elliptical with an area $\pi \varepsilon_x$ and characterized by Twiss parameters A_{Tx1} and B_{Tx1}. Then the beam at the exit of the cell is described by a phase-space ellipse of area $\pi \varepsilon_x$ characterized by Twiss parameters

$$A_{Tx2} = f[(x|x), (x|a), (a|x), (a|a), A_{Tx1}, B_{Tx1}], \qquad (6.3a)$$

$$B_{Tx2} = g[(x|x), (x|a), (a|x), (a|a), A_{Tx1}, B_{Tx1}], \qquad (6.3b)$$

where f and g are functions defined by the optical properties of the cell. As a side condition here, it must be taken into account that according to Eq. (1.19b) the determinant of the previous transfer matrix equals 1:

$$(x|x)(a|a) - (x|a)(a|x) = 1. \qquad (6.3c)$$

Thus there are three equations [Eqs. (6.3a)–(6.3c)] to determine the four unknowns $(x|x)$, $(x|a)$, $(a|x)$, and $(a|a)$, which leaves one condition for us to choose freely. For this condition we can use the periodicity of single-particle trajectories over the repetition length l_s. This l_s is the length along the optic axis necessary for an individual particle to move once around the phase-space ellipse. The length l_s normally is not identical to the cell length l_0 of the optical structure. One way to determine l_s is to define a so-called "phase shift" (Courant and Snyder, 1958)

$$\psi_x = \int_z^{z+l_0} \frac{\varepsilon_x\, ds}{R_x^2} = \int_z^{z+l_0} \frac{ds}{B_{Tx}}, \tag{6.3d}$$

with $R_x = \sqrt{\varepsilon_x B_{Tx}}$ taken from Eq. (5.10a).

To understand the meaning of the phase shift ψ_x we assume an arbitrary phase ellipse given in Eq. (5.7b) as $x^2 + (xA_T + aB_T)^2 = \varepsilon_x$ and recognize that by replacing x and a by

$$x = -u\sqrt{B_T}, \qquad a = (v + uA_T)/\sqrt{B_T}, \tag{6.4}$$

(Wilson, 1977) we find $u^2 + v^2 = \varepsilon_x$. The x, a ellipse thus has turned into an u, v circle (Fig. 6.2), with the consequence that any point of this circle can be characterized by

$$w = u + iv = \sqrt{\varepsilon}\, e^{i\sigma}.$$

Here, σ is the angle by which a particle has advanced in moving around the u, v circle by going from z_1 to z_2. As is shown later, this angle σ is

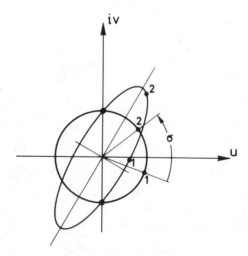

Fig. 6.2. An x, a phase-space ellipse is transformed into an u, v circle. The points marked 1 or 2, characterizing a particle at times t_1 and t_2, are shown both on the x, a ellipse and on the u, v circle. Note that the so-called phase advance is the angle σ.

identical to the phase shift ψ of Eq. (6.3d) for $z_2 = z_1 + l_0$, i.e., in case one has proceeded by a full cell of length l_0.

The u coordinate of a point on the u, v circle is the real part of $\sqrt{\varepsilon} \exp(i\sigma)$, according to the above definition of w. With $x = -u\sqrt{B_T}$ according to Eq. (6.4) and $R = \sqrt{\varepsilon B_T}$ according to Eq. (5.10a), this relation becomes the real part of

$$x = R\,e^{i\sigma}, \tag{6.5a}$$

and we find by differentiating Eq. (6.5a) with respect to z:

$$x' = (R' + iR\sigma')\,e^{i\sigma}, \tag{6.5b}$$

$$x'' = [(R'' - R\sigma'^2) + i(2R'\sigma' + R\sigma'')]\,e^{i\sigma}. \tag{6.5c}$$

Introducing these x and x'' into Eq. (6.1a) yields $R'' + k_x^2 R - R\sigma'^2 + i(2R'\sigma' + R\sigma'') = 0$, which transforms to the well-known (Steffen, 1965) envelope equation [Eq. (5.20)],

$$R'' = k_x^2 R - (\varepsilon_x^2 / R^3) = 0,$$

if we define $\sigma = \int \varepsilon_x / R^2$, as was done in Eq. (6.3d) for one cell, so that

$$\sigma' = \varepsilon_x / R^2 \quad \text{and} \quad \sigma'' = -2\varepsilon_x R' / R^3.$$

If Eq. (6.5a) is a solution of Hill's equation [Eq. (6.1a)], then any linear combination must also be a solution. Thus, we can write

$$x(z) = R(pe^{i\sigma} + qe^{-i\sigma}) = f(R, \sigma), \tag{6.6a}$$

with the first derivative of this $x(z)$ being $x'(z) = dx/dz$, or

$$x'(z) = R'(pe^{i\sigma} + qe^{-i\sigma}) + (i\sigma'R)(pe^{i\sigma} - qe^{-i\sigma}) = g(R, \sigma, R', \sigma'). \tag{6.6b}$$

Denoting the quantities x and $x' \approx a$ as well as R and σ at the beginning ($z = z_1$) and at the end ($z = z_2$) of a cell by indices 1 and 2, we can always express their relations in the familiar form of transfer matrices

$$\begin{pmatrix} x_2 \\ a_2 \end{pmatrix} = \begin{pmatrix} (x|x) & (x|a) \\ (a|x) & (a|a) \end{pmatrix} \begin{pmatrix} x_1 \\ a_1 \end{pmatrix}. \tag{6.7}$$

Replacing x_1, a_1 and x_2, a_2 in Eq. (6.7) by $f_1 = f(R_1, \sigma_1)$, $g_1 = g(R_1, \sigma_1, R_1', \sigma_1')$ and $f_2 = f(R_2, \sigma_2)$, $g_2 = g(R_2, \sigma_2, R_2', \sigma_2')$, we find two complex equations. Splitting these equations into real and imaginary parts yields four equations with the four unknowns $(x|x)$, $(x|a)$, $(a|x)$, and $(a|a)$, the elements of the transfer matrix of Eq. (6.7). With $\sigma' = \varepsilon_x / R^2$, $\sigma'' = -2\varepsilon_x R'/R^3$ as derived above and with $e^{i\sigma} = \cos \sigma + i \sin \sigma$, we finally find for $\sigma = \sigma_2 - \sigma_1$,

$$
\begin{pmatrix} (x|x) & (x|a) \\ (a|x) & (a|a) \end{pmatrix}
$$

$$
= \begin{pmatrix} R_2 \cos \sigma \dfrac{R_1'}{\varepsilon_x} \sin \sigma & \dfrac{R_1 R_2 \sin \sigma}{\varepsilon} \\ \dfrac{A}{-R_1 R_2} & \dfrac{R_1}{R_2} \cos \sigma + \dfrac{R_1 R_2'}{\varepsilon_x} \sin \sigma \end{pmatrix} \tag{6.8a}
$$

with

$$
A = (\varepsilon_x + R_1 R_1' R_2 R_2' / \varepsilon_x) \sin \sigma + (R_1 R_1' - R_2 R_2') \cos \sigma.
$$

Replacing R_1, R_2 and R_1', R_2' by the expression of Eqs. (5.10a) transforms Eq. (6.8a) (Brown, 1981) to

$$
\begin{pmatrix} (x|x) & (x|a) \\ (a|x) & (a|a) \end{pmatrix}
$$

$$
= \begin{pmatrix} \sqrt{\dfrac{B_{T2}}{B_{T1}}} (\cos \sigma + A_{T1} \sin \sigma) & \sqrt{B_{T1} B_{T2}} \sin \sigma \\ \dfrac{(1 + A_{T1} A_{T2}) \sin \sigma + (A_{T2} - A_{T1}) \cos \sigma}{-\sqrt{B_{T1} B_{T2}}} & \sqrt{\dfrac{B_{T1}}{B_{T2}}} (\cos \sigma - A_{T2} \sin \sigma) \end{pmatrix}. \tag{6.8b}
$$

For known coefficients of the left-hand transfer matrix, we thus find

$$
\sin \sigma = \frac{(x|a)}{\sqrt{B_{T1} B_{T2}}}, \tag{6.9a}
$$

$$
\tan \sigma = \frac{(x|a)}{(x|x) B_{T1} - (x|a) A_{T1}} = \frac{(x|a)}{(a|a) B_{T2} + (x|a) A_{T2}}. \tag{6.9b}
$$

At this point we should recall that there are many possibilities to pass a beam through a periodic optical structure. Each of these possibilities is characterized by a phase shift σ or an initial phase-space ellipse, i.e., by A_{T1} and B_{T1} in Eq. (6.8b).

6.1.1 Systems with Postulated Optical Properties

To illustrate the meaning of Eqs. (6.8a) and (6.8b) it is useful to discuss optical systems in which certain elements of the corresponding transfer matrix disappear (Brown, 1981).

6.1.1.1 Point-to-Point Focusing Systems

For the special case $(x|a) = 0$ in which (see Section 1.3.2.1) an object–image relation exists between the profile planes at z_1 and z_2, Eq. (6.9a) postulates $\sin \sigma = 0$ or with $n = 0, 1, 2, 3, \ldots$,

$$\sigma = n\pi, \tag{6.10}$$

and $\cos \sigma = \pm 1$. In this case, Eq. (6.8b) transforms to

$$\begin{pmatrix} (x|x) & 0 \\ (a|x) & (a|a) \end{pmatrix} = \begin{pmatrix} \sqrt{\dfrac{B_{T2}}{B_{T1}}} & 0 \\ \dfrac{A_{T1} - A_{T2}}{\sqrt{B_{T1}B_{T2}}} & \sqrt{\dfrac{B_{T1}}{B_{T2}}} \end{pmatrix}, \tag{6.11a}$$

which postulates $A_{T2} - A_{T1} = (a|x)(x|x)B_{T1} = (a|x)(a|a)B_{T2}$. For a telescopic system for which both $(x|a)$ and $(a|x)$ vanish simultaneously (see Section 1.3.2.2) Eq. (6.10) yields

$$A_{T1} = A_{T2}.$$

Thus, Eq. (6.11a) becomes

$$\begin{pmatrix} (x|x) & 0 \\ 0 & (a|a) \end{pmatrix} = \begin{pmatrix} M & 0 \\ 0 & 1/M \end{pmatrix}, \tag{6.11b}$$

with $M = \sqrt{B_{T2}/B_{T1}}$, in which case Eq. (5.13) reads

$$\begin{pmatrix} B_{T2} \\ A_{T2} \\ C_{T2} \end{pmatrix} = \begin{pmatrix} M^2 & 0 & 0 \\ 0 & 1 & 0 \\ 0 & 0 & 1/M^2 \end{pmatrix} \begin{pmatrix} B_{T1} \\ A_{T1} \\ C_{T1} \end{pmatrix}.$$

6.1.1.2 Point-to-Parallel Focusing Systems

For the special case $(a|a) = 0$, i.e., for a point-to-parallel focusing system (see Section 1.3.2.4), Eq. (6.9b) reads

$$\tan \sigma = 1/A_{T2}, \tag{6.12}$$

independent of the choice of B_{T1} or B_{T2}. In this case, Eq. (6.8b) transforms to

$$\begin{pmatrix} (x|x) & (x|a) \\ (a|x) & 0 \end{pmatrix} = \begin{pmatrix} (A_{T1} + A_{T2}) \sqrt{\dfrac{B_{T2}/B_{T1}}{1 + A_{T2}^2}} & \sqrt{\dfrac{B_{T1}B_{T2}}{1 + A_{T2}^2}} \\ -\sqrt{\dfrac{1 + A_{T2}^2}{B_{T1}B_{T2}}} & 0 \end{pmatrix}. \quad (6.13a)$$

If the system under investigation is not only point-to-parallel $[(a|a) = 0]$, but also parallel-to-point focusing $[(x|x) = 0]$, as described in Section 1.3.2.4, Eq. (6.13a) yields

$$A_{T1} = -A_{T2},$$

or $\tan \sigma = 1/A_{T2} = -1/A_{T1}$. In this case, Eq. (6.13a) reads

$$\begin{pmatrix} 0 & (x|a) \\ (a|x) & 0 \end{pmatrix} = \begin{pmatrix} 0 & f \\ -1/f & 0 \end{pmatrix}, \quad (6.13b)$$

with $f^2 = B_{T1}B_{T2}/(1 + A_{T1}^2)$. For $A_{T1} = 0$, this transfer matrix was already found in Eq. (5.29) as one that connects two beam waists, in which case Eq. (5.13) becomes

$$\begin{pmatrix} A_{T2} \\ B_{T2} \\ C_{T2} \end{pmatrix} = \begin{pmatrix} 0 & 0 & f^2 \\ 0 & -1 & 0 \\ 1/f^2 & 0 & 0 \end{pmatrix} \begin{pmatrix} A_{T1} \\ B_{T1} \\ C_{T1} \end{pmatrix}.$$

6.1.1.3 Optical Systems with a Phase Shift of $\sigma = (n + 1/2)\pi$

As another special case, we may postulate

$$\sigma = (n + 1/2)\pi,$$

or $\cos \sigma = 0$ and $\sin \sigma = +1$, with $n = 0, 1, 2, 3, \ldots$. In this case, Eq. (6.8b) transforms to

$$\begin{pmatrix} (x|x) & (x|a) \\ (a|x) & (a|a) \end{pmatrix} = \begin{pmatrix} A_{T1} \sqrt{\dfrac{B_{T2}}{B_{T2}}} & \sqrt{B_{T1}B_{T2}} \\ \dfrac{1 + A_{T1}A_{T2}}{-\sqrt{B_{T1}B_{T2}}} & -A_{T2} \sqrt{\dfrac{B_{T1}}{B_{T2}}} \end{pmatrix}. \quad (6.14)$$

6.1.2 Systems with Matched Beams

If the beam ellipses are identical at the entrance and at the exit of a cell, i.e., if

$$R = R_1 = R_2, \qquad R' = R_1' = R_2', \qquad B_T = B_{T1} = B_{T2}, \qquad A_T = A_{T1} = A_{T2},$$

we speak of a matched beam and of unit cells (Brown, 1981). In this case,

Eqs. (6.8) simplify to

$$
\begin{pmatrix} (x|x) & (x|a) \\ (a|x) & (a|a) \end{pmatrix} = \begin{pmatrix} \dfrac{RR'}{\varepsilon_x} \cos\mu \sin\mu & \dfrac{R^2}{\varepsilon_x} \sin\mu \\ -\left(\dfrac{\varepsilon_x}{R^2} + \dfrac{R'^2}{\varepsilon_x}\right)\sin\mu & \cos\mu + \dfrac{RR'}{\varepsilon_x}\sin\mu \end{pmatrix}
$$

$$
= \begin{pmatrix} \cos\mu + A_T \sin\mu & B_T \sin\mu \\ -C_T \sin\mu & \cos\mu - A_T \sin\mu \end{pmatrix}, \tag{6.15}
$$

where σ is renamed μ, and C_T is used for $(1 + A_T^2)/B_T$, according to Eq. (5.8). Note here that the trace of the transfer matrix of Eq. (6.15) is

$$
(x|x) + (a|a) = 2\cos\mu. \tag{6.16}
$$

Therefore μ is independent of A_T and B_T and consequently of the reference point z, which is not true, however, for the whole matrix. Note also that a stable motion of particles through a multitude of unit cells can only be expected for $|(x|x) + (a|a)| < 2$ so that $|\cos\mu|$ in Eq. (6.16) stays below 1.

For a unit cell as described by Eqs. (6.15), we find

$$
\frac{(x|x) - (a|a)}{(x|a)} = \frac{2A_T}{B_T}, \qquad \frac{(a|x)}{(x|a)} = \frac{1 + A_T^2}{-B_T^2},
$$

which can be combined to

$$
A_T = \frac{(x|x) - (a|a)}{\sqrt{4 - [(x|x) + (a|a)]^2}} = \frac{(x|x) - (a|a)}{2\sin\mu} \tag{6.17a}
$$

$$
B_T = \frac{2(x|a)}{\sqrt{4 - [(x|x) + (a|a)]^2}} = \frac{(x|a)}{\sin\mu}, \tag{6.17b}
$$

with $4\sin^2\mu = 4 - [(x|x) + (a|a)]^2$, according to Eq. (6.16).

6.1.2.1 Unit Cells with Beam Waists at Entrance and Exit

For a particle beam that has a waist $[A_{Ti} = 0]$ at the entrance ($z = z_i$) to a cell, the matching condition postulates a waist $[A_{T(i+2)} = 0]$ also at the exit ($z = z_{i+2}$). In this case the transfer matrix of Eq. (6.15) reads

$$
\begin{pmatrix} (x|x) & (x|a) \\ (a|x) & (a|a) \end{pmatrix} = \begin{pmatrix} \cos\mu & B_T \sin\mu \\ -B_T^{-1}\sin\mu & \cos\mu \end{pmatrix}.
$$

With $(x|x)(a|a) - (x|a)(a|x) = 1$ according to Liouville's theorem [Eq. (1.19b)], Eqs. (6.17a) and (6.17b) can be rewritten as

$$
A_T = (x|x) - (a|a) = 0, \tag{6.18a}
$$

$$
B_T = \sqrt{-(x|a)/(a|x)}, \tag{6.18b}
$$

with $\sin^2 \mu = -(x|a)(a|x)$, or

$$\cos \mu = (x|x). \tag{6.18c}$$

6.1.2.2 Mirror-Symmetric Unit Cells

For a mathematical investigation, a mirror-symmetric unit cell advantageously is split into a first and a second half. The transfer matrix of the first half transporting the beam from z_i to z_{i+1} is most generally

$$\begin{pmatrix} x_{i+1} \\ a_{i+1} \end{pmatrix} = \begin{pmatrix} (x_1|x) & (x_1|a) \\ (a_1|x) & (a_1|a) \end{pmatrix} \begin{pmatrix} x_i \\ a_i \end{pmatrix}. \tag{6.19a}$$

This relation $\mathbf{X}_{i+1} = T_x \mathbf{X}_i$ can also be read as $\mathbf{X}_i = T_x^{-1} \mathbf{X}_{i+1}$, or explicitly, with $T_x T_x^{-1} = 1$, as

$$\begin{pmatrix} x_i \\ a_i \end{pmatrix} = \begin{pmatrix} (a_1|a) & -(x_1|a) \\ -(a_1|x) & (x_1|x) \end{pmatrix} \begin{pmatrix} x_{i+1} \\ a_{i+1} \end{pmatrix}. \tag{6.19b}$$

Looking at the first half of a mirror-symmetric system in reversed direction is the same as looking at the second half in the forward direction. Thus, Eq. (6.19b) describes the second half of the unit cell under investigation if the direction of the optic axis is reversed. This changes the signs of the elements in the second row and the second column since clearly $\sin \alpha = a$ changes its sign when z changes its direction. Consequently, the transfer matrix of the second half of the unit cell reads $\mathbf{X}_{i+2} = T_{xr} \mathbf{X}_{i+1}$, or explicitly,

$$\begin{pmatrix} x_{i+2} \\ a_{i+2} \end{pmatrix} = \begin{pmatrix} (a_1|a) & (x_1|a) \\ (a_1|x) & (x_1|x) \end{pmatrix} \begin{pmatrix} x_{i+1} \\ a_{i+1} \end{pmatrix}. \tag{6.19c}$$

Combining $\mathbf{X}_{i+1} = T_x \mathbf{X}_i$ of Eq. (6.19a) and $\mathbf{X}_{i+2} = T_{xr} \mathbf{X}_{i+1}$ of Eq. (6.19c), we find the transformation of the full unit cell, i.e., from z_i to z_{i+2}, as $\mathbf{X}_{i+2} = T_{xr} T_x \mathbf{X}_i$,

$$\begin{pmatrix} x_{i+2} \\ a_{i+2} \end{pmatrix} = \begin{pmatrix} (x|x) & (x|a) \\ (a|x) & (a|a) \end{pmatrix} \begin{pmatrix} x_i \\ a_i \end{pmatrix}$$

$$= \begin{pmatrix} (x_1|x)(a_1|a) + (x_1|a)(a_1|x) & 2(x_1|a)(a_1|a) \\ 2(x_1|x)(a_1|x) & (x_1|x)(a_1|a) + (x_1|a)(a_1|x) \end{pmatrix} \begin{pmatrix} x_i \\ a_i \end{pmatrix}. \tag{6.20}$$

Equation (6.20) states that for a mirror-symmetric unit cell, $(x|x)$ equals $(a|a)$, which is the condition found in Eq. (6.18a) for a unit cell that has beam waists at its entrance and exit. By introducing the $(x|x)$ of Eq. (6.20) into Eq. (6.18c), we then find the difference of the phase shifts $\sigma_{x(i+2)} - \sigma_{xi} = \mu$ which we will call the phase advance per unit cell as

$$\cos \mu = (x_1|x)(a_1|a) + (x_1|a)(a_1|x). \tag{6.21}$$

Together with $(x_1|x)(a_1|a) - (x_1|a)(a_1|x) = 1$ taken from Eq. (1.19b) it thus follows that

$$2(x_1|x)(a_1|a) = \cos \mu + 1, \tag{6.22a}$$

$$2(x_1|a)(a_1|x) = \cos \mu - 1, \tag{6.22b}$$

and consequently $\sin^2 \mu = -4(x_1|x)(x_1|a)(a_1|x)(a_1|a)$. Determining $(x_1|x)$ and $(a_1|x)$ from Eqs. (6.22a) and (6.22b) and introducing these terms into by Eq. (6.20), we find from Eq. (6.18b) with $\varepsilon_x B_T = R_{xi}^2 = R_{x(i+2)}^2$, i.e., equal half-diameters of the beam waists at z_i and z_{i+2}:

$$R_{xi}^2 = R_{x(i+2)}^2 = \varepsilon_x \frac{(x_1|a)}{(x_1|x)} \sqrt{\frac{1 + \cos \mu}{1 - \cos \mu}} = \varepsilon_x \sqrt{\frac{-(x_1|a)(a_1|a)}{(x_1|x)(a_1|x)}}. \tag{6.23a}$$

Because of the mirror symmetry of the unit cell, there must be a beam maximum at z_{i+1} in the middle between the beam waists, i.e., after the first half of the unit cells. At this position, we find $R_{x(i+1)}^2 = (x_1|x)^2 R_{xi}^2 + (x_1|a)^2 \varepsilon_x^2 / R_{xi}^2$, according to Eq. (5.15a) or explicitly with Eq. (6.23a),

$$R_{x(i+1)}^2 = 2\varepsilon_x \frac{(x_1|x)(x_1|a)}{\sin \mu} = \varepsilon_x \sqrt{\frac{-(x_1|x)(x_1|a)}{(a_1|x)(a_1|a)}}, \tag{6.23b}$$

with the so-called "beat factor" $R_{x(i+1)}/R_{xi}$ being

$$\frac{R_{x(i+1)}^2}{R_{xi}^2} = \frac{2(x_1|x)^2}{1 + \cos \mu} = \frac{(x_1|x)}{(a_1|a)}. \tag{6.23c}$$

To match a particle beam with a given ε_x to some mirror-symmetric unit cell, we must thus fulfill two conditions:

(1) we must vary one parameter of the unit cell such that Eqs. (6.22a) and (6.22b) are fulfilled for some choice of the phase advance μ;

(2) we must determine the size of the maximum beam diameter $R_{x(i+1)}$ from Eq. (6.23b) and check whether this diameter is small enough that the beam can pass through the apertures of the given optical elements; if this is not the case, we must vary a second parameter of the unit cell and start over with trying to fulfill condition 1.

Note that for $\cos \mu = 0$, the phase advance per unit cell is $(n + 1/2)\pi$, which is equivalent to requiring $(x|x) = (a|a) = 0$ according to Section 6.1.1.3, since for the unit cell discussed above we had postulated $A_{T1} = A_{T2} = 0$.

6.1.2.2.1 Periodic Arrays of Focusing Lenses

A special case for a series of mirror-symmetric cells could be a periodic array of thin lenses of focal lengths f separated by distances $2d$. Half a unit cell of such a periodic array, extending from profile plane i to profile plane $i + 1$ in Fig. 6.3, thus consists of a distance d followed by the first half of one of the lenses. Since this half lens has the focal length $2f$, the transfer matrix for the half unit cell becomes

$$T = \begin{pmatrix} 1 & d \\ -1/2f & 1 - (d/2f) \end{pmatrix} = \begin{pmatrix} 1 & 0 \\ -1/2f & 1 \end{pmatrix} \begin{pmatrix} 1 & d \\ 0 & 1 \end{pmatrix}. \quad (6.24)$$

For a fixed phase advance μ, Eq. (6.21), or one of the Eqs. (6.22a) or (6.22b) reads $\cos \mu = 1 - d/f$, which yields

$$f = d/(1 - \cos \mu). \quad (6.25a)$$

The beam extrema in this case are found from Eqs. (6.23a) and (6.23b) as

$$R_{xi}^2 = \varepsilon_x d \sqrt{\frac{1 + \cos \mu}{1 - \cos \mu}} = \varepsilon_x \sqrt{d(2f - d)}, \quad (6.25b)$$

$$R_{x(i+1)}^2 = \frac{2\varepsilon_x d}{\sin \mu} = 2\varepsilon_x f \sqrt{\frac{d}{2f - d}}, \quad (6.25c)$$

with $\sin^2 \mu = [2 - (d/f)]d/f$.

As an example, from light optics we may assume that an object of 0.5 mm diameter must be observed through a tube 500 mm long. If this observation must be done in a wavelength region for which no glass-fiber system exists, we may use an array of unit cells each containing only one lens. In this case it is advantageous to have an object–image relation, i.e., a phase shift of $n\pi$ over two unit cells, which means between the entrance of the ith and the exit of the $(i + 1)$ unit cell. Per unit cell thus the phase advance μ should be $\pi/2$ or $(n + 1/2)\pi$. Consequently, we find from Eqs. (6.25a)-(6.25c):

$$f = d, \qquad R_{xi}^2 = \varepsilon_x d, \qquad R_{x(i+1)}^2 = 2\varepsilon_x d.$$

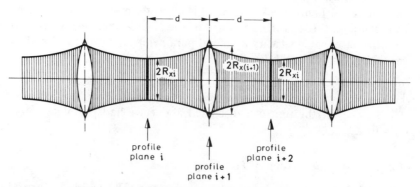

Fig. 6.3. A periodic array of focusing lenses with a matched beam. Note that beam waists exist at i and $i + 2$, i.e., in the middle between lenses.

If the lens diameters must stay below 5 mm, we may choose $R_{x(i+1)} < 2$ mm. For an arbitrary $d = 25$ mm, which requires 10 lenses of focal lengths of 25 mm each, we find $\varepsilon_x = R^2_{x(i+1)}/2d < 0.08$ mm. Thus, only those rays can pass through the system which leave the ±0.25 mm object with maximum angles of inclination of $\alpha_{max} = \pm\varepsilon_x/0.25 = \pm0.32$ or about $\pm18°$. To match this light beam to the transport system we can use a first lens to form an image of size $2R_{xi} = 2\sqrt{\varepsilon_x d} \approx 2.83$ mm, which is about 5.7 times larger than the object. The full system should thus consist of a first matching lens, a lens chanel of 10 lenses, and a final eyepiece.

6.1.2.2.2   Periodic Arrays of Alternating Focusing and Defocusing Lenses

In strong focusing particle accelerators or beam guidance lines, focusing and defocusing quadrupoles are used with two consecutive quadrupoles, usually of opposite type. Thus, the system can be characterized by FDFD $\cdots$ or F0D0F0D0 $\cdots$ with 0 standing for intermediate field-free regions. In such a lens arrangement, a given particle is always further away from the optic axis in a focusing lens than it is in a defocusing lens (Fig. 6.4). Since the deflection a particle experiences in a lens is proportional to the distance this particle is away from the optic axis, the deflection toward the optic axis always overcompensates the deflection away from it in an array of equally strong lenses. This causes a quadrupole lens channel to be overall focusing in the x and y directions simultaneously, although the individual quadrupole singlets are strongly astigmatic.

Let us now consider a periodic array of thin quadrupole lenses of alternating focal lengths $+f$ and $-f$ separated by distances d. As shown in Fig. 6.5, one cell of such an array consists of one-half of a defocusing thin lens at the front and back end (with these half lenses each having a focal length $-2f$) plus a focusing lens of focal length $+f$ in the middle between the defocusing lenses. In such a mirror-symmetric unit cell, the beam maxima must be at the positions of the focusing lenses with beam minima at the positions of the defocusing lenses. The first half of such a unit cell reaching from the center of the defocusing to the center of the focusing lens is thus described by a transfer matrix:

$$T = \begin{pmatrix} 1 + d/2f & d \\ -d/4f^2 & 1 - d/2f \end{pmatrix} = \begin{pmatrix} 1 & 0 \\ -1/2f & 1 \end{pmatrix} \begin{pmatrix} 1 & d \\ 0 & 1 \end{pmatrix} \begin{pmatrix} 1 & 0 \\ 1/2f & 1 \end{pmatrix}. \quad (6.26)$$

For a fixed phase advance μ, we find from Eq. (6.22b) $\cos \mu = 1 - d^2/2f^2$, or

$$f = d/\sqrt{2(1 - \cos \mu)}. \quad (6.27a)$$

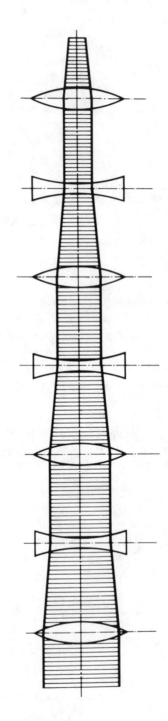

Fig. 6.4. A periodic array of alternating quadrupole lenses. Note that the particle trajectories shown traverse the focusing lenses at a greater distance from the optic axis compared to the defocusing lenses. Thus, the overall action of such a periodic quadrupole channel is focusing in both the x and y directions.

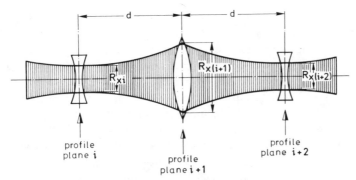

Fig. 6.5. A periodic array of focusing and defocusing lenses with a matched beam. Note that beam waists exist at positions of defocusing lenses.

With the matrix elements of Eq. (6.26), the beam extrema are found in this case from Eqs. (6.23a) and (6.23b) and

$$R_{xi}^2 = \frac{\varepsilon_x d}{1 + \sqrt{(1 - \cos \mu)/2}} \sqrt{\frac{1 + \cos \mu}{1 - \cos \mu}} = 2\varepsilon_x f \sqrt{\frac{2f - d}{2f + d}}, \quad (6.27b)$$

$$R_{x(i+1)}^2 = \frac{2\varepsilon_x d[1 + \sqrt{(1 - \cos \mu)/2}]}{\sin \mu} = 2\varepsilon_x f \sqrt{\frac{2f + d}{2f - d}}. \quad (6.27c)$$

Note that because of the symmetric arrangement of the quadrupole lenses, the beam minima in the x direction are at those places at which we find beam maxima in the y direction and vice versa.

As an example, assume a thin-lens quadrupole channel of limited diameter. With a fixed d, a numerical evaluation of Eqs. (6.27a) and (6.27c) shows that $R_{x(i+1)}$ becomes minimal at around $\mu = 77°$. For this μ, Eqs. (6.27a)–(6.27c) yield

$$f \approx 0.803d, \qquad R_{xi} \approx 0.88\sqrt{\varepsilon_x d}, \qquad R_{x(i+1)} \approx 1.82\sqrt{\varepsilon_x d}, \quad (6.28)$$

so that the "beat factor" is about 2.07. The minimum for $R_{x(i+1)}$ is very flat so that very good results are also obtained for $\mu \approx 77°$. For $\mu = 90°$, for instance, the maximum beam diameter is only about 2% larger than for $\mu = 77°$.

For the special case of $\mu = (n + \frac{1}{2})\pi$, Eqs. (6.27a)–(6.27c) yield with $n = 0, 1, 2, 3, \ldots$,

$$f = d/\sqrt{2}, \qquad R_{xi}^2 = \sqrt{2}\varepsilon_x d/(1 + \sqrt{2}), \qquad R_{x(i+1)}^2 = \sqrt{2}\varepsilon_x d(1 + \sqrt{2}).$$

The beat factor of such a periodic array would thus be $R_{x(i+1)}/R_{xi} = 1 + \sqrt{2} \approx 2.4$.

An example of such a periodic array are the cells of a ring accelerator for which the phase advance normally stays below $\pi/2$. Besides acceleration gaps, such an accelerator may contain deflecting magnets of alternating gradients, i.e., $dB_y/dx \neq 0$. In the y direction here the pole-gap distances $2G_0$ in the inhomogeneous magnets limit the acceptable phase-space area $\varepsilon_y = y_{00}\beta_{00}$ of a particle beam. For a distance of $d = 6$ m between the focusing elements, maximal apertures of $y_m = \pm.02$ m, and an assumed $\mu \approx 77°$, Eq. (6.28) yields, after exchanging ε_x for ε_y, the maximum phase-space area ε_y of a particle beam that can be transmitted by such an accelerator, as $\varepsilon_y \approx 0.02^2/(6 \times 1.82^2) \approx 20$ mm mrad. From the magnitude of the beat factor of about 2.1, we further conclude that the focusing lenses must be about twice as wide as the defocusing ones, which permits a considerably cheaper design of each second inhomogeneous magnet in such an accelerator. Because of Eq. (5.3b), $n\varepsilon_y$ is a constant, with n the refractive index. Since n increases with increasing particle energy $-eV$ according to Eq. (2.24), the largest value of ε_y must be transmitted at the beginning of the acceleration cycle. Thus, it is advantageous to introduce the particles to the periodic structure with the highest possible kinetic energy.

6.2 RINGS OF UNIT CELLS

When optical elements are arranged in a ring, the particle beam travels through identical structures during every turn. Because of its repetitive occurrence small disturbances of particle trajectories in this case can add to large effects.

6.2.1 Lateral Beam Deviations

If a particle beam travels through a ring structure of n cells of lengths l_0, we often use a quantity that describes how often a particle has moved around the phase ellipse per turn in the ring. This quantity is called *the tune* (see, for instance, Guignard, 1976):

$$Q = \frac{1}{2\pi} \int_z^{z+nl_0} \frac{dz}{B_T}. \tag{6.29a}$$

For a ring that consists of N matched unit cells, in each of which [Eq. (6.3d)] the phase advance is $\mu = dz/B_T$, Eq. (6.29a) can be rewritten as

$$Q = N\mu/2\pi, \tag{6.29b}$$

since the phase shift per unit cell is independent of the reference point z where a unit cell starts and the previous one ends.

If one element in the ring is imperfect, a particle trajectory deviates from its design trajectory at this position. If this trajectory passes the imperfection

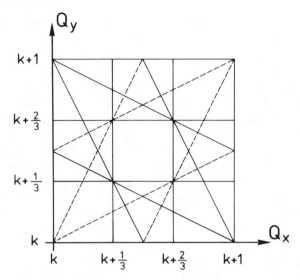

Fig. 6.6. Ring stability diagram showing lines of instability.

in each turn at exactly the same position, the trajectory deviation is in the same direction for each turn. Consequently, such a particle will quickly be lost.

Advantageously, we thus choose the x and y tunes, Q_x and Q_y, never to be integers, since in this case the particle under consideration passes through the imperfect element each time with different x and y coordinates. In other words, we try to avoid the condition

$$aQ_x + bQ_y = \text{integer}. \qquad (6.30)$$

Consequently (Guignard, 1976), only points far from any of the lines shown in Fig. 6.6 are stable working points for a ring structure.

6.2.2 Longitudinal Beam Deviations

Besides lateral deviations, longitudinal deviations from an ideal particle motion are also of interest in a ring structure. Assuming particles of equal mass–charge ratios, the time for one revolution in the ring generally depends on their energy–charge ratios. However, a ring can always be designed such that the flight time around the ring becomes independent of the energy–charge ratio of the particles (see Section 4.4), and the more energetic, faster particles travel along slightly longer trajectories. In this case, we speak of an isochronous ring (Schaffer, 1981), in which originally separate bunches of particles remain separated.

When this condition of isochronicity is not fulfilled, the distances between bunches of particles disappear after several revolutions in the ring. However, if acceleration gaps are placed at certain positions in the ring, the voltage across such acceleration gaps can be made to rise slightly while a bunch of particles passes. Thus, the velocities of the first particles in a bunch increase more than the velocities of the last particles. When these particles reach the next acceleration gap in a ring, the initially first particles arrive too early, i.e., at a time of low acceleration voltage, and gain less energy so that they arrive relatively late at the next acceleration gap where they consequently gain more energy, etc. This property is called phase focusing and ensures that on average all particles gain the same energy but that an individual particle moves from the front to the end of a bunch and back, having the relative highest energy when it is at the front and the relative lowest energy when it is at the end of a bunch. Similar to a x, a or y, b phase space, there is also a Δt, Δk phase-space volume that changes its shape but not its volume similarly to Fig. 5.1.

With increasing energy during an acceleration cycle, the more energetic particles arrive at a second acceleration gap shorter and shorter times ahead of the less energetic ones. Eventually, the isochronous condition is reached, and thereafter more energetic particles arrive at the next acceleration gap even later than particles of lower energy. The only way to conserve the phase-focusing property in the Δt, Δk phase space after the isochronous condition has been passed, is to change the phase of the accelerating voltage so that the first particles in a bunch gain more energy than the last ones. Since there is no phase focusing at all when the isochronous condition is fulfilled, considerable losses in particle intensity must be expected when the particles pass through the corresponding transition energy. Consequently, ring accelerators are normally designed such that their transition energy is higher than the maximal desired particle energy.

REFERENCES

Bovet, C., Gouivan, R., Gumowski, I., and Reich, K. H. (1970). CERN/MPS-SI/Int. DL/70/4.

Brown, K. L. (1981). *Nucl. Instrum. Methods* **187**, 51.

Brown, K. L. and Howry, S. K. (1970). SLAC Rep. No. 91.

Courant, E. D., and Snyder, H. S. (1958). *Ann. Phys.* (*New York*) **3**, 48.

Guignard, G. (1976). CERN Rep. No. 76-06, 77-10.

Livingood, J. (1961) "Principles of Cyclic Particle Accelerators." van Nostrand Co. Inc., Princeton.

Schaffer, G. (1981). "IKOR-Study", KFAJ Rep., KFA Juelich, W-Germany.

Steffen, K. G. (1965). "High Energy Beam Optics." Wiley (Interscience), New York.

Wilson, E. J. N. (1977). CERN-Rep. No. 77-07.

Fringing Fields

The optical properties of quadrupole lenses and of sector fields have been discussed in Chapters 3 and 4. In both cases, the electrostatic field strengths and the magnetic flux densities were assumed to start and end abruptly. Since across an abrupt boundary Maxwell's equations would not be fulfilled, the results could only be rough approximations. In this chapter we shall determine particle trajectories in realistic fringing fields. Fortunately, the results of Chapters 3 and 4 do not become obsolete y these calculations. Fields with such abrupt boundaries give almost the same result as the real field distributions if the positions of the abrupt boundaries are chosen appropriately, normally not coinciding with the electrode or pole-face boundaries (Bucherer, 1909).

7.1 PARTICLE TRAJECTORIES IN FRINGING FIELDS OF DIPOLE MAGNETS

For all discussions of fringing field effects, we shall choose a ξ, η, ζ-coordinate system (Fig. 7.1), which is fixed to the boundary of the magnet

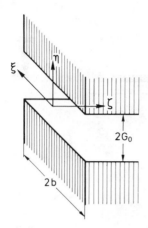

Fig. 7.1. A coordinate system in the fringing field region of a homogeneous deflecting magnet.

pole pieces. Here, η is a coordinate perpendicular to the plane of deflection so that $\eta = 0$ describes the midplane between the magnet poles; ξ is a coordinate parallel to and ζ a coordinate perpendicular to the magnet boundary, with both ξ and ζ lying in the plane $\eta = 0$. The point $\xi = \eta = \zeta = 0$ is chosen to be in the middle of the magnet boundary of Fig. 7.1.

In this ξ, η, ζ-coordinate system, the equations of motion read for a particle of mass m and charge (ze),

$$m\ddot{\xi} = (ze)[\dot{\eta}B_\zeta - \dot{\zeta}B_\eta], \qquad m\ddot{\eta} = (ze)[\dot{\zeta}B_\xi - \dot{\xi}B_\zeta]. \qquad (7.1)$$

If the projection of the particle trajectory on the $(\eta = 0)$ plane forms the angle ε with the ζ axis (Fig. 7.2), and if this trajectory is inclined at an angle β' with respect to the $(\eta = 0)$ plane we find from Eq. (7.1),

$$\sin \varepsilon(\zeta) = \frac{\dot{\xi}_a + \int \dot{\xi} \, d\zeta}{v \cos \beta'} = \sin \varepsilon_a - \frac{1}{\cos \beta'} \int_{\zeta_a}^{\zeta} \left(B_\eta - B_\zeta \frac{\tan \beta'}{\cos \varepsilon} \right) \frac{d\zeta}{B_0\rho}, \qquad (7.2a)$$

$$\sin \beta'(\zeta) = \frac{\dot{\eta}_a + \int \ddot{\eta} \, d\zeta}{v} = \sin \beta'_a - \int_{\zeta_a}^{\zeta} (B_\zeta \tan \varepsilon - B_\xi) \frac{d\zeta}{B_0\rho}. \qquad (7.2b)$$

Here, use is made of $\dot{\xi} = v \sin \varepsilon \cos \beta'$, $\dot{\eta} = v \sin \beta'$, and $\dot{\zeta} = v \cos \varepsilon \cos \beta'$, with $v = ds/dt$ the particle velocity. Furthermore, here $mv/(ze) = B_0\rho = \chi_B$ is the magnetic rigidity of the particle under consideration, and the index a denotes a position of the particle trajectory well outside the magnetic field. Note here also that Eq. (7.2a) can be transformed to

$$\tan \varepsilon(\zeta) = \tan \varepsilon_a - \frac{1}{\cos^3 \varepsilon_a \cos \beta'} \int_{\zeta_a}^{\zeta} \left(B_\eta - B_\zeta \frac{\tan \beta'}{\cos \varepsilon} \right) \frac{d\zeta}{B_0\rho} + \cdots, \qquad (7.2c)$$

because $\sin \varepsilon = \sin \varepsilon_a + \Delta$ transforms to $\tan \varepsilon = \tan \varepsilon_a + (\Delta/\cos^3 \varepsilon_a) + \cdots$, if terms with $\Delta^2, \Delta^3, \ldots$, are neglected.

7.1.1 Positions of Effective Field Boundaries in Dipole Magnets

For a reference particle of mass m and charge $(z_0 e)$, which moves along a radius of curvature ρ_0 in the plane of symmetry ($\eta = 0$; $\dot{\eta} = \tan \beta' = 0$), Eq (7.2b) vanishes, and Eq. (7.2a) simplifies to

$$\sin \varepsilon(\zeta) = \sin \varepsilon_a - \int_{\zeta_a}^{\zeta} \frac{B_\eta(0, 0, \zeta)}{B_0 \rho_0} \, d\zeta, \tag{7.3a}$$

$$\tan \varepsilon(\zeta) = \tan \varepsilon_a - \frac{1}{\cos^3 \varepsilon_a} \int_{\zeta_a}^{\zeta} \frac{B_\eta(0, 0, \zeta)}{B_0 \rho_0} \, d\zeta. \tag{7.3b}$$

The corresponding particle trajectory is straight at ζ_a, where B_η vanishes, and is a circle of radius ρ_0 at some ζ_b, where B_η equals B_0. Between these two ζ values, the curvature of the trajectory increases monotonically from 0 to $1/\rho_0$ as B_η increases from 0 to B_0 (Fig. 7.2).

We may now extrapolate the circular part of the trajectory backward until at some $\zeta = \zeta^*$, it forms the same angle ε_a with the ζ axis as the real trajectory did at $\zeta = \zeta_a$. The point $\zeta = \zeta^*$ we call the position of the *effective field boundary* in which an *effective flux density distribution* $B_\eta^*(0, 0, \zeta)$ is assumed to change abruptly from 0 to B_0 as indicated in Fig. 7.2 so that $B_\eta^*(0, 0, \zeta < \zeta^*) = 0$ and $B_\eta^*(0, 0, \zeta \geqslant \zeta^*) = B_0$. Extrapolating also the straight part of the real trajectory up to $\zeta = \zeta^*$, we find the so-called *effective trajectory*, which is straight for $\zeta < \zeta^*$, experiences a parallel shift $\Delta \xi$ at $\zeta = \zeta^*$, and is circular for $\zeta \geqslant \zeta^*$.

To define the position ζ^* of the effective field boundary we must evaluate Eq. (7.3a) once for the distribution of the real flux density $B_\eta(0, 0, \zeta)$ and once for the distribution of an "effective flux density distribution" $B_\eta^*(0, 0, \zeta)$. If $\sin \varepsilon_b$ of Eq. (7.3a) must the same for the two flux density distributions, we find $B_0(\zeta_b - \zeta^*) = \int B_\eta(0, 0, \zeta) \, d\zeta$, or

$$\zeta^* = \zeta_b - \int_{\zeta_a}^{\zeta_b} \frac{B_\eta(0, 0, \zeta)}{B_0 G_0} \, d\zeta = I_1 G_0, \tag{7.4a}$$

where $2G_0$ is the width of the magnet air gap. Analogously, we find a displacement in the ξ direction,

$$\xi_b - \xi_a = \int_{\zeta_a}^{\zeta_b} \tan \varepsilon \, d\zeta,$$

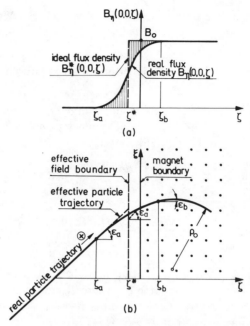

Fig. 7.2. A particle in the fringing field of a magnetic dipole field. (a) The distribution of the magnetic flux density $B_\zeta(0, 0, \zeta)$ in the fringing field region is shown together with the so-called *effective flux density distribution* $B_\zeta^*(0, 0, \zeta)$. Note that the two shaded areas are equally large. (b) The *real trajectory* in the fringing field region is shown together with the so-called *effective trajectory*, indicated by dashed lines. The effective trajectory is straight up to the effective field boundary at $\zeta = \zeta^*$ and continues along a circle of radius ρ_0 from there on. Furthermore, at the effective field boundary, the effective trajectory, experiences a parallel shift $\Delta\xi$, which moves the trajectory a little toward the center of curvature of the optic axis. Note that ε_a is positive if for the field-free region, the normal to the field boundary is further away from the center of curvature of the optic axis than the particle trajectory.

with $\tan \varepsilon$ determined from Eq. (7.3b) for the real $B_\eta(0, 0, \zeta)$ or the effective $B_\eta^*(0, 0, \zeta)$ flux density distribution. The difference between these two displacements, i.e., the distance $\Delta\xi$ by which all particle trajectories and thus also the optic axis should be shifted in the ξ-direction when they cross the effective field boundary is

$$\Delta\xi_b = \left[\frac{(\zeta_b - \zeta^*)^2}{2} - \int\limits_{\zeta_a}^{\zeta_b}\!\!\int \frac{B_\eta(0, 0, \zeta)}{B_0} \, d\zeta \, d\zeta \right] \frac{1}{\rho_0 \cos^3 \varepsilon_a} = (I_2 G_0) \frac{G_0/\rho_0}{\cos^3 \varepsilon_a}.$$

(7.4b)

Here again, $2G_0$ is the width of the magnet air gap, and $1/\rho_0$ is the curvature of the optic axis. As before ζ_a and ζ_b must be chosen well outside and well inside the magnet so that $B_\eta(0, 0, \zeta_a) = 0$ and $B_\eta(0, 0, \zeta_b) = B_0$. Consequently, no differences arise in Eqs. (7.4a) and (7.4b) if ζ_a is varied in a region where B_η vanishes and/or if ζ_b is varied in a region where B_η equals B_0.

Note here that a fulfilled Eq. (7.4a) is equivalent to the postulate that the two shaded areas in Fig. 7.2 are of equal size. Note further that $\Delta\xi_b$ of Eq. (7.4b) is not zero if Eq. (7.4a) is fulfilled. Note finally that both I_1 and I_2 of Eqs. (7.4) can be unknown at the design stage for an optical system. However, when the optical design postulates the effective field boundaries to be at certain positions in space, we must know ζ^* and $\Delta\xi$ before we fabricate the magnetic sector field: ζ^* determines the distance between the already fixed effective field boundary and the position of the pole-shoe contour of the magnet under construction, and $\Delta\xi$ determines the distance by which the finally fabricated magnetic sector field must be shifted, so that the real and the effective particle trajectories coincide in the main field region.

Looking at measured or calculated flux density distributions as in Fig. 7.3, we find that the flux density drops from 100% to about 10% of B_0 over a length comparable to $2G_0$, the width of the magnet air gap. However, the distance $\zeta^* - \zeta_a$ is often much larger than $\zeta_b - \zeta^*$. In fact, the long tail of $B_\eta(0, 0, \zeta)$ in most cases has an extension comparable to the physical dimension of the magnet yoke. Furthermore, this part of the fringing field is strongly affected by any small piece of iron in this region, such as an accidentally misplaced screwdriver. For this reason, high-performance magnets are normally equipped with fringing field shunts (Fig. 7.4), which cut off the long tail of the flux density distribution (Figs. 7.3 and 7.5) and limit the fringing field to a region of approximately G_0 or $2G_0$.

Fringing field shunts consist mainly of iron pieces, (S_{1a} and S_{1b} in Fig. 7.4), which may be connected magnetically outside of the drawing plane shown. The pieces S_{2a} and S_{2b} in Fig. 7.4 are unnecessary in case the pieces

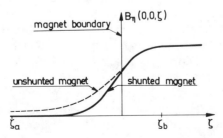

Fig. 7.3. The distribution of $B_\eta(0, 0, \zeta)$, the η component of the magnetic flux density for the case of an unshunted and a shunted magnet (see also Fig. 7.4). Note in the first case the long tail of the distribution, which can easily be distorted by any piece of steel.

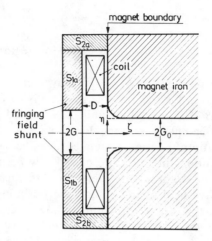

Fig. 7.4. A fringing field shunt for a magnet; G, D and G_0 are of about equal magnitude. The coils should be as far removed from the midplane $\eta = 0$ as possible. The origin of the coordinate system is placed at the position of the magnet boundary (see also Fig. 7.1). Note the multiple shamfered magnet pole tips, which, compared to sharply pointed ones, cause considerably reduced saturation effects. Single or multiple shamfers should approximate a rounded edge with a radius of curvature of about $0.8G_0$ or larger so as to approximate a shape that would postulate B_η to be constant along the iron surface for infinitely permeable steel (Rogowski, 1923). For the later-quoted approximate numerical fringing field calculations of Eqs. (7.6) and (7.11) or Figs. 7.5, 7.6, and 7.10, however, pointed pole tips are assumed (dashed lines) because of the simpler mathematical treatment.

S_{1a} and S_{1b} are perfectly symmetric to the plane $\eta = 0$. The existence of the pieces S_{2a} and S_{2b}, however, is desirable if the fabrication and position tolerances destroy the perfect symmetry of the pieces S_{1a} and S_{1b}. In this case, the pieces S_{2a} and S_{2b} ensure that the magnetic scalar potential of the pieces S_{1a} and S_{1b} is at the magnetic scalar potential of the midpart of the magnet yoke, i.e., a part at $\eta = 0$. However, since the pieces S_{2a} and S_{2b} must never carry a strong magnetic flux, their cross sections can be small. It is of great importance that the relative flux density distribution $B_\eta(\zeta)/B_0$ (Figs. 7.2 or 7.3) throughout the fringing field be identical for all values of B_0, i.e., that no appreciable saturation effects of the magnet poles can be observed. If we choose $D > 2G_0$ or at least $D > G_0$, such saturation effects normally are so small that they can be neglected even in the neighborhood of the tips of the shunts. The tips of the magnet poles, however, in all cases should be phased off or rounded, as indicated in Fig. 7.4 (see also Hübner and Wollnik, 1970).

In order to find the values of ζ^* and $\Delta\xi$ from Eqs. (7.4a) and (7.4b), the detailed distribution of $B_\eta(0, 0, \zeta)$ must be known from field measurements.

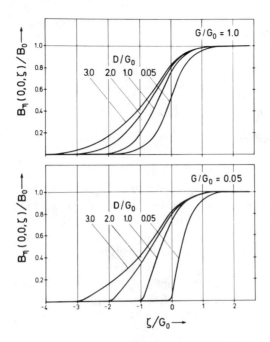

Fig. 7.5. Fringing field distributions $B_\eta(0, 0, \zeta)$ for different fringing field shunt geometries, i.e., different choices of D/G_0 and G/G_0 (Fig. 7.4). These distributions are calculated (Herzog, 1940) under the assumption of pointed pole tips.

During the design state, however, such measurements are usually not available, and we use numerical solutions of Poisson's equation for a realistic magnet design or, at least, solutions of Laplace's equation for an idealized magnet made of iron with infinite permeability.

Under the assumption that the coils as well as the iron pieces S_{2a}, S_{2b} of Fig. 7.4 are far from the plane of midsymmetry ($\eta = 0$), the component $B_\eta(0, 0, \zeta)$ of the magnetic flux density $\mathbf{B}$, can be determined by conformal mapping. For mathematical simplicity, we assume here sharply pointed pole tips (see the dashed pole contour in Fig. 7.4) so that we can use the Schwartz–Christoffel conformal mapping formula, as for any electrostatic electrode arrangement (Fig. 7.12). The fringing field distributions thus obtained are plotted in Fig. 7.5 for different geometries of the fringing-field shunts, i.e., for different choices of D/G_0 and G/G_0 (see Fig. 7.4 and Herzog, 1935a). Note that for small shunt apertures $2G$, the field $B_\eta(0, 0, \zeta)$ is almost cut off at $\zeta = D$, whereas for larger values of G, the field penetrates a little through the wider aperture. Introducing such idealized fringing field distributions into the integral of Eq. (7.4a), Herzog (1935a,b; 1940) calculated I_1 and thus ζ^* for different ratios D/G_0 and G/G_0. His result can be simplified to

$$I_1\pi = \ln\frac{4G_0^2}{q_1q_2} - \frac{G}{G_0}\ln\frac{q_1}{q_2} - \frac{D}{G_0}\arccos\frac{D^2+G^2-G_0^2}{q_1q_2}, \quad (7.5a)$$

where q_1 and q_2 are abbreviations for

$$q_1^2 = D^2 + (G_0+G)^2, \qquad q_2^2 = D^2 + (G_0-G)^2.$$

In detail, $\arccos(D^2+G_0^2+G^2)/(q_1q_2)$ describes the angle θ_1 in Fig. 7.12. The resultant I_1 is shown graphically in Fig. 7.6a for different geometries of the fringing field shunt characterized by D/G_0 and G/G_0 (see Fig. 7.4). Note that I equals about -0.42 for $D = G = G_0$ so that in this case the effective field boundary lies $0.42G_0$ outside the magnet boundary. Analogously, as in Eq. (7.4a) we can also introduce idealized fringing field distributions into the integral of Eq. (7.4b); however, no comparably simple function to Eq. (7.5a) can be obtained. The result of a numerical evaluation of Eq. (7.4b) is plotted in Fig. 7.6b as function of D/G_0 and G/G_0. Note that I_2 equals about -0.29 for $D = G = G_0$. In a rough approximation I_2 can be expressed as

$$I_2 \approx -[0.15 + 0.04(D/G_0)^2 + 0.04(G/G_0) + 0.05(G/G_0)^2]. \quad (7.5b)$$

By rounding the pole tips in Fig. 7.4 by large radii of curvature or by large shamfers, any magnet can be designed such that the effective field boundary coincides with the physical boundary of the magnet poles ($\zeta^* = 0$). This simplifies the alignment of the magnet system. However, this normally causes the shape of the effective field boundary, i.e., ζ^* as function of ξ, to deviate more from the iron contour of the magnet pole, an effect that cannot easily be compensated for by other measures. Thus, it is normally advantageous to design the magnet for optical performance and tolerate the problems caused by the fact that ζ^* lies outside the boundary of the magnet poles.

7.1.2 Fringing Field Shunts for Homogeneous and Inhomogeneous Dipole Magnets

For homogeneous magnets, the width of the air gap $2G_0$ is constant for all values of ξ (Fig. 7.1). Choosing constant values for G and D (see Fig. 7.4), fringing field shielding arrangements result for sector fields, as indicated in Fig. 7.7. For inhomogeneous magnets having inclined planar or conical

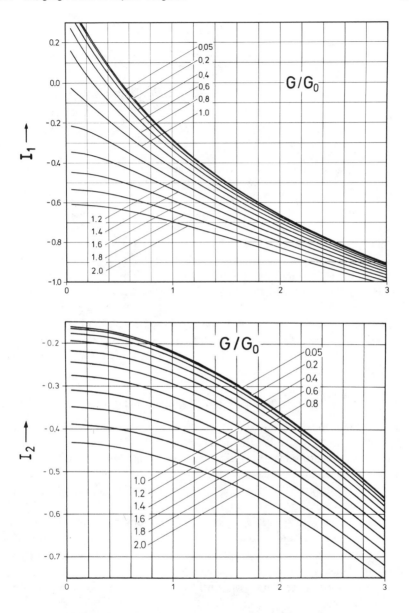

Fig. 7.6. (a) The quantity $\zeta^*/G_0 = I_1$ is plotted as determined from Eq. (7.4a). (b) The quantity I_2 as determined from Eq. (7.4b) with $B_\eta(0, 0, \zeta)$ is calculated according to the Schwarz–Christoffel conformal mapping formula. In both cases, pointed pole tips or electrodes are assumed (Fig. 7.4), and the influence of the coils of a magnet are neglected. Note that (a) was calculated from Eq. (7.5a), whereas (b) had to be calculated in a more involved manner.

effective field
boundary

magnet boundary

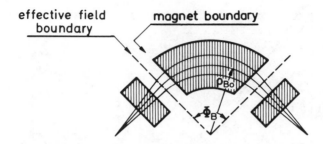

Fig. 7.7. Fringing field shunts are shown for a homogeneous magnetic sector field. Note that the effective field boundaries and the magnet boundaries are parallel.

pole faces as in Figs. 4.9 and 4.16, the air gap increases or decreases linearly with ξ. Thus, it seems reasonable to change D and G of Fig. 7.4 accordingly, which results in the shielding arrangement shown in Fig. 7.8. In these cases, the inhomogeneity should be roughly preserved throughout the fringing field. The integrals of Eqs. (7.4a) and (7.4b) and later of Eq. (7.9) require knowledge of $B_{\eta}(0, 0, \zeta)$, the measured η-component of the magnetic flux density distribution along the ζ axis. For a start, however, the numerical values of Figs. 7.6 and later of Fig. 7.10, determined for an idealized homogeneous magnet, can be used.

7.1.3   y-Focusing in Fringing Fields of Dipole Magnets

For particle optical calculations we normally use an x, y, z-coordinate system with the origin fixed at the intersection of the effective field boundary and the incoming still-straight optic axis, the z axis. Here y and the η axes coincide, whereas in the case where the optic axis is not perpendicular to the magnet boundary, the x and z axes are rotated by an angle $\varepsilon_a = \varepsilon'$, compared to the ξ and ζ axes (Fig. 7.9).

Because of symmetry, all magnet field lines are perpendicular to the $(\eta = y = 0)$ plane. Charged particles that traverse the fringing field of a magnet not in this plane of symmetry between the pole faces experience ξ and ζ components of **B** and thus, according to Eq. (7.1b) or (7.2b), forces in the direction of y. In detail, we find to first order in ξ or η for straight field boundaries;

$$B_{\xi}(\xi, \eta\, \zeta) = 0,$$
$$B_{\eta}(\xi, \eta, \zeta) = B_{\eta}(0, 0, \zeta) + \cdots,$$
$$B_{\zeta}(\xi, \eta, \zeta) = \eta(\partial B_{\zeta}/\partial \eta)_{0,0,\zeta} + \cdots = \eta(\partial B_{\eta}/\partial \zeta)_{0,0,\zeta} + \cdots.$$

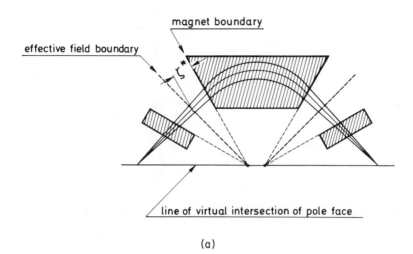

(a)

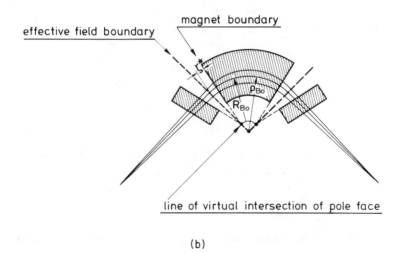

(b)

Fig. 7.8. Fringing field shielding shunts are shown for (a) a wedge magnet and for (b) a radially inhomogeneous magnetic sector field, respectively. Note that the effective field boundaries are inclined to the magnet boundaries so that they would intersect at that ξ coordinate, where the prolonged pole faces would touch (Figs. 4.9 and 4.16).

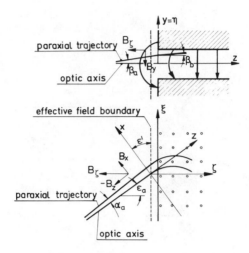

Fig. 7.9. Spatial distribution of the magnetic flux density in the fringing field of a deflecting magnet projected on the planes $\xi = 0$ and $\eta = 0$.

Here, we have used curl $\mathbf{B} = 0$, or $(\partial B_\zeta/\partial \eta) = (\partial B_\eta/\partial \zeta)$. Introducing this $B_\eta(\xi, \eta, \zeta)$ into Eq. (7.2b), we find

$$\sin \beta'(\zeta) = \sin \beta'_a + \int_{\zeta_a}^{\zeta} \eta \, \frac{\tan \varepsilon}{B_0 \rho_0} \left(\frac{\partial B_\eta}{\partial \zeta} \right)_{0,0,\zeta} d\zeta + \cdots,$$

$$= \sin \beta'_a + \int_{\zeta_a}^{\zeta} \frac{\eta}{B_0 \rho_0} \left[\tan \varepsilon_a - \frac{1}{\cos^3 \varepsilon_a} \right.$$

$$\left. \times \int_{\zeta_a}^{\zeta} \frac{B_{\eta(0,0,\zeta)}}{B_0 \rho_0} \, d\zeta \right] \left(\frac{\partial B_\eta}{\partial \zeta} \right)_{0,0,\zeta} d\zeta + \cdots. \qquad (7.6a)$$

Neglecting the last term in Eq. (7.6a), replacing η by the initial η_a, and integrating from ζ_a to ζ_b, Herzog (1955) found with $B_\eta(0, 0, \zeta_a) = 0$ and $B_\eta(0, 0, \zeta_b) = B_0$,

$$\sin \beta'_b - \sin \beta'_a \approx \eta_a (\tan \varepsilon_a / \rho_0), \qquad (7.6b)$$

which astonishingly enough does not depend on the detailed fringing field distribution.

Equation (7.6b) describes the deflection of a thin y-focusing lens placed at the entrance effective field boundary. At the exit effective field boundary, an analogous thin focusing lens must be assumed. According to Eqs. (7.6b) and (1.4), the focusing powers of these thin lenses equal (Fig. 1.6),

$$1/f'_y = -\tan \varepsilon'/\rho_0, \qquad 1/f''_y = -\tan \varepsilon''/\rho_0. \qquad (7.7)$$

Here, ε' and ε'' describe angles at which the optic axis of a beam passes through the effective field boundaries at the entrance and exit of the sector field (see also Fig. 4.8). Looking in more deail at the above derivation (see Wollnik, 1964, or Wollnik and Ewald, 1965), we must determine η in Eq. (7.6a) as function of ζ,

$$\eta - \eta_a = \int_{\zeta_a}^{\zeta} \frac{\tan \beta'}{\cos \varepsilon} d\zeta \approx \int_{\zeta_a}^{\zeta} \frac{\sin \beta'}{\cos \varepsilon} d\zeta,$$

so that Eq. (7.6a) becomes to first order in η and β',

$$\sin \beta'_b - \sin \beta'_a = \int_{\zeta_a}^{\zeta_b} \left[\eta_a + (\zeta - \zeta^*) \frac{\sin \beta_a}{\cos \varepsilon_a} - \frac{\eta_a \tan \varepsilon_a}{\cos \varepsilon_a} \int_{\zeta_a}^{\zeta} \frac{B_{\eta(0,0,\zeta)}}{B_0 \rho_0} d\zeta \right]$$

$$\times \left[\tan \varepsilon_a - \frac{1}{\cos^3 \varepsilon_a} \int_{\zeta_a}^{\zeta} \frac{B_{\eta(0,0,\zeta)}}{B_0 \rho_0} d\zeta \right]$$

$$\times \left(\frac{\partial B_\eta}{\partial \zeta} \right)_{0,0,\zeta} \left(\frac{d\zeta}{B_0 \rho_0} \right) + \cdots . \tag{7.8a}$$

With $B_\eta(0, 0, \zeta_a) = 0$, $B_\eta(0, 0, \zeta_b) = B_0$, $(\partial B_\eta / \partial \zeta)_{0,0,\zeta_a} = (\partial B_\eta / \partial \zeta)_{0,0,\zeta_b} = 0$, and $\int (\zeta - \zeta^*)(\partial B_\eta / \partial \zeta) d\zeta = 0$, when integrated from ζ_a to ζ_b, the Eq. (7.8a) yields

$$\sin \beta' - \sin \beta'_a = -\frac{\eta_0}{\rho_0} \left[\tan \varepsilon_a + \frac{1}{\cos \varepsilon_a} (1 + 2 \tan^2 \varepsilon_a) I_3 \frac{G_0}{\rho_0} \right] + \cdots , \tag{7.8b}$$

with

$$I_3 = \int_{\zeta_a}^{\zeta_b} \left[\int_{\zeta_a}^{\zeta} \frac{B_{\eta(0,0,\zeta)}}{B_0^2 G_0} d\zeta \right] \left(\frac{\partial B_\eta}{\partial \zeta} \right)_{0,0,\zeta} d\zeta$$

$$= \int_{\zeta_a}^{\zeta_b} \frac{B_{\eta(0,0,\zeta)}}{B_0 G_0} \left[1 - \frac{B_{\eta(0,0,\zeta)}}{B_0} \right] d\zeta,$$

as we find by partial integration. Equation (7.8b) deviates from Eq. (7.6b) by the term that contains the fringing field integral I_3. Note again that $2G_0$ is the width of the magnetic air gap, whereas ρ_0 is the radius of curvature of the optic axis. The focusing powers $1/f'_y$ and $1/f''_y$ of the thin fringing field lenses thus finally read

$$\frac{1}{f'_y} = \frac{1}{\rho_0} \left[\tan \varepsilon' + \frac{1 + \sin^2 \varepsilon'}{\cos^3 \varepsilon'} I'_3 \frac{G_0}{\rho_0} \right], \tag{7.9a}$$

$$\frac{1}{f_y''} = \frac{1}{\rho_0}\left[\tan \varepsilon'' + \frac{1 + \sin^2 \varepsilon''}{\cos^3 \varepsilon''} I_3'' \frac{G_0}{\rho_0}\right]. \tag{7.9b}$$

The quantities I_3' and I_3'' are fringing field integrals for the entrance and the exit sides of a magnet, as defined under Eq. (7.8b).

The integrals I_3 of Eq. (7.9) can also be approximated by calculating $B_\eta(0, 0, \zeta)$ for a shielded magnet having sharply pointed pole tips (Fig. 7.4) and being built from iron of infinite permeability. As pointed out by Wollnik (1964) an integral like I_3 had already been evaluated by Herzog (1940) for the analogous fringing field of electrostatic deflectors [see Eq. (7.14)], yielding

$$I_3\pi = 2\ln\frac{2G_0}{q_1} - \frac{G_0}{D}\arccos\left(1 - \frac{2D^2}{q_1^2}\right), \tag{7.10}$$

with q_1 defined following Eq. (7.5), and $\theta_3 = \arccos\left[1 - (2D^2/q_1^2)\right]$ given in Fig. 7.12. This result is shown graphically in Fig. 7.10, where I_3 is plotted for different geometries of fringing field shunts, i.e., for different choices

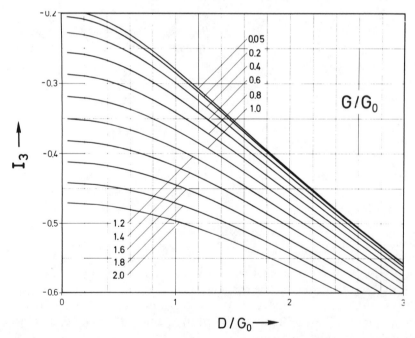

Fig. 7.10. The integral I_3, which describes an additional fringing field defocusing action [Eqs. (7.9a) or (7.9b)] is plotted for different geometries of fringing field shunts, i.e., for different choices of D/G_0 and G/G_0 (Fig. 7.4). For simplicity in these calculations, pointed pole tips or electrodes are assumed, and the influence of the coils in a magnet is neglected.

of D/G_0 and G/G_0 (Fig. 7.4). Note that I_3 is always negative, indicating an always-defocusing action in y direction. Note also that I_3 equals about -0.37 for $D = G = G_0$.

7.1.4 Transfer Matrices of Realistic Magnetic Sector Fields

When ϕ_B is the angle of deflection of a magnetic sector field, i.e., the angle between the effective field boundaries (Figs. 7.8 and 7.9), we find the transfer matrices of the realistic magnetic sector field as a combination of transfer matrices describing the actions of the fringing fields and of the main sector field:

$$
\begin{pmatrix} 1 & 0 & 0 & 0 \\ -1/f''_x & 1 & 0 & 0 \\ 0 & 0 & 1 & 0 \\ 0 & 0 & 0 & 1 \end{pmatrix}
\begin{pmatrix} (x|x) & (x|a) & (x|\delta_K) & (x|\delta_m) \\ (a|x) & (a|a) & (x|\delta_K) & (a|\delta_m) \\ 0 & 0 & 1 & 0 \\ 0 & 0 & 0 & 1 \end{pmatrix}
\begin{pmatrix} 1 & 0 & 0 & 0 \\ -1/f'_x & 1 & 0 & 0 \\ 0 & 0 & 1 & 0 \\ 0 & 0 & 0 & 1 \end{pmatrix},
$$

$$(7.11a)$$

$$
\begin{pmatrix} 1 & 0 \\ -1/f''_y & 1 \end{pmatrix}
\begin{pmatrix} (y|y) & (y|b) \\ (b|y) & (b|b) \end{pmatrix}
\begin{pmatrix} 1 & 0 \\ -1/f'_y & 1 \end{pmatrix}.
$$

$$(7.11b)$$

Here, the elements of the middle transfer matrices must be taken from Eqs. (4.8), (4.38), and (4.61).

7.2 PARTICLE TRAJECTORIES IN FRINGING FIELDS OF ELECTROSTATIC DEFLECTORS

Particle trajectories in the fringing field $E_\zeta(0, 0, \zeta)$ of an electrostatic deflector are found analogously to the magnetic case. An electrostatic sector field is always radially inhomogeneous. According to Wollnik (1964) or Wollnik and Ewald (1965), we should thus design fringing field shunts as indicated in Fig. 7.11 analogous to Fig. 7.8b. For a spherical or toroidal condensor, the shunt aperture $2G$ should furthermore follow the y curvature of the electrodes (see R_{E0} in Fig. 4.17). Advantageously, the potentials of these shunts are chosen to be identical to the middle equipotential surface of the deflector $(V_1 + V_2)/2$. If this potential is not the ground potential $V = 0$ to which the charged particles had been accelerated long before they entered the electrostatic deflector, a second diaphragm, which is at ground potential, should be placed before the fringing field shunt (Fig. 7.11).

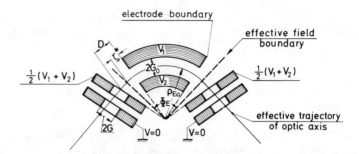

Fig. 7.11. Fringing field shunts for an electrostatic sector field formed by electrodes at potentials V_1 and V_2. For $V_1 + V_2 = 0$, the outer fringing field shunts can be omitted.

7.2.1 Electrostatic Deflectors Biased Symmetrically to Ground

Consider a symmetrically grounded electrostatic parallel-plate deflector (Fig. 7.12), the electrodes of which are at potentials V_1 and V_2, with

$$V_1 + V_2 = 0. \qquad (7.12)$$

An ion of charge $(z_0 e)$, mass m_0, and kinetic energy $K_0 = -(z_0 e) V_0$ that moves in the plane of symmetry ($\eta = 0$) in such a parallel-plate condensor

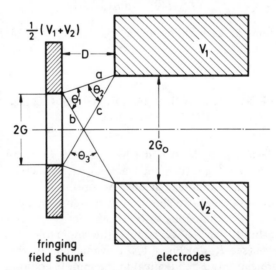

Fig. 7.12. The geometry of the fringing field shunt for an electrostatic deflector, which is analogous to the magnetic case of Fig. 7.4 [Wollnik and Ewald (1965)]. Note that because of $a^2 + b^2 - 2(ab) \cos \theta_1 = c^2$, we find $\cos \theta_1 = (D^2 + G_0^2 + G^2)/(q_1 q_2)$, with q_1 and q_2 given below Eq. (7.5), and that because of $b^2 + c^2 - 2(bc) \cos \theta_3 = \arccos(q_1 \theta_3) = a^2$, we find $\cos \theta_3 = 1 - 2D^2/q_1^2$.

experiences the deflecting force

$$m_0 \ddot{\xi} = (z_0 e) E_\xi(0, 0, \zeta).$$

Here we can also define an effective field boundary at $\zeta = \zeta^*$, with

$$\zeta^* = \zeta_b - \int_{\zeta_a}^{\zeta_b} \frac{E_\xi(0, 0, \zeta)}{E_0} \, d\zeta = I_1 G_0, \qquad (7.13a)$$

and a shift of all particle trajectories including the optic axis at this boundary,

$$\Delta\xi = \left[\frac{(\zeta_b - \zeta^*)^2}{2} - \int\!\!\int_{\zeta_a}^{\zeta_b} \frac{E_\xi(0, 0, \zeta)}{E_0} \, d\zeta \, d\zeta \right] \frac{1}{\rho_0} = (I_2 G_0) \frac{G_0}{\rho_0}. \qquad (7.13b)$$

Here, $2G_0$ is the electrode gap, ρ_0 the radius of curvature of the optic axis, and E_0 the electrostatic field strength at the optic axis. The integrals I_1 and I_2 are identical to those of Eqs. (7.4a) and (7.4b) and thus numerically also given by Eqs. (7.5a) and (7.5b) or graphically in Fig. 7.6, where in both cases the geometry of the fringing field shunts, i.e., D/G_0 and G/G_0, must be taken from Fig. 7.12. As in Eq. (7.4), the magnitudes of ζ_a and ζ_b must be chosen well outside the parallel-plate deflector and well inside, respectively, resulting in $E_\xi(0, 0, \zeta_a) = 0$ and $E_\xi(0, 0, \zeta_b) = E_0$.

An effective field boundary, as determined by Eq. (7.13a), produces most but not all observed fringing field focusing effects. In the case of a particle beam that enters and leaves a magnetic sector field perpendicularly ($\varepsilon' = \varepsilon'' = 0$), a focusing action was found in the y direction but none in the x direction. For an electrostatic sector field, such a focusing action exists in the x direction but none in the y direction. In detail, Herzog (1940) found the action of thin lenses at the entrance and exit boundaries that have the focusing powers [see Matsuda (1971) for the factor 2]:

$$1/f_x' = -2 I_3'(G_0/\rho_0^2), \qquad 1/f_x'' = -2 I_3''(G_0/\rho_0^2), \qquad (7.14)$$

with

$$I_3 = \int_{\zeta_a}^{\zeta_b} \frac{E_\xi(0, 0, \zeta)}{G_0 E_0} \left[1 - \frac{E_\xi(0, 0, \zeta)}{E_0} \right] d\zeta.$$

Here also $2G_0$ is the electrode gap, ρ_0 the radius of curvature of the optic axis, and E_0 the electrostatic field strength at the optic axis, whereas ζ_a and ζ_b must again be chosen well outside and well inside the electrostatic field so that $E_\xi(0, 0, \zeta_a) = 0$ and $E_\xi(0, 0, \zeta_b) = E_0$. The integral of Eq. (7.14) is identical to that in Eq. (7.8b) and thus numerically also given by Eq. (7.10) or graphically in Fig. 7.10, where the geometry of the fringing field shunt, i.e., the magnitude of D/G_0 and G/G_0, must be taken from Fig. 7.12.

7.2.2 Electrostatic Deflectors Biased Asymmetrically to Ground

Consider now an electrostatic deflector the electrodes of which are at potentials V_1 and V_2, with

$$V_1 + V_2 \neq 0. \tag{7.15}$$

From the previous explanations, it is already known that under these circumstances, we should use two fringing field shunts, as indicated in Fig. 7.11. When the apertures $2G$ in both shunts are small, we have a good approximation of a dipole sheet (Fig. 2.5) formed by the two shunts at potentials 0 and $(V_1 + V_2)/2$.

If a particle of rest mass m_{00} and charge $(z_0 e)$ arrives at the left side of a dipole sheet with an energy K_1 and moves to the other side of the dipole sheet, this particle gains the energy $\Delta K = -(z_0 e)(V_1 + V_2)/2$. If the particle under consideration arrived at the left side of the dipole sheet inclined (Fig. 2.5) under angles α_1 and β_1, this particle will leave the right side of the dipole sheet inclined under angles α_2 and β_2, where according to Eq. (2.24), we find for a planar particle motion $a_i = \sin \alpha_i$ and $b_i = \sin \beta_i$, with $i = 1, 2$,

$$\frac{a_2}{a_1} = \frac{b_2}{b_1} = \sqrt{\frac{K_2(1 + \eta_2)}{K_1(1 + \eta_1)}} = \sqrt{\left(1 + \frac{\Delta K}{K_1}\right)\left(1 + \frac{\Delta K}{K_1 + 2m_{00}c^2}\right)}$$

$$= (1 + \kappa)\left(1 + \frac{\kappa \eta_1}{1 + \eta_1}\right),$$

and $K_2 = K_1 + \Delta K = K_1(1 + \kappa)$, as well as $\eta_1 = K_1/(m_{00}c^2)$ and $\eta_2 = K_2/(m_{00}c^2)$ [see also Eq. (2.3)].

For relativistically slow particles [$K_1, K_2 \ll 931$ MeV/u or $\eta_1 \approx \eta_2 \ll 1$], these relations simplify to

$$a_2/a_1 = b_2/b_1 = \sqrt{K_2/K_1} = \sqrt{1 + \kappa}. \tag{7.16a}$$

A dipole sheet also changes the relative energy deviation of an arbitrary particle from a reference particle that can move along the optic axis. Consider a particle that had the energy $K_1(1 + \delta_{K1})$ before and the energy $K_2(1 + \delta_{K2}) = K_1(1 + \delta_{K1}) + \Delta K$ behind the dipole sheet. With $K_2 = K_1 + \Delta K$, we thus find

$$\delta_{K1} = \delta_{K2}[1 + (\Delta K/K_1)] = \delta_{K2}(1 + \kappa), \tag{7.16b}$$

and the transfer matrices of a dipole sheet read

$$\begin{pmatrix} x_2 \\ a_2 \\ \delta_{K2} \\ \delta_{m2} \end{pmatrix} = \begin{pmatrix} 1 & 0 & 0 & 0 \\ 0 & 1/\sqrt{1+\kappa} & 0 & 0 \\ 0 & 0 & 0 & 1/(1+\kappa) \\ 0 & 0 & 1 & 0 \end{pmatrix} \begin{pmatrix} x_1 \\ a_1 \\ \delta_{K1} \\ \delta_{m1} \end{pmatrix}, \tag{7.17a}$$

$$\begin{pmatrix} y_2 \\ b_2 \end{pmatrix} = \begin{pmatrix} 1 & 0 \\ 0 & 1/\sqrt{1+\kappa} \end{pmatrix} \begin{pmatrix} y_1 \\ b_1 \end{pmatrix}. \tag{7.17b}$$

There are two further effects that change the focusing properties of an unsymmetrically grounded deflector compared to a symmetrically grounded one. First, the position $\zeta^* = I_1 G_0$ of the effective field boundary is changed to $\zeta^*_{new} = \zeta^* + \Delta\zeta^* = (I_1 + \Delta I_1)G_0$ and second, the fringing field focusing action is changed.

Herzog (1940) found the shift of the effective field boundary $\Delta\zeta^* = \Delta I_1 G_0$ with respect to $\zeta^* = I_1 G_0$ determined from Eq. (7.5a) or Fig. 7.6a as

$$\Delta\zeta^* = -G_0 \frac{(z_0 e)(V_1 + V_2)}{2\pi K_1} \ln 2, \tag{7.18}$$

for both ends of the sector field. This shift of the effective field boundary does not depend on the special shape of the fringing field shunt arrangement. Normally, $(V_1 + V_2)$ is smaller than $(V_1 - V_2)$ with $\rho_0(V_1 - V_2)/2G = 2K_1/(z_0 e)$ so that the quantity $\Delta\zeta^*$ stays smaller than $0.44 G_0^2/\rho_0$, a value obtained from Eq. (7.18) if $V_1 + V_2$ is replaced by $V_1 - V_2$. This shift is so small that for most cases it can be neglected. Herzog (1940) also found the focusing powers of the fringing field lenses of an unsymmetrically grounded elecrostatic sector field as

$$1/f'_x = (G_0/\rho_0^2)(-2I'_3 + I'_4), \qquad 1/f''_x = (G_0/\rho_0^2)(-2I''_3 + I''_4). \tag{7.19}$$

The quantities I'_3 and I''_3 describe the same defocusing fringing field integrals as those in Eqs. (7.10) and (7.14) and Fig. 7.10 for a symmetrically grounded electrostatic deflector, whereas the quantities I'_4 and I''_4 describe additional focusing fringing field integrals,

$$I_4\pi = \left[\frac{(z_0 e)(V_1 + V_2)}{4K_1}\right]^2 \frac{G_0}{D} \arccos\left(1 - \frac{2D^2}{q_1^2}\right). \tag{7.20}$$

Here, q_1 is defined following Eq. (7.5), and $\theta_3 = \arccos[1 - (2D^2/q_1^2)]$ is given in Fig. 7.12. For $D = G = G_0$, for example, we read $\tan(\theta_3/2) = 0.5$ or $\theta_3 \approx 53°$ from Fig. 7.12, which causes I_4 to be approximately $0.295[(z_0 e)(V_1 + V_2)/(4K_1)]^2$.

7.2.3 Transfer Matrices of Realistic Electrostatic Sector Fields

When ϕ_E is the angle of deflection of an electrostatic sector field, i.e., the angle between the effective field boundaries (Fig. 7.11), the transfer matrices of the realistic sector field are given by a combination of transfer matrices describing the actions of the fringing fields and of the main sector field. Contrary to Eq. (7.11) for the magnetic case, the possibly accelerating or decelerating effects of dipole sheets at the entrance or exit of the sector field must be taken into account. Thus one finds

$$
\begin{pmatrix}
1 & 0 & 0 & 0 \\
-1/f_x'' & \sqrt{1+\kappa} & 0 & 0 \\
0 & 0 & 1+\kappa & 0 \\
0 & 0 & 0 & 1
\end{pmatrix}
\begin{pmatrix}
(x|x) & (x|a) & (x|\delta_K) & (x|\delta_m) \\
(a|x) & (a|a) & (a|\delta_K) & (a|\delta_m) \\
0 & 0 & 1 & 0 \\
0 & 0 & 0 & 1
\end{pmatrix}
$$

$$
\times
\begin{pmatrix}
1 & 0 \\
-1/f_x' & 1/\sqrt{1+\kappa} \\
0 & 0 \\
0 & 0
\end{pmatrix}
\begin{pmatrix}
0 & 0 \\
0 & 0 \\
1/(1+\kappa) & 0 \\
0 & 1
\end{pmatrix},
\qquad (7.21a)
$$

$$
\begin{pmatrix}
1 & 0 \\
-1/f_y'' & \sqrt{1+\kappa}
\end{pmatrix}
\begin{pmatrix}
(y|y) & (y|b) \\
(b|y) & (b|b)
\end{pmatrix}
\begin{pmatrix}
1 & 0 \\
-1/f_y' & 1/\sqrt{1+\kappa}
\end{pmatrix}.
\qquad (7.21b)
$$

The elements of the middle transfer matrices must be taken here from Eqs. (4.61a) and (4.61b) whereas the f_x', f_x'' are given in Eq. (7.19). The quantity $\kappa = \Delta K / K_1$ characterizes the amount by which the midpotential of the electrostatic deflector deviates from ground [Eq. (7.15)], with K_1 being the particle energy before the sector field. For an electrostatic sector field biased symmetrically to ground, κ vanishes.

7.3 PARTICLE TRAJECTORIES IN FRINGING FIELDS OF QUADRUPOLE LENSES

The magnetic flux density or the electrostatic field strength in a quadruple lens is zero along the optic axis $x = y = 0$ [Eqs. (3.1) and (3.3)]. The field gradients g_B and g_E, however, must somehow change along the optic axis from zero outside the quadrupole to fixed values g_{B0} and g_{E0} in the main field region (Fig. 7.13) of a magnetic or electrostatic quadrupole, respectively. Analogous to Fig. 7.2, we here can define an *effective gradient distribution* g_B^* and g_E^* along the z axis. This gradient distribution rises abruptly from zero to g_{B0} and g_{E0} at effective field boundaries $z = z^*$ so that $g_B^*(z < z^*) = g_E^*(z < z^*) = 0$ as well as $g_B^*(z > z^*) = g_{B0}$ and $g_E^*(z > z^*) = g_{E0}$. Consequently, the magnetic flux density in a real magnetic quadrupole follows,

$$
B_x(x, y, z) = -g_B y + \cdots,
$$

$$
B_y(x, y, z) = g_B x + \cdots,
$$

$$
B_z(x, y, z) = 0,
$$

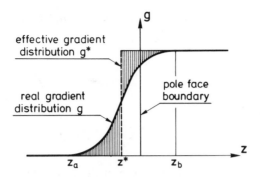

Fig. 7.13. The distribution g of the gradient in the fringing field of a quadrupole is shown together with the effective gradient distribution g^*.

or the electrostatic field strength in a real electrostatic quadrupole,

$$E_x(x, y, z) = -g_E x + \cdots,$$

$$E_y(x, y, z) = g_E y + \cdots,$$

$$E_z(x, y, z) = 0.$$

The forces acting on a particle of mass m_0, charge $z_0 e$, and velocity v_0 can be described by

$$m_0 \ddot{x} = (z_0 e) x g, \qquad m_0 \ddot{y} = (z_0 e) y g,$$

in both cases. Here, g represents g_B and g_E / v_0 in the magnetic and electrostatic cases, respectively. Integrating these relations over the time t, we find with $dz \approx dt v_0$, as well as with the magnetic and electrostatic rigidities, $(m_0 v_0)/(z_0 e) = B_0 \rho_{B0} = \chi_{B0}$ and $(m_0 v_0^2)/(z_0 e) = E_0 \rho_{E0} = \chi_{E0}$, which both [see also Eqs. (2.11) and (2.14)] are abbreviated by χ_0:

$$\dot{x}(z) \approx \dot{x}(z_a) - x_1 \int_{z_a}^{z} \frac{g \, dz}{\chi_0}, \qquad (7.22a)$$

$$\dot{y}(x) \approx \dot{y}(z_a) + y_1 \int_{z_a}^{z} \frac{g \, dz}{\chi_0}. \qquad (7.22b)$$

Here it is assumed that the particle does not change its x or y value appreciably while traversing the fringing field. To define the position of an effective field boundary, we must evaluate Eqs. (7.22a) and (7.22b) for a particle once crossing the real fringing field with $g(z \neq 0)$ and once crossing the effective fringing field with $g^*(z < z^*) = 0$ and $g^*(z > z^*) = g_0$ [here, g_0 represents either g_{B0} or g_{E0}/v_0] and postulate that at some z_b, the $\dot{x}(z_b)$

and $\dot{y}(z_b)$ values are identical. In this case, we find from both Eqs. (7.22a) and (7.22b),

$$(z_b - z^*)g_0 = \int_{z_a}^{z_b} g \, dz, \qquad (7.23)$$

or

$$z^* = z_b - \int_{z_a}^{z_b} \frac{g}{g_0} \, dz = I_{1q} G_0, \qquad (7.24)$$

where $2G_0$ denotes the aperture width of the quadrupole.

For an evaluation of Eq. (7.24), the gradient distributions g_B or g_E should be determined for each individual quadrupole as function of z. However, an experimental evaluation requires great effort, and a theoretical calculation of the integral I_{1q} is difficult, since we would have to integrate over a solution of the exact three-dimensional Laplace equation. Fortunately, the position of the effective gradient boundary does not need to be known with utmost accuracy for a quadrupole lens, since according to Eqs. (3.9a) or (3.9b), the focusing power $1/f$ of a quadrupole lens equals $k \sin(kw)$. The change in focusing power due to a variation of the length w can thus be compensated for by a variation of k, i.e., by a slightly varied pole-tip flux density or a slightly varied electrode potential. For most cases, it is thus sufficient to replace the integral I_{1q} in Eq. (7.23) by the integral I_1 of Eq. (7.4) and Fig. 7.6a (see also Matsuda and Wollnik, 1972).

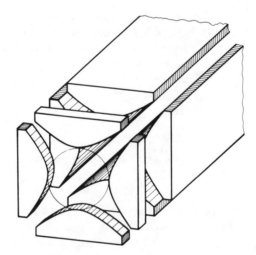

Fig. 7.14. A quadrupole fringing field shunt. In many cases, the aperture in the fringing field shunt can be simplified to be a round hole, a square, or a rectangle.

In the fringing field region, the quadrupole gradient is very sensitive to any disturbances, as is the fringing field distribution of a sector field. In order to eliminate this sensitivity, we should limit the fringing field region also in a quadrupole by some fringing field shunt, as indicated in Fig. 7.14 or as described by Steffen (1965).

REFERENCES

Bucherer, A. H. (1909). *Ann. Phys.* (*Leipzig*) **28**, 513.
Enge, H. (1967). *In* "Focusing of Charged Particles," A. Septier (ed.). Academic, New York.
Enge, H. (1975). Int. Rep. ICI-3038-2/75, Industrial Coils Inc., Boston.
Herzog, R. (1935a). *Arch. Elektrotech.* (*Berlin*) **29**, 790.
Herzog, R. (1935b). *Z. Phys.* **97**, 596.
Herzog, R. (1940). *Z. Phys.* **41**, 18.
Herzog, R. (1950). *Acta Phys. Austriaca* **4**, 431.
Herzog, R. (1955). *Z. Naturforsch.* **10A**, 886.
Hübner, H. and Wollnik, H., (1970). *Nucl. Instrum. Methods* **86**, 141.
Matsuda, H. (1971). *Nucl. Instrum. Methods* **91**, 637.
Matsuda, H. and Wollnik, H. (1972). *Nucl. Instrum. Methods* **103**, 117.
Rauscher, M. (1970). Thesis, University of Giessen.
Rogowski, W. (1923). *Arch. Elektrotechnnik* (*Berlin*) **12**, 1.
Steffen, K. G. (1965). "High Energy Beam Optics." (Interscience), New York.
Wollnik, H. (1964). Thesis, Techn. Hochsch, München.
Wollnik, H. and Ewald, A. (1965). *Nucl. Instrum. Methods* **36**, 93.

8

Image Aberrations

Trajectories of charged particles relative to an optic axis have been described throughout this book in terms of a Gaussian or first-order approximation. This description is adequate for narrow particle beams; however, for realistic beams, deviations from this first-order theory are often observed: the aberrations of second, third, and higher order. Normally, these aberrations are only of interest in image-profile planes where the particle beam is concentrated to a small cross section. For such image-profile planes, however, a correction of aberrations is highly desirable.

To determine the magnitudes of aberrations of an optical system is rather involved. For the design of high-performance optical systems, on the other hand, it is sufficient to know the origin of aberrations and to understand their interdependence. For this reason, the detailed derivation of image aberrations is outlined only in the Appendix to this chapter.

8.1 SYSTEMATICS OF IMAGE ABERRATIONS

The geometry of an optical system is defined by the geometry of its optic axis. This optic axis is straight in field-free regions or quadrupoles (see Fig.

3.1) and circular, with ρ_0 the radius of curvature, in radially inhomogeneous sector fields (see Figs. 4.1, 4.16, and 4.17). The electrostatic potential in all these fields is constant along the optic axis (Chapters 3 and 4), as is the magnitude of the electric or magnetic field E_0 or B_0. Thus, a particle can move along the optic axis if it has a rest mass m_{00}, energy K_0, and charge $(z_0 e)$, or, in other words, rigidities $\chi_0 = E_0 \rho_{E0}$ or $\chi_0 = B_0 \rho_{B0}$.

To describe the trajectory of an arbitrary particle, we must define its energy–charge as well as its mass–charge ratios:

$$\frac{K}{(ze)} = \frac{K_0}{(z_0 e)}(1 + \delta_K), \qquad \frac{m_0}{(ze)} = \frac{m_{00}}{(z_0 e)}(1 + \delta_m)$$

[Eqs. (2.15), (2.16b) and (2.16c)]. According to Eq. (2.20d), the rigidity of such a particle is $\chi = \chi_0(1 + \Delta)$, with

$$\Delta = (\delta_K + \hat{\delta}_K)\frac{(1 + 2\eta_0)^2 + h}{2(1 + \eta_0)(1 + 2\eta_0)} + (\delta_m)\frac{(1 + 2\eta_0) - h}{2(1 + \eta_0)(1 + 2\eta_0)}$$

$$+ (\delta_K + \hat{\delta}_K - \delta_m)^2 \frac{h[1 - 4\eta_0(1 + \eta_0)] - (1 + 2\eta_0)^2}{8(1 + \eta_0)^2(1 + 2\eta_0)^2} + \cdots, \qquad (8.1)$$

where, as defined in Eqs. (2.19) or (4.59), a magnetic field is characterized by $h = 0$ and an electrostatic field by $h = 1$. Assume now that such an arbitrary particle had started from a point $|x_0| < x_{00}$, $|y_0| < y_{00}$ under angles $|\alpha_0| < \alpha_{00}$, $|\beta_0| < \beta_{00}$, as indicated in Fig. 8.1. Tracing this particle through a given optical system from z_0 to z_e, we find the final position x_e, y_e at which this particle arrives under angles α_e, β_e. Furthermore, we find the time t_e this particle is delayed at z_e compared to the reference particle, if

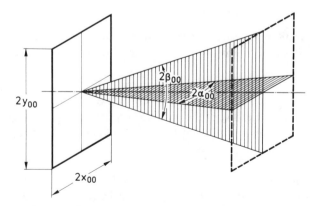

Fig. 8.1. The size $2x_{00}2y_{00}$ of an object is indicated together with angles of inclination $2\alpha_{00}$ and $2\beta_{00}$, at which particles emerge from the center $x_0 = y_0 = 0$ of the object.

it was delayed by the time t_0 at z_0. These x_e, a_e, t_e, y_e, b_e are functions of x_0, a_0, t_0, y_0, b_0, as well as of two of the small quantities Δ, δ_K, δ_m. So as to not complicate the formulas discussed here, we shall in most cases restrict ourselves to particles of constant mass ($m_0 = m_{00}$) and charge ($z = z_0$) and keep only the energy ($K \neq K_0$) variable. Thus, we have $\delta_m = 0$ and only $\delta_K \neq 0$, which we shall abbreviate as $\delta \neq 0$.

At z_0 and z_e we may now define general position vectors with the components $r_{i,0}$ and $r_{k,e}$, with $i, k \in \{1, 2, 3, 4, 5, 6\}$. Explicitly, these vectors read as follows:

$$r_{1,0} = x_0, \qquad r_{4,0} = \delta_0, \qquad r_{1,e} = x_e, \qquad r_{4,e} = \delta_e,$$

$$r_{2,0} = a_0, \qquad r_{5,0} = y_0, \qquad r_{2,e} = a_e, \qquad r_{5,e} = y_e,$$

$$r_{3,0} = t_0, \qquad r_{6,0} = b_0, \qquad r_{3,e} = t_e, \qquad r_{6,e} = b_e.$$

Note here that according to Eqs. (2.22a) and (2.22b), we had defined a and b as

$$a = \frac{(\sin \alpha)\sqrt{(1 + \delta)(1 + \eta)}}{\sqrt{1 + \tan^2 \beta \cos^2 \alpha}} = \frac{p_x}{p_0}, \qquad b = \frac{(\sin \beta)\sqrt{(1 + \delta)(1 + \eta)}}{\sqrt{1 + \tan^2 \alpha \cos^2 \beta}} = \frac{p_y}{p_0}.$$

Very generally, the relation between $r_{k,0}$ and $r_{i,e}$ is given by

$$r_{i,e} = \sum_{k=1}^{6} r_{k,0}\left\{ (r_i|r_k) + \sum_{l=1}^{6} \frac{r_{l,0}}{2}\left\{ (r_i|r_kr_l) + \sum_{m=1}^{6} \frac{r_{m,0}}{3}\{(r_i|r_kr_lr_m) + \cdots\}\right\}\right\}, \quad (8.2)$$

with $i, k, l, m, n, \ldots$, all ranging from 1 to 6. Here, $(r_i|r_k)$, $(r_i|r_kr_l)$, $(r_i|r_kr_lr_m)$, $\ldots$, are elements of first, second, third, $\ldots$, order. For each value of i, we thus have $6 \cdot 6$ elements of first order, $6 \cdot 6^2$ elements of second order, $6 \cdot 6^3$ elements of third order, etc. However, many of these elements are zero for most cases of interest.

(1) In time-independent fields, none of the coefficients can explicitly depend on the time t. Except $(r_3|r_3) = 1$, one thus finds

$$(r_i| \cdots r_3^{n_3} \cdots) = 0. \tag{8.3a}$$

(2) If there are no accelerating fields in the optical system, the energy cannot vary, so the rigidity of the particle under investigation stays constant, and we have $\delta_0 = \delta_e = \delta$. Except for $(r_4|r_4) = 1$, we thus find

$$(r_4| \cdots) = 0. \tag{8.3b}$$

If the particles of a beam are all accelerated by the same potential ΔV, Eq. (8.3b) is also valid except for $(r_4|r_4) = 1 + \kappa$ [Eq. (7.17)].

(3) If the optical elements exhibit a twofold symmetry, i.e., if all fields have a common plane of deflection ($y = 0$), all deflections in the x direction

must be symmetric, and all deflections in the y direction must be antisymmetric with respect to ($y = 0$). Consequently, we find

$$(r_i | \cdots r_5^{n_5} r_6^{n_6} \cdots)_{\text{odd}} = 0, \qquad \text{for} \quad i = 1, 2, 3, 4, \qquad (8.3c)$$

$$(r_i | \cdots r_5^{n_5} r_6^{n_6} \cdots)_{\text{even}} = 0, \qquad \text{for} \quad i = 5, 6. \qquad (8.3d)$$

Because of Eqs. (8.3a)–(8.3d), quite a few coefficients of Eq. (8.2) vanish. Rewriting Eq. (8.2) as

$$r_{i,1} = [\text{first order}] + \tfrac{1}{2}[\text{second order}] + \tfrac{1}{6}[\text{third order}] + \cdots,$$

we thus find, with $r_{i,1} \in \{x_1, a_1, t_1\}$ and $r_i \in \{x, a, t\}$,

$[\text{first order}] = (r_i|x)x_0 + (r_i|a)a_0 + (r_i|\delta)\delta_0,$

$\begin{aligned}[\text{second order}] = \ & (r_i|xx)x_0^2 + 2(r_i|xa)x_0a_0 + (r_i|aa)a_0^2 \\ & + (r_i|yy)y_0^2 + 2(r_i|yb)y_0b_0 + (r_i|bb)b_0^2 \\ & + 2(r_i|x\delta)x_0\delta_0 + 2(r_i|a\delta)a_0\delta_0 + (r_i^2|\delta\delta)\delta_0^2,\end{aligned}$

$\begin{aligned}[\text{third order}] = \ & (r_i|xxx)x_0^3 + 3(r_i|xxa)x_0^2a_0 + 3(r_i|xaa)x_0a_0^2 + (r_i|aaa)a^3 \\ & + 3(r_i|xyy)x_0y_0^2 + 6(r_i|xyb)x_0y_0b_0 + 3(r_i|xbb)x_0b_0^2 \\ & + 3(r_i|ayy)a_0y_0^2 + 6(r_i|ayb)a_0y_0b_0 + 3(r_i|abb)a_0b_0^2 \\ & + 3(r_i|xx\delta)x_0^2\delta_0 + 6(r_i|xa\delta)x_0a_0\delta_0 + 3(r_i|aa\delta)a_0^2\delta_0 \\ & + 3(r_i|yy\delta)y_0^2\delta_0 + 6(r_i|yb\delta)y_0b_0\delta_0 + 3(r_i|bb\delta)b_0^2\delta_0 \\ & + 3(r_i|x\delta\delta)x_0\delta_0^2 + 3(r_i|a\delta\delta)a_0\delta_0^2 + (r_i|\delta\delta\delta)\delta_0^3,\end{aligned}$

and with $r_{i,1} \in \{y_1, b_1\}$ and $r_i \in \{y, b\}$,

$[\text{first order}] = (r_i|y)y_0 + (r_i|b)b_0,$

$\begin{aligned}[\text{second order}] = \ & 2(r_i|yx)y_0 + 2(r_i|ya)y_0a_0 + 2(r_i|bx)b_0x_0 \\ & + 2(r_i|ba)b_0a_0 + 2(r_i|y\delta)y_0\delta_0 + 2(r_i|b\delta)b_0\delta_0,\end{aligned}$

$\begin{aligned}[\text{third order}] = \ & (r_i|yyy)y_0^3 + 3(r_i|yyb)y_0^2b_0 + 3(r_i|ybb)y_0b_0^2 + (r_i|bbb)b_0^3 \\ & + 3(r_i|yxx)y_0x_0^2 + 6(r_i|yxa)y_0x_0a_0 + 3(r_i|yaa)y_0a_0^2 \\ & + 3(r_i|bxx)b_0x_0^2 + 6(r_i|bxa)b_0x_0a_p + 3(r_i|baa)b_0a_0^2 \\ & + 6(r_i|yx\delta)y_0x_0\delta_0 + 6(r_i|ya)y_0a_0\delta_0 + 6(r_i|ba\delta)b_0a_0\delta_0 \\ & + 3(r_i|y\delta\delta)y_0\delta_0^2 + 3(r_i|b\delta\delta)b_0\delta_0^2,\end{aligned}$

where the first-order terms are as defined in Table 8.1, and the second-order terms are as defined in Table 8.2. To determine image aberrations of complex systems, we can furthermore describe the relations between higher order position vectors at z_{i+1} and z_i of a section of an optical system by transfer matrices. The overall transfer matrix of this optical system is then found as the product of these individual transfer matrices.

TABLE 8.1

Coefficients of First-Order Aberrations for a Sector Field[a]

Symbol	Characteristics of aberrations of first order	Maximal size of aberration
$(x\vert x)$	First order aberration due to x_0 deviation at object. In an image profile plane for the x direction, the magnitude of $(x\vert x)$ describes the x magnification in an image profile plane	$2x_{00}(x\vert x)$
$(x\vert a)$	a_0 Aperture aberration of first order. In an image profile plane for the x direction, the quantity $(x\vert a)$ vanishes	$2a_{00}(x\vert a)$
$(x\vert\delta)$	Energy over charge dispersion	$2\delta_0(x\vert\delta)$
$(y\vert y)$	First-order aberration due to y_0 deviation at object. In an image profile plane for the y direction, the magnitude of $(y\vert y)$ describes the y magnification	$2y_{00}(y\vert y)$
$(y\vert b)$	b_0 Aperture aberration of first order. In an image profile plane for the y direction, the quantity $(y\vert b)$ vanishes	$2b_{00}(y\vert b)$

[a] The size of the aberrations for a beam assumes an object of size $\pm x_{00}$ and $\pm y_{00}$, where particles leave under maximal angles relative to the optic axis $\pm\alpha_{00} \approx \pm a_{00}$ and $\pm\beta_{00} \approx \pm b_{00}$ from each point of this object.

General transfer matrices of second order are shown in Tables 8.3 and 8.4, for which the corresponding position vectors read at z_i:

$$\mathbf{X}_i = x_i,\, a_i,\, t_i,\, \delta_i,\, x_i^2,\, x_i a_i,\, a_i^2,\, y_i^2,\, y_i b_i,\, b_i^2,\, x_i\delta_i,\, a_i\delta_i,\, \delta_i^2,$$

$$\mathbf{Y}_i = y_i,\, b_i,\, y_i x_i,\, b_i x_i,\, y_i a_i,\, b_i a_i,\, y_i\delta_i,\, b_i\delta_i.$$

These coefficients of second order were calculated for the main fields by Hintenberger and Koenig (1957), Ewald and Liebl (1955, 1957), Boerboom et al. (1959), and Wollnik (1964, 1965). In the form of transfer matrices, the same results were obtained by Brown et al. (1964) and Takeshita (1966). For the fringing fields, these coefficients were determined by Wollnik (1964) and Wollnik and Ewald (1965), whereas the results of the main and of the fringing fields were combined in the form of transfer matrices by Wollnik (1967a, b) and Enge (1967).

These aberration coefficients allow to determine how much a trajectory deviates from its first-order solution at a position z_i. However, normally the goal of the calculation of image aberrations is to determine the width of a particle beam at z_i. Starting at z_0 with a beam filling an upright parallelogram of area $4x_{00}a_{00}$, the range of a_0 and b_0 varies for different x_0 and y_0. Thus, the image aberrations at z_i cause the corresponding particles to be spread out over a region Δx_i, which is unsymmetric with respect to particles with $a_0 = b_0 = 0$. Starting at z_0 with a beam filling an upright rectangle of area $4x_{00}a_{00}$ and amending the optical system at z_0 by a virtual thin object lens of appropriate focal length l_x (see Section 5.23), the trajectories in the optical system are identical though the range of a_0 and b_0 is identical for all x_0 and y_0. The range Δx_i over which the particles are spread out at z_i is the same in both cases; however, in the second case the trajectories with $a_0 = b_0 = 0$ are found in the middle of Δx_i and out of the middle in the first case. For this reason the so-called normalized aberrations calculated for an optical system, which is amended by a virtual thin object lens, allows for a more direct interpretation of the beam widening of an optical system.

TABLE 8.2

Coefficients of Second-Order Aberrations for a Sector Field[a]

Symbol	Characteristics of aberrations of second order	Maximal size of aberration
$(x\|xx)$	Aberration due to x_0 deviation $[x_0^2 < x_{00}^2]$	$x_{00}^2(x\|xx)/2$
$(x\|xa)$	Mixed xa aberration $[x_0^2 a_0^2 < x_{00}^2 a_{00}^2]$	$\pm x_{00}a_{00}(x\|xa)$
$(x\|aa)$	a-Aperture aberration $[a_0^2 < a_{00}^2]$	$a_{00}^2(x\|aa)/2$
$(x\|yy)$	Aberration due to y_0 deviation $[y_0^2 < y_{00}^2]$	$y_{00}^2(x\|yy)/2$
$(x\|yb)$	Mixed yb aberration $[y_0^2 b_0^2 < y_{00}^2 b_{00}^2]$	$\pm y_{00}b_{00}(x\|yb)$
$(x\|bb)$	b-Aperture aberration $[b_0^2 < b_{00}^2]$	$b_{00}^2(x\|bb)/2$
*$(x\|x\delta)$	Mixed x chromatic aberration $[x_0^2\delta^2 < x_{00}^2\delta_0^2]$	$\pm x_{00}\delta_0(x\|x\delta)$
*$(x\|a\delta)$	Mixed a chromatic aberration $[a_0^2\delta^2 < a_{00}^2\delta_0^2]$	$\pm a_{00}\delta_0(x\|a\delta)$
$(x\|\delta\delta)$	Quadratic chromatic aberration $[\delta^2 < \delta_0^2]$	$\delta_0^2(x\|\delta\delta)/2$
$(y\|yx)$	Mixed yx aberration $[y_0^2 x_0^2 < y_{00}^2 x_{00}^2]$	$\pm y_{00}x_{00}(y\|yx)$
$(y\|bx)$	Mixed bx aberration $[b_0^2 x_0^2 < b_{00}^2 x_{00}^2]$	$\pm b_{00}x_{00}(y\|bx)$
$(y\|ya)$	Mixed ya aberration $[y_0^2 a_0^2 < y_{00}^2 a_{00}^2]$	$\pm y_{00}a_{00}(y\|ya)$
$(y\|ba)$	Mixed ba aperture aberration $[b_0^2 a_0^2 < b_{00}^2 a_{00}^2]$	$\pm b_{00}a_{00}(y\|ba)$
*$(y\|y\delta)$	Mixed y chromatic aberration $[y_0^2\delta^2 < y_{00}^2\delta_0^2]$	$\pm y_{00}\delta_0(y\|y\delta)$
*$(y\|b\delta)$	Mixed b chromatic aberration $[b_0^2\delta^2 < b_{00}^2\delta_0^2]$	$\pm b_{00}\delta_0(y\|b\delta)$

[a] The magnitudes of the aberrations assume an object as indicated in Fig. 8.1. Note that in the x direction, no aberrations exist that are proportional to xy, xb, ay, ab, $y\delta$, $b\delta$, but that aberrations proportional to y^2, yb, b^2 do occur. Note that in the y direction, no aberrations appear that are proportional to x^2, xa, a^2, y^2, yb, b^2, $x\delta$, $a\delta$, δ^2. For the case of a quadrupole lens, only the chromatic aberrations $(x\|x\delta)$, $(x\|a\delta)$, $(y\|y\delta)$, $(y\|b\delta)$, which are marked by an asterisk, persist whereas all other aberrations vanish.

8.2 ORIGIN OF IMAGE ABERRATIONS

There are geometric and chromatic aberrations. If r_{l0}, r_{m0}, r_{n0} represent x_0, a_0, y_0, b_0, we speak of geometric aberrations, which exist for any optical system. If one of the expressions r_l, r_m, r_n represents δ, we speak of chromatic

TABLE 8.3

Second-Order x-Transfer Matrix for a Sector Field [a]

$(x\mid x)$	$(x\mid a)$	0	$(x\mid\delta)$	$(x\mid xx)$	$(x\mid xa)$	$(x\mid aa)$	$(x\mid yy)$	$(x\mid yb)$	$(x\mid bb)$	$(x\mid x\delta)$	$(x\mid a\delta)$	$(x\mid\delta\delta)$
$(a\mid x)$	$(a\mid a)$	0	$(a\mid\delta)$	$(a\mid xx)$	$(a\mid xa)$	$(a\mid aa)$	$(a\mid yy)$	$(a\mid yb)$	$(a\mid bb)$	$(a\mid x\delta)$	$(a\mid a\delta)$	$(a\mid\delta\delta)$
$(t\mid x)$	$(t\mid a)$	1	$(t\mid\delta)$	$(t\mid xx)$	$(t\mid xa)$	$(t\mid aa)$	$(t\mid yy)$	$(t\mid yb)$	$(t\mid bb)$	$(t\mid x\delta)$	$(t\mid a\delta)$	$(t\mid\delta\delta)$
0	0	0	1	0	0	0	0	0	0	0	0	0
0	0	0	0	$(x\mid x)^2$	$2(x\mid x)(x\mid a)$	$(x\mid a)^2$	0	0	0	$2(x\mid x)(x\mid\delta)$	$2(x\mid a)(x\mid\delta)$	$(x\mid\delta)^2$
0	0	0	0	$(x\mid x)(a\mid x)$	$2(x\mid a)(a\mid x)-1$	$(x\mid a)(a\mid a)$	0	0	0	$G(x\mid\delta)$	$G(a\mid\delta)$	$(x\mid\delta)(a\mid\delta)$
0	0	0	0	$(a\mid x)^2$	$2(a\mid x)(a\mid a)$	$(a\mid a)^2$	0	0	0	$2(a\mid x)(a\mid\delta)$	$2(a\mid a)(a\mid\delta)$	$(a\mid\delta)^2$
0	0	0	0	0	0	0	$(y\mid y)^2$	$2(y\mid y)(y\mid b)$	$(y\mid b)^2$	0	0	0
0	0	0	0	0	0	0	$(y\mid y)(b\mid y)$	$2(y\mid b)(b\mid y)-1$	$(y\mid b)(b\mid b)$	0	0	0
0	0	0	0	0	0	0	$(b\mid y)^2$	$2(b\mid y)(b\mid b)$	$(b\mid b)^2$	0	0	0
0	0	0	0	0	0	0	0	0	0	$(x\mid x)$	$(x\mid a)$	$(x\mid\delta)$
0	0	0	0	0	0	0	0	0	0	$(a\mid x)$	$(a\mid a)$	$(a\mid\delta)$
0	0	0	0	0	0	0	0	0	0	0	0	1

[a] In this x-transfer matrix the abbreviations $G(x,\delta) = (x\mid x)(a\mid\delta) + (a\mid x)(x\mid\delta)$ and $G(a\mid\delta) = (x\mid a)(a\mid\delta) + (a\mid a)(x\mid\delta)$ are used. Note that the elements of rows 4–13 do not contain any new information and can readily be calculated once the first four elements of the first three rows are known.

TABLE 8.4

Second-Order y-Transfer Matrix for a Sector Field[a]

$(y\|y)$	$(y\|b)$	$(y\|yx)$	$(y\|ya)$	$(y\|bx)$	$(y\|ba)$	$(y\|y\delta)$	$(y\|b\delta)$
$(b\|y)$	$(b\|b)$	$(b\|yx)$	$(b\|ya)$	$(b\|bx)$	$(b\|ba)$	$(b\|y\delta)$	$(b\|b\delta)$
0	0	$(y\|y)(x\|x)$	$(y\|y)(x\|a)$	$(y\|b)(x\|x)$	$(y\|b)(x\|a)$	$(y\|y)(x\|\delta)$	$(y\|b)(x\|\delta)$
0	0	$(y\|y)(a\|x)$	$(y\|y)(a\|a)$	$(y\|b)(a\|x)$	$(y\|b)(a\|a)$	$(y\|y)(a\|\delta)$	$(y\|b)(a\|\delta)$
0	0	$(b\|y)(x\|x)$	$(b\|y)(x\|a)$	$(b\|b)(x\|x)$	$(b\|b)(x\|a)$	$(b\|y)(x\|\delta)$	$(b\|b)(x\|\delta)$
0	0	$(b\|y)(a\|x)$	$(b\|y)(a\|a)$	$(b\|b)(a\|x)$	$(b\|b)(a\|a)$	$(b\|y)(a\|\delta)$	$(b\|b)(a\|\delta)$
0	0	0	0	0	0	$(y\|y)$	$(y\|b)$
0	0	0	0	0	0	$(b\|y)$	$(b\|b)$

[a] Note that the elements of the rows 3–8 do not contain any new information and can readily be calculated once the first two elements of the first two rows are known.

aberrations, which are only present when the beam contains particles of different magnetic and electrostatic rigidities.

8.2.1   Geometric Aberrations

8.2.1.1   Geometric Aberrations in Fields with Straight Optic Axes

Quadrupoles or rotationally symmetric lenses have straight optic axes around which they exhibit a four-fold symmetry* to which higher multipole fields are added in the case of imperfectly formed electrodes or pole faces. In such systems, only geometric aberrations of odd order can exist, since all electromagnetic forces are antisymmetric with respect to x and y. Thus, only those coefficients $(r_k | \cdots x^i a^j \cdots y^m b^n \cdots)$ of Eq. (8.1) are nonzero, for which $i + j$ are even (or odd) and $m + n$ are odd (or even) numbers for $r_k = x$ (or y).

Consequently, particle trajectories originating at z_0 from a point $P_0(x_0, y_0 = 0)$ under angles $\alpha_0 \approx a_0$, $\beta_0 = 0$ finally arrive in the profile plane z at a point $P(x, y = 0)$ under angles $\alpha, \beta = 0$, i.e., with r_{k1} being x_1 or $a_1 \approx \alpha_1$:

$$r_{k1}(z) = (r_k|x)x_0 + (r_k|a)a_0 + \tfrac{1}{6}[(r_k|xxx)x_0^3 + 3(r_k|xxa)x_0^2 a_0$$
$$+ 3(r_k|xaa)x_0 a_0^2 + (r_k|aaa)a_0^3] + \cdots. \qquad (8.4)$$

Here, all coefficients $(r_k | \cdots)$ vary with z. Analogously, the same Eq. (8.4) is valid for r_{k1} being y_1 or $b_1 \approx \beta_1$ if all quantities x, a are exchanged for y, b with $x_0 = 0$, $y_0 \neq 0$ and $a_0 = 0$, $b_0 \neq 0$.

In the case of a point source, i.e., $x_0 = y_0 = 0$, Eq. (8.4) simplifies considerably, so that we find for $r_k = x$ or $r_k = a$,

$$x(z) = (x|a)a_0 + \tfrac{1}{6}(x|aaa)a_0^3 + \cdots,$$

$$a(z) = (a|a)a_0 + \tfrac{1}{6}(a|aaa)a_0^3 + \cdots.$$

If there is a field-free region behind z_n, we find that at $z = z_n + \Delta z$, the particle under investigation deviates from the optic axis by

$$x(z_n + \Delta z) = [(x_n|a) + \Delta z(a_n|a)]a_0 + \tfrac{1}{6}[(x_n|aaa) + \Delta z(a_n|aaa)]a_0^3 + \cdots.$$
$$(8.5a)$$

---

* In this context, some authors speak of a system with two planes of symmetry instead of a fourfold symmetry. For an ideal quadrupole, these two statements are identical. For a quadrupole with imperfectly formed pole faces or electrodes, which have two planes of symmetry, however, second-order aberrations do not vanish completely, as would be the case for a quadrupole that has a fourfold symmetry (Halbach, 1969).

For small angles $\alpha_0 \approx a_0$, the term with $a_0^3 \approx \alpha_0^3$ can be neglected, and we find that $x_n(z_n + \Delta z)$ vanishes for all a_0 if Δz equals $-(x_n|a)/a_n|a)$. At this $z_n + \Delta z$, Eq. (8.5a) can be rewritten as

$$x_n(z_n + \Delta z) = \left[x_n|aaa) - \frac{(x_n|a)}{(a_n|a)}(a_n|aaa) \right] \frac{a_0^3}{6} + \cdots . \tag{8.5b}$$

For $\Delta z = 0$, the case where the original z_n has been chosen to denote the focus position, i.e., $(x_n|a) = 0$, Eq. (8.5b) simplifies to

$$x_n(z_n) = \tfrac{1}{6}(x_n|aaa)a_0^3 + \cdots . \tag{8.5c}$$

In this case, the third-order aperture aberration $(x_n|aaa)a_0^3/6$ fully describes the width of the particle beam. So long as terms proportional to a_0^3 can be neglected, x_n is zero.

A point source emits particles under many angles $\alpha_0 \approx a_0$ with $|a_0| < a_{00}$. Knowing that particles with small a_0 are focused at z_n according to Eq. (8.5c), we may now look for particles of larger a_0. The first statement that can be made is that according to Eq. (8.5c), the deviation x_n increases rapidly with a_0 so that the particle density drops quickly with increasing x_n. Assuming that the source emits equally many particles for each $|a_0| < a_{00}$, 50% are emitted with angles below $0.5\alpha_{00} \approx 0.5a_{00}$, and 90% are emitted with angles below $0.9\alpha_{00} \approx 0.9a_{00}$. Consequently, we can expect 50 or 90% of the particles to stay within

$$x_n(50\%) = \tfrac{1}{6}(x_n|aaa)(0.5a_{00})^3 + \cdots \approx \tfrac{1}{6}(0.125)(x_n|aaa)a_{00}^3 + \cdots ,$$

$$x_n(90\%) = \tfrac{1}{6}(x_n|aaa)(0.9a_{00})^3 + \cdots \approx \tfrac{1}{6}(0.729)(x_n|aaa)a_{00}^3 + \cdots .$$

Note that these considerations apply only for a true point source, in which case the beam diameter at z_n is dominated by the aberration $(x_n|aaa)a_{00}^3/6$. In the case of a nonnegligibly small source $|x_0| < x_{00}$, even those particles which started at angles $\alpha_0 = 0$ cause a certain beamwidth at z_n. Thus, 50 or 90% of the initial particles arrive in a region of $\pm x_n$, smaller than the previously determined $\pm 0.125/6$ or $\pm 0.729/6$ times $(x_n|aaa)a_{00}^3$.

In order to concentrate all particles of a beam to a small region of x_n, the optical system should be modified such that the higher order coefficients of Eqs. (8.4) or (8.5a)–(8.5c) decrease in magnitude. Alternatively, we can optimize the focus spot of a given optical system by counterbalancing part of the aperture aberration $[(x_n|aaa) + \Delta z(a_n|aaa)]a_0^3$ of Eq. (8.5a) by a first-order defocusing action (Fig. 8.2). For this purpose, we assume that in Eq. (8.5a), the coefficient $(x_n|a)$ vanishes; i.e., there is a first-order focus at z_n. In order to determine the overall beam diameter at that focus, replace a_0 in Eq. (8.5a) by the maximal a_{00}. In this way, we find also that the

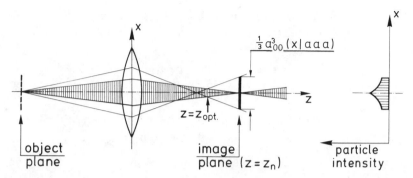

Fig. 8.2. Particle trajectories of equal rigidity in a lens with a straight optic axis. Note that trajectories that are only slightly inclined to the optic axis are focused in the "image plane," whereas trajectories that are more inclined are focused at a different location due to the nonvanishing aberration $2(x|aaa)\alpha_0^3/6$. Note that the smallest possible beam diameter is only about $(x|aaa)a^3/9$, which we find from Eq. (8.6c).

corresponding particles are focused at

$$\Delta z_{opt} = -\frac{(x_n|aaa)a_{00}^2 + \cdots}{6(a_n|a) + (a_n|aaa)a_{00}^2 + \cdots}.$$

Introducing the Δz_{opt} thus found into Eq. (8.5b) for $-(x_n|a)/(a_n|a)$ yields

$$x(z_n + \Delta z_{opt}) = \frac{(x_n|aaa)(a_{00}^2 - a_0^2)a_0 + \cdots}{6 + (a_n|aaa)a_{00}^2/(a_n|a) + \cdots}. \qquad (8.6a)$$

This $x(z_n + \Delta z_{opt})$ vanishes for $a_0 = 0$ as well as for $a_0 = a_{00}$ and has a maximum in between at that position where dx/da_0 vanishes, i.e., for $a_0^2 = a_{00}^2/3$. In this case, the $x(z_n + \Delta z_{opt})$ of Eq. (8.6a) becomes

$$x_{max}(z_n + \Delta z_{opt}) = \frac{(x_n|aaa)a_{00}^3(2/3\sqrt{3}) + \cdots}{6 + (a_n|aaa)a_{00}^2/(a_n|a) + \cdots}. \qquad (8.6b)$$

For $(a_n|aaa)a_{00}^2 \ll (a_n|a)$, which usually is the case, Eq. (8.6b) can be rewritten with $2/3\sqrt{3} \approx 0.385$, as

$$x_{max}(z_n + \Delta z_{opt}) \approx \tfrac{1}{6}(0.385)(x_n|aaa)a_{00}^3 + \cdots. \qquad (8.6c)$$

Thus, the beam diameter at z_n is almost three times wider than the beam diameter at $z_n + \Delta z_{opt}$, as is indicated in Fig. 8.2.

8.2.1.2   Geometric Aberrations in Fields with Curved Optic Axes

Deflecting fields have a curved optic axis and are characterized by a twofold symmetry with respect to the plane $y = 0$ in Fig. 4.1. As in the case of quadrupoles, here also higher multipole fields are added to the resulting field distribution to allow for imperfectly formed electrodes or pole faces. For systems characterized by a two-fold symmetry, all electromagnetic forces in the x direction are symmetric and in the y direction antisymmetric with respect to $y = 0$. Thus, in Eq. (8.1) the only nonvanishing coefficients $(r_k| \cdots y^i b^j \cdots)$ are those for which k equals 1, 2, 3, and $i + j$ is even, as well as those for which k equals 5, 6 and $i + j$ is odd [Eqs. (8.2c) and (8.2d)]. All these coefficients are listed in Table 8.2.

Trajectories of monoenergetic particles starting from a point source at z_0 under angles $\pm\alpha_0 \approx \pm a_0$ and $\pm\beta_0 \approx \pm b_0$ thus finally have distances $x(z)$ from the optic axis with

$$x(z) = \pm(x|a)a_0 + \tfrac{1}{2}[(x|aa)a_0^2 + (x|bb)b_0^2 + \cdots]$$
$$\pm \tfrac{1}{6}[(x|aaa)a_0^3 + 3(x|abb)a_0 b_0^2 + \cdots], \tag{8.7}$$

where all coefficients $(x|r_k r_n)$ are functions of z. Since a_0 can have positive or negative values, some terms have plus and minus signs, indicating that they cause a symmetrical beam broadening so that the magnitude of the corresponding terms finally must be counted twice.

Because of the terms proportional to a_0^2 and b_0^2 in Eq. (8.7), the beam is always unsymmetric to the optic axis, with this effect most noticeable at the position where $(x|a)$ vanishes, i.e., at the x image. To illustrate the magnitude and the origin of second-order image aberrations, particle trajectories are shown in a homogeneous magnet in Fig. 8.3. An image is found at point A in Fig. 8.3 for trajectories that are only slightly inclined to the optic axis ($\beta_0 \approx b_0 = 0$, $\alpha_0 \approx a_0 \ll 1$). Trajectories with larger $\alpha_0 \approx a_0$ meet at point B. The distance between A and B is taken from Fig. 8.3 to be

$$AB = 2(\rho_0 - \rho_0 \cos \alpha_0), \tag{8.8a}$$

with ρ_0 denoting the radius of the circular optic axis. The distance AB is mainly the α-aperture aberration $(x|aa)a_0^2/2$, with $a_0 \approx \alpha_0$ and $(x|aa)/2$ the corresponding coefficient of second order. Because $a_0 \approx \alpha_0$ varies from zero to a maximal $a_{00} \approx \alpha_{00}$ in Eq. (8.8a), the aperture aberration $(x|aa)a_0^2/2$ causes an intensity distribution that has the unsymmetric shape, indicated in the middle part of Fig. 8.3 by the thick line from A to B.

Particle trajectories (dotted lines in Fig. 8.3) that leave the source S inclined ($\alpha_0 = 0$, $\beta_0 \neq 0$) to the plane of symmetry ($\eta = 0$) cross the surface $\eta = 0$ at points C' and C''. When projected onto the plane of symmetry, both C' and C'' fall on C. The deviation AC in the upper part of Fig. 8.3 results from the fact that particles with $\beta = 0$ move along a helix where the projection of this helix on the plane of symmetry ($\eta = 0$) is a circle, the radius ρ of which is determined by the balance between the centripetal force $(ze)v_{\parallel}B_0$ and the centrifugal force $mv_{\parallel}^2/\rho$. Here, (ze) is the charge of

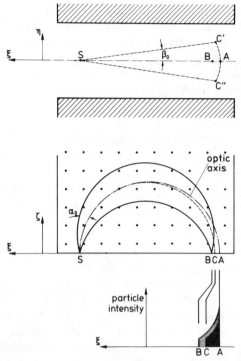

Fig. 8.3. Particle trajectories in a homogeneous magnet are shown relative to a circular optic axis. Also shown is the intensity distribution that must be expected in the image plane. Heavy lines indicate trajectories in the midplane ($\eta = 0$) as well as the corresponding intensity distribution. Dotted lines indicate trajectories inclined to this midplane as well as the corresponding intensity distribution. Note that B as well as C', C'' are at the same side of A.

the particle and B_0 the main field flux density, $v = |\mathbf{v}|$ is the particle velocity and $v_{\parallel} = v \cos \beta_0$ the velocity component parallel to the plane $\eta = 0$. With $\rho_0 = mv/(ze)B_0$ and $\rho = mv_{\parallel}/(ze)B_0 = \rho_0 \cos \beta_0$, we find $AC = 2(\rho_0 - \rho)$, or

$$AC = 2(\rho_0 - \rho_0 \cos \beta_0). \tag{8.8b}$$

The distance AC is mainly the β-aperture aberration $(x|bb)b_0^2/2$, with $b_0 \approx \beta_0$ and $(x|bb)/2$ the corresponding aberration coefficients of second order. Because $b_0 \approx \beta_0$ in Eq. (8.8b) varies from 0 to a maximal $b_{00} \approx \beta_{00}$, the aperture aberration $(x|bb)b_0^2/2$ causes an unsymmetric intensity distribution, indicated in Fig. 8.3 by the dotted line from A to C. The distance in η direction between C' and C'' may not be desirable; normally, however, it is of no importance in a particle spectrometer, since the particle intensity is usually recorded by a slit parallel to the η axis.

At this point it should be mentioned that according to Eqs. (8.8a) and (8.8b), the coefficients $(x|aa)$ and $(x|bb)$ are of equal magnitude for a 180° magnetic sector field. For sector fields, in which the source and detector of charged particles are outside the main field region, the coefficients $(x|aa)$ and $(x|bb)$ normally are not equal but have the same signs.

8.2.2 Chromatic Aberrations

Chromatic aberrations are observed when we investigate trajectories of particles of different rigidities $\chi = \chi_0(1 + \Delta)$, i.e., of different energy–charge $(K/ze) = K_0(1 + \delta_K)/(z_0 e)$ and mass–charge $(m_0/ze) = m_{00}(1 + \delta_m)/(z_0 e)$ ratios. For the sake of simplicity, we shall restrict ourselves to particles of constant mass $(m_0 = m_{00})$ and charge $(z = z_0)$ and keep only the energy $(K \neq K_0)$ variable. Thus, we have $\delta_m = 0$ and $\delta_K \neq 0$, which we abbreviate as $\delta \neq 0$. Consequently, Eq. (8.1) transforms to

$$\Delta = \frac{(1 + 2\eta_0)^2 + h}{2(1 + \eta_0)(1 + 2\eta_0)}\delta + \frac{h[1 - 4\eta_0(1 + \eta_0)] - (1 + 2\eta_0)^2}{8(1 + \eta_0)^2(1 + 2\eta_0)^2}\delta^2 + \cdots,$$

(8.9)

where h equals 0 and 1 for magnetic and electrostatic fields, respectively [Eq. (2.19) or (4.59)]. Note that according to Eq. (8.9), there is a fixed relation between Δ and δ so that a Δ achromaticity can easily be transformed to a δ achromaticity and vice versa.

8.2.2.1 Chromatic Aberrations in Fields with Straight Optic Axes

For rotational symmetric and for quadrupole lenses, i.e., for systems with a straight optic axis, the only second-order aberrations are chromatic ones. Among the other aberrations for sector fields these aberrations are also listed in Table 8.2 but marked with an asterisk. For the case of a point source, their effect is illustrated in Fig. 8.4. For reference particles $(m_{00}, z_0 e, K_0)$, the lens under consideration has a focal length f_0 so that these particles are focused at z_n (shaded area of Fig. 8.4). For arbitrary particles $[m_{00}, z_0 e, K = K_0(1 + \delta)]$, the lens has a focal length $f = f_0(1 + \delta + \cdots)$. For $\delta > 0$, the corresponding particles are thus focused downstream from z_n. For an arbitrary particle that started with $x_0 = y_0 = b_0 = 0$, the distance $x(z)$ from the optic axis is found from Eq. (8.2) as

$$x(z) = \pm(x|a)a_0 \pm (x|a\delta)a_0\delta_0 \pm \tfrac{1}{2}(x|a\delta\delta)a_0\delta_0^2 + \cdots,$$

with $(x|a)$, $(x|a\delta)$, $(x|a\delta\delta)$, etc., all functions of z. Since both a_0 and δ can take up positive and negative values, the $\pm$ signs again indicate that the

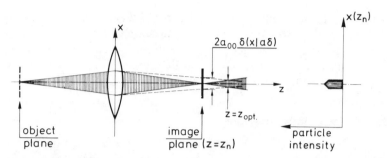

Fig. 8.4. Particle trajectories of equal mass and charge but two different energies (heavy lines and dotted lines) in a lens with a straight optic axis. Note that the dotted lines are not focused at $z = z_n$. Note also that the narrowest beam diameter is found at a path coordinate $z_{opt} = z_n + (x|a\delta)\delta_0/(a|a)$.

corresponding terms cause a symmetric beam broadening and that the overall beam widening is twice the sum of all terms.

Note here that the terms $(x|a\delta)a_0\delta_0$ and $(x|a\delta\delta)a_0\delta_0^2$ can in no way be corrected as long as the optic axis remains straight. This fact had been known to Isaac Newton when he decided that with lenses made from only one type of glass, the chromatic aberrations of an astronomical lens telescope cannot be corrected. For this reason, he constructed mirror-type telescopes. Lens telescopes became feasible only much later when glasses were known with refractive indices that varied with the wavelength of light.

8.2.2.2 Chromatic Aberrations in Fields with Curved Axes

In a sector field, particles of different rigidities are deflected differently and at the same time are focused after different flight distances. This situation is described by the chromatic aberrations listed in Table 8.2. For the case of a point source, their effect is illustrated in Fig. 8.5 for a 90° homogeneous sector magnet with particles of rigidities χ_0 and $\chi = \chi_0(1 + \Delta)$. These particles move along trajectories of radii ρ_0 and $\rho = \rho_0(1 + \Delta) = \rho_0[1 + \delta_K[(1 + 2\eta_0) + h/(1 + 2\eta_0)]/2(1 + \eta_0) + \cdots]$ according to Eq. (8.9b) with $\delta_m = 0$. As shown in Fig. 8.5, these particles are finally focused at points A and C, respectively. In detail, the deviation $x(z)$ of an arbitrary particle from the optic axis and the corresponding angle of inclination $a(z) \approx \alpha(z)$ are found from Eq. (8.2) for particles that had started from a point source $(x_0 = y_0 = 0)$, as

$$x(z) = \pm(x|a)a_0 \pm (x|\delta)\delta$$

$$+ \tfrac{1}{2}[(x|aa)a_0^2 + (x|bb)b_0^2 \pm 2(x|a\delta)a_0\delta_0 + (x|\delta\delta)\delta_0^2 + \cdots], \quad (8.10a)$$

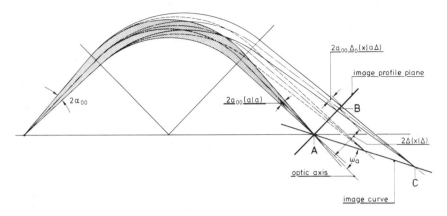

Fig. 8.5. A magnetic sector field focusing ions of equal mass and charge but three different energies K_0, $K_0(1 + \delta)$, and $K_0(1 + 2\delta)$ on different points along an image curve. This image curve crosses the optic axis with an angle ω_a defined in Eq. (8.13a) (see also Fig. 4.1).

$$a(z) = \pm(a|a)a_0 \pm (a|\delta)\delta$$
$$+ \tfrac{1}{2}[(a|aa)a_0^2 + (a|bb)b_0^2 \pm 2(a|a\delta)a_0\delta_0 + (a|\delta\delta)\delta_0^2 + \cdots], \quad (8.10b)$$

where all coefficients of Eqs. (8.10a) and (8.10b) are functions of z. For $\delta = 0$, the first-order image occurs at point A in Fig. 8.5, i.e., at that z_n where $(x_n|a)$ vanishes. Trajectories of particles with $\delta_0 \neq 0$ but $\alpha_0 = 0$ are displaced by a distance x_{n0} and inclined by $a_{n0} \approx \alpha_{n0}$,

$$x_{n0}(\delta_0, \alpha_0 = 0) = \pm(x_n|\delta)\delta_0 + (x_n|\delta\delta)\delta_0^2/2 + \cdots, \quad (8.11a)$$

$$a_{n0}(\delta_0, \alpha_0 = 0) = \pm(a_n|\delta)\delta_0 + (a_n|\delta\delta)\delta_0^2/2 + \cdots. \quad (8.11b)$$

Trajectories of particles with $\delta_0 \neq 0$ and $a_0 \neq 0$ are displaced by x_n and inclined by a_n because of $(x_n/a) = 0$,

$$x_n(\delta_0, a_0) = \pm(x_n|\delta)\delta_0 \pm (x_n|a\delta)a_0\delta_0 + (x_n|\delta\delta)\delta_0^2/2$$
$$\pm (x_n|a\delta\delta)a_0\delta_0^2/2 + \cdots, \quad (8.12a)$$

$$a_n(\delta_0, a_0) = (a|a)a_0 \pm (a_n|\delta)\delta_0 \pm (a_n|a\delta)a_0\delta_0 + (a_n|\delta\delta)\delta_0^2/2$$
$$\pm (a_n|a\delta\delta)a_0\delta_0^2/2 + \cdots. \quad (8.12b)$$

In case there is a field-free region around z_n, we find the side-beam particles ($\delta \neq 0$) focused at point C in Fig. 8.5 with the coordinates x_{n0} and $\Delta z = BC = z_n - z_{n0} = (x_n - x_{n0})/(a_n - a_{n0})$, or explicitly,

$$z - z_n = \frac{(x_n|a\delta)a_0\delta_0 + (x_n|a\delta\delta)a_0\delta_0^2/2 + \cdots}{(a_n|a)a_0 + (a_n|a\delta)a_0\delta_0 + \cdots}.$$

Thus, an image curve is established with $x^* = dx_{n0}/d(\rho_0\delta_0)$, $z^* = d(z - z_n)/d(\rho_0\delta_0)$, and

$$\tan \omega_\alpha = z^*/x^*, \qquad r_\alpha = (x^{*2} + z^{*2})^{3/2}/x^*z^{**} - x^{**}z^*$$

where the angle ω_α is shown in Fig. 8.5. In detail, we find

$$\cotan \omega_\alpha = \frac{(x_n|x)(x_n|a\delta) + \cdots}{(x_n|\delta) + \cdots}, \tag{8.13a}$$

$$r_\alpha = \frac{[(x_n|\delta)^2 + (x_n|x)^2(x_n|a\delta)^2]^{3/2}}{(x_n|x)\{(x_n|\delta)(x_n|a\delta\delta) - (x_n|a\delta)[(x_n|\delta\delta) + 2(x_n|\delta)(x_n|x)(a_n|a\delta)]\}}, \tag{8.13b}$$

with $(x_n|x) = 1/(a_n|a)$ because of $(x_n|a) = 0$. For $(x_n|a\delta) = 0$, we find $\omega_\alpha = \pi/2$ from Eq. (8.13a), showing that in this case the image plane is perpendicular to the optic axis. Furthermore, the radius of curvature of the image curve is determined from Eq. (8.13b) as $r_\alpha = (x_n|\delta)^2/[(x_n|x)(x_n|a\delta\delta)]$.

8.3 RELATIONS BETWEEN COEFFICIENTS OF EQUATION (8.2) DUE TO THE CONDITION OF SYMPLECTICITY

The motion of charged particles in electromagnetic fields is most generally described by the Lorentz equation [Eq. (2.6)]. In order to make this Lorentz equation more easily applicable to any coordinate system, Eq. (2.6) can be rewritten as the Lagrange equations [Eq. (2.31b)]. These three differential equations of second order can also be transformed (see, for instance, Goldstein, 1980) into six differential equations of first order, the so-called canonical equations. The word "canonical" here characterizes the fact that these equations describe a directed particle motion.

8.3.1 Canonical Transformations with Time Being the Independent Variable

Instead of the Cartesian x, y, z-coordinate system, let us introduce "phase-space coordinates," in which a particle is characterized at any time T by its position (x, y, z) and the components of its generalized momentum $\partial L/\partial v$ [Eq. (2.35)]:

$$x, \qquad p_x = \partial L/\partial \dot{x}, \tag{8.14a}$$

$$y, \qquad p_y = \partial L/\partial \dot{y}, \tag{8.14b}$$

$$z, \qquad p_z = \partial L/\partial \dot{z}. \tag{8.14c}$$

Here, $L = m_0 c^2 (1 + 2\eta) - (ze) V + (ze) vA$ is the Lagrangian of Eq. (2.32) and $\dot{x}, \dot{y}, \dot{z}$ are the time derivatives of x, y, z. At this time let us define the so-called Hamiltonian,

$$H = \dot{x} p_x + \dot{y} p_y + \dot{z} p_z - L(x, \dot{x}, y, \dot{y}, z, \dot{z}), \qquad (8.15a)$$

[see Eq. (2.33)] in time-independent fields is the sum of the kinetic and potential energy of a particle under consideration. The total derivative of this Hamiltonian is determined as

$$dH = \left[\left(\dot{x}\, dp_x - \frac{\partial L}{\partial x}\, dx \right) + \left(p_x - \frac{\partial L}{\partial \dot{x}} \right) d\dot{x} \right]$$
$$+ \left[\left(\dot{y}\, dp_y - \frac{\partial L}{\partial y}\, dy \right) + \left(p_y - \frac{\partial L}{\partial \dot{y}} \right) d\dot{y} \right]$$
$$+ \left[\left(\dot{z}\, dp_z - \frac{\partial L}{\partial z}\, dz \right) + \left(p_z - \frac{\partial L}{\partial \dot{z}} \right) d\dot{z} \right] + \frac{\partial H}{\partial T}\, dT. \qquad (8.15b)$$

Because of the definition of p_x, p_y, p_z in Eqs. (8.14a)–(8.14c), the second terms within each bracket of Eq. (8.15b) vanish. By comparing coefficients, we thus find the canonical equations,

$$\partial H / \partial p_x = \dot{x}, \qquad \partial H / \partial x = -\partial L / \partial x = -\dot{p}_x, \qquad (8.16a)$$
$$\partial H / \partial p_y = \dot{y}, \qquad \partial H / \partial y = -\partial L / \partial y = -\dot{p}_y, \qquad (8.16b)$$
$$\partial H / \partial p_z = \dot{z}, \qquad \partial H / \partial z = -\partial L / \partial z = -\dot{p}_z, \qquad (8.16c)$$
$$\partial H / \partial T = -\partial L / \partial T. \qquad (8.16d)$$

In a matrix notation, Eqs. (8.16) can also be written as

$$\begin{pmatrix} dx/dT \\ dp_x/dT \\ dy/dT \\ dp_y/dT \\ dz/dT \\ dp_z/dT \end{pmatrix} = \begin{pmatrix} 0 & 1 & 0 & 0 & 0 & 0 \\ -1 & 0 & 0 & 0 & 0 & 0 \\ 0 & 0 & 0 & 1 & 0 & 0 \\ 0 & 0 & -1 & 0 & 0 & 0 \\ 0 & 0 & 0 & 0 & 0 & 1 \\ 0 & 0 & 0 & 0 & -1 & 0 \end{pmatrix} \begin{pmatrix} \partial H/\partial x \\ \partial H/\partial p_x \\ \partial H/\partial y \\ \partial H/\partial p_y \\ \partial H/\partial z \\ \partial H/\partial p_z \end{pmatrix}, \qquad (8.17a)$$

with the side condition of Eq. (8.16d). If J denotes the unit-type matrix of Eq. (8.17a), we can also write Eq. (8.17a) with the canonically conjugate coordinates $\mathbf{r} = (x, p_x, y, p_y, z, p_z)$ as

$$\mathbf{r} = J(\partial H / \partial \mathbf{r}). \qquad (8.17b)$$

8.3.2 Canonical Transformations with an Independent Position Variable

Instead of the coordinates of Eqs. (8.14a)–(8.14c), with time T the independent variable, we may also use the curvilinear coordinates for Eq.

(8.2) in which an arbitrary trajectory is described relative to the optic axis and where the coordinate z measured along this optic axis is the independent variable. We shall rewrite these coordinates for particles of given mass m_{00} and charge $(z_0 e)$ as

$$x, \qquad\qquad p_x = p_0 a, \tag{8.18a}$$

$$y, \qquad\qquad p_y = p_0 b, \tag{8.18b}$$

$$t = T - T_0, \qquad -\delta K_0 = K_0 - K. \tag{8.18c}$$

Here, p_0 is the momentum of a reference particle, and $K - K_0$ is the difference of the initial kinetic energies of an arbitrary and a reference particle measured at the first standard profile plane of an optical system. This difference $K - K_0$ is identical to the difference $E - E_0$ of the corresponding total energies at any position z measured along the optic axis. Finally, T and T_0 are flight times through the optical system under consideration for an arbitrary particle moving along its trajectory and for a reference particle moving along the optic axis.

In time-independent fields, we may express the total energy E as the Hamiltonian [Eq. (2.33)]:

$$E = H(x, p_x, y, p_y, z, p_z, T).$$

Analogously, we can express p_z as a function F of the other variables

$$p_z = -F(x, p_x, y, p_y, T, E, z).$$

From these two relations we find the total derivatives

$$dE = -K_0 \, d\delta = \left(\frac{\partial H}{\partial x} dx + \frac{\partial H}{\partial p_x} dp_x \right) + \left(\frac{\partial H}{\partial y} dy + \frac{\partial H}{\partial p_y} \right)$$

$$+ \left(\frac{\partial H}{\partial z} dz + \frac{\partial H}{\partial p_z} dp_z \right) + \frac{\partial H}{\partial T} dT, \tag{8.19a}$$

$$-dp_z = \left(\frac{\partial F}{\partial z} dx + \frac{\partial F}{\partial p_x} dp_x \right) + \left(\frac{\partial F}{\partial y} dy + \frac{\partial F}{\partial p_y} dp_y \right)$$

$$+ \left(\frac{\partial F}{\partial T} dT + \frac{\partial F}{\partial E} dE \right) + \frac{\partial F}{\partial z} dz. \tag{8.19b}$$

At this point, we may transform Eqs. (8.16) slightly and obtain

$$\frac{dx}{dz} = \dot{x} \frac{dT}{dz} = \frac{\partial H / \partial p_x}{\partial H / \partial p_z}, \qquad \frac{dp_x}{dz} = \dot{p}_x \frac{dT}{dz} = -\frac{\partial H / \partial x}{\partial H / \partial p_z},$$

$$\frac{dy}{dz} = \dot{y} \frac{dT}{dz} = \frac{\partial H / \partial p_y}{\partial H / \partial p_z}, \qquad \frac{dp_x}{dz} = \dot{p}_y \frac{dT}{dz} = -\frac{\partial H / \partial y}{\partial H / \partial p_y},$$

so that Eq. (8.19a) becomes with $\dot{z} = \partial H/\partial p_z$:

$$-dp_z = \frac{\left(\dfrac{\partial H}{\partial x}dx + \dfrac{\partial H}{\partial p_x}dp_x\right) + \left(\dfrac{\partial H}{\partial y}dy + \dfrac{\partial H}{\partial p_y}dp_y\right) + \left(\dfrac{\partial H}{\partial T}dT - dE\right) + \dfrac{\partial H}{\partial z}dz}{\dfrac{\partial H}{\partial p_z}}.$$

(8.19c)

Comparing the coefficients of Eqs. (8.19b) and (8.19c) yields with Eqs. (8.16) and Eqs. (8.18),

$$\frac{\partial F}{\partial p_x} = \frac{\partial F}{p_0\,\partial a} = \frac{\dot{x}}{\dot{z}} = \frac{dx}{dz}, \qquad \frac{\partial F}{\partial x} = -\frac{\dot{p}_x}{\dot{z}} = -\frac{dp_x}{dz} = -p_0\frac{da}{dz}, \qquad (8.20a)$$

$$\frac{\partial F}{\partial p_y} = \frac{\partial F}{p_0\,\partial b} = \frac{\dot{y}}{\dot{z}} = \frac{dy}{dz}, \qquad \frac{\partial F}{\partial y} = -\frac{\dot{p}_y}{\dot{z}} = -\frac{dp_y}{dz} = -p_0\frac{db}{dz}, \qquad (8.20b)$$

$$\frac{\partial F}{\partial E} = -\frac{\partial F}{K_0\,\partial\delta} = -\frac{1}{\dot{z}} = -\frac{dT}{dz}, \qquad \frac{\partial F}{\partial T} = \frac{\partial H/\partial T}{\dot{z}} = -\frac{dE}{dz} = -K_0\frac{d\delta}{dz}, \qquad (8.20c)$$

with $dT = dt$. If we replace H, T, z, p_z of Eqs. (8.14a)-(8.14c) by F, z, t, $K_0\delta$ of Eqs. (8.18a)-(8.18c) we find that Eqs. (8.20) and (8.16) are identical. Thus, the variables x, ap_0, y, bp_0, t, $K_0\delta$ of Eqs. (8.18) are canonically conjugate, and F is the corresponding Hamiltonian.

Similarly, since the canonical equations [Eqs. (8.16a)-(8.16c)] were written in the matrix notation of Eqs. (8.17), we can rewrite Eqs. (8.20a)-(8.20c) as

$$\begin{pmatrix} dx/dz \\ da/dz \\ dy/dz \\ db/dz \\ dt/dz \\ db/d \end{pmatrix} = \begin{pmatrix} 0 & p_0^{-1} & 0 & 0 & 0 & 0 \\ -p_0^{-1} & 0 & 0 & 0 & 0 & 0 \\ 0 & 0 & 0 & p_0^{-1} & 0 & 0 \\ 0 & 0 & -p_0^{-1} & 0 & 0 & 0 \\ 0 & 0 & 0 & 0 & 0 & K_0^{-1} \\ 0 & 0 & 0 & 0 & -K_0^{-1} & 0 \end{pmatrix} \begin{pmatrix} \partial F/\partial x \\ \partial F/\partial a \\ \partial F/\partial t \\ \partial F/\partial b \\ \partial F/\partial t \\ \partial F/\partial\delta \end{pmatrix}. \quad (8.21a)$$

With J denoting the matrix of Eq. (8.21a), we can also write Eq. (8.21a) with $\mathbf{r} = (x, a, y, b, t, \delta)$ as

$$\mathbf{r} = J(\partial F/\partial \mathbf{r}). \qquad (8.21b)$$

8.3.3 The Condition of Symplecticity

For an arbitrary optical system, we now may describe a particle trajectory by position vectors

$$\mathbf{r} = (r_1 = x, r_2 = a, r_3 = y, r_4 = b, r_5 = t, r_6 = \delta)$$

in a six-dimensional space and determine $r_e = \mathbf{r}(z_e)$ if $r_0 = \mathbf{r}(z_0)$ is known.

Very generally then the relation between r_{k0} and r_{le}, with $k, l = 1, 2, 3, 4, 5, 6$, can be characterized by

$$\frac{\partial r_{le}}{\partial z} = \sum_k \frac{\partial r_{le}}{\partial r_{k0}} \frac{\partial r_{k0}}{\partial z}.$$

Explicitly, this relation reads

$$
\begin{pmatrix}
\dfrac{\partial r_{1e}}{\partial z} \\[6pt]
\dfrac{\partial r_{2e}}{\partial z} \\[6pt]
\dfrac{\partial r_{3e}}{\partial z} \\[6pt]
\dfrac{\partial r_{4e}}{\partial z} \\[6pt]
\dfrac{\partial r_{5e}}{\partial z} \\[6pt]
\dfrac{\partial r_{6e}}{\partial z}
\end{pmatrix}
=
\begin{pmatrix}
\dfrac{\partial r_{1e}}{\partial r_{10}} & \dfrac{\partial r_{1e}}{\partial r_{20}} & \dfrac{\partial r_{1e}}{\partial r_{30}} & \dfrac{\partial r_{1e}}{\partial r_{40}} & \dfrac{\partial r_{1e}}{\partial r_{50}} & \dfrac{\partial r_{1e}}{\partial r_{60}} \\[6pt]
\dfrac{\partial r_{2e}}{\partial r_{10}} & \dfrac{\partial r_{2e}}{\partial r_{20}} & \dfrac{\partial r_{2e}}{\partial r_{30}} & \dfrac{\partial r_{2e}}{\partial r_{40}} & \dfrac{\partial r_{2e}}{\partial r_{50}} & \dfrac{\partial r_{2e}}{\partial r_{60}} \\[6pt]
\dfrac{\partial r_{3e}}{\partial r_{10}} & \dfrac{\partial r_{3e}}{\partial r_{20}} & \dfrac{\partial r_{3e}}{\partial r_{30}} & \dfrac{\partial r_{3e}}{\partial r_{40}} & \dfrac{\partial r_{3e}}{\partial r_{50}} & \dfrac{\partial r_{3e}}{\partial r_{60}} \\[6pt]
\dfrac{\partial r_{4e}}{\partial r_{10}} & \dfrac{\partial r_{4e}}{\partial r_{20}} & \dfrac{\partial r_{4e}}{\partial r_{30}} & \dfrac{\partial r_{4e}}{\partial r_{40}} & \dfrac{\partial r_{4e}}{\partial r_{50}} & \dfrac{\partial r_{4e}}{\partial r_{60}} \\[6pt]
\dfrac{\partial r_{5e}}{\partial r_{10}} & \dfrac{\partial r_{5e}}{\partial r_{20}} & \dfrac{\partial r_{5e}}{\partial r_{30}} & \dfrac{\partial r_{5e}}{\partial r_{40}} & \dfrac{\partial r_{5e}}{\partial r_{50}} & \dfrac{\partial r_{5e}}{\partial r_{60}} \\[6pt]
\dfrac{\partial r_{6e}}{\partial r_{10}} & \dfrac{\partial r_{6e}}{\partial r_{20}} & \dfrac{\partial r_{6e}}{\partial r_{30}} & \dfrac{\partial r_{6e}}{\partial r_{40}} & \dfrac{\partial r_{6e}}{\partial r_{50}} & \dfrac{\partial r_{6e}}{\partial r_{60}}
\end{pmatrix}
\begin{pmatrix}
\dfrac{\partial r_{10}}{\partial z} \\[6pt]
\dfrac{\partial r_{20}}{\partial z} \\[6pt]
\dfrac{\partial r_{30}}{\partial z} \\[6pt]
\dfrac{\partial r_{40}}{\partial z} \\[6pt]
\dfrac{\partial r_{50}}{\partial z} \\[6pt]
\dfrac{\partial r_{60}}{\partial z}
\end{pmatrix}
\qquad (8.22a)
$$

where the derivatives vary with varying positions z_0 and z_e. Equation (8.22a) can also be written in an abbreviated form as

$$\partial \mathbf{r}_e / \partial z = A(\partial \mathbf{r}_0 / \partial z). \qquad (8.22b)$$

Here, A is the matrix of Eq. (8.22a), the so-called Jacobi matrix. According to Eq. (8.21), we may now replace $\partial \mathbf{r}_0 / \partial z$ by $J(z_0) \, \partial F / \partial \mathbf{r}_0$, yielding

$$\frac{\partial \mathbf{r}_e}{\partial z} = A \frac{\partial \mathbf{r}_0}{\partial z} = A J(z_0) \frac{\partial F}{\partial \mathbf{r}_0}, \qquad (8.23)$$

where $J(z_0)$ is the matrix J of Eq. (8.21a) at $z = z_0$. The expression $\partial F / \partial \mathbf{r}_0$ in Eq. (8.23) can also be written explicitly for any $l = 1, 2, 3, 4, 5, 6$ as

$$\frac{\partial F}{\partial r_{l0}} = \sum_k \frac{\partial F}{\partial r_{ke}} \frac{\partial r_{ke}}{\partial r_{l0}},$$

which is identical to stating $\partial F / \partial \mathbf{r}_0 = A'(\partial F / \partial \mathbf{r}_e)$. Conversely, we may replace $\partial \mathbf{r}_e / \partial z$ in the left-hand side of Eq. (8.23) by $J_e(\partial F / \partial \mathbf{r}_e)$, again according to Eq. (8.21a) at $z = z_0$. Thus, we find (Berz *et al.*, 1986) the most important so-called symplectic condition* (Thirring, 1977; Dragt, 1982)

$$J_e = A J_0 A'. \qquad (8.24)$$

---

* The validity of Eq. (8.24) is identical to postulating the existence of an "eikonal" from which, for instance, Glaser (1956) derived properties of rotationally symmetric lenses or Plies and Rose (1971) properties of magnetic sector fields.

From this relation, Liouville's theorem is easily derived, since we read from Eq. (8.21a),

$$J_e = p_{0e}^{-4} K_{0e}^{-2}, \qquad J_0 = p_{00}^{-4} K_{00}^{-4},$$

so that Eq. (8.24) yields

$$|A| = \pm |J_e||J_0| = \pm \frac{p_{00}^2 K_{00}}{p_{0e}^2 K_{0e}}. \qquad (8.25a)$$

If there were no accelerations between z_0 and z_e, Eq. (8.25a) simplifies to

$$|A| = \pm 1. \qquad (8.25b)$$

Although plus/minus signs appear in Eqs. (8.25a) and (8.25b) only the plus signs are applicable, since there are cases of $|A| = +1$ (for instance, the transformation in a field-free region) and since $|A|$ cannot vary discontinuously from $+1$ to -1 in a real optical system.

Knowing $d\mathbf{r}_0$ over all particles in a volume at z_0, we can determine $d\mathbf{r}_e$ in a volume at z_e as $d\mathbf{r}_e = |A| \, d\mathbf{r}_0$ because of the integration rules in a multidimensional space (see, for instance, Stromberg, 1981); hence we find

$$\int dr_e \Big/ \int dr_0 = |A| = p_{00}^2 K_{00} / p_{0e}^2 K_{0e}.$$

This is Liouville's theorem for the case of not quite canonical coordinates x, a, y, b, t, δ. For the case of truly canonical coordinates $\bar{r} = (x, p_0 a, y, p_0 b, t, \delta K_0)$, as defined in Eqs. (8.18a)–(8.18c), we find $|A| = 1$, or

$$\int d\bar{r}_e = \int d\bar{r}_0 = \text{const},$$

i.e., the phase-space areas at z_0 and z_e are equal in area [Eq (5.1)].

8.3.3.1   Relations between Coefficients of First Order

Using only the linear terms in Eq. (8.2), we can determine the Jacobi matrix of Eq. (8.22a) as the first-order transfer matrix

$$A = \begin{pmatrix} (x|x) & (x|a) & 0 & 0 & 0 & (x|\delta) \\ (a|x) & (a|a) & 0 & 0 & 0 & (a|\delta) \\ 0 & 0 & (y|y) & (y|b) & 0 & 0 \\ 0 & 0 & (b|y) & (b|b) & 0 & 0 \\ (t|x) & (t|a) & 0 & 0 & 1 & (t|\delta) \\ 0 & 0 & 0 & 0 & 0 & (1+\kappa)^{-1} \end{pmatrix}. \qquad (8.26a)$$

It is assumed here that the reference particles had the energies K_{00} and $K_{0e} = K_{00}(1 + \kappa)$ at z_0 and z_e [see also Eq. (7.16b)]. Introducing the matrix of Eq. (8.26a) into Eq. (8.24), we find

$$
\begin{pmatrix}
0 & p_{0e}^{-1} & 0 & 0 & 0 & 0 \\
-p_{0e}^{-1} & 0 & 0 & 0 & 0 & 0 \\
0 & 0 & 0 & p_{0e}^{-1} & 0 & 0 \\
0 & 0 & -p_{0e}^{-1} & 0 & 0 & 0 \\
0 & 0 & 0 & 0 & 0 & K_{0e}^{-1} \\
0 & 0 & 0 & 0 & K_{e0}^{-1} & 0
\end{pmatrix}
$$

$$
=
\begin{pmatrix}
-(x|a)p_{00}^{-1} & (x|x)p_{00}^{-1} & 0 & 0 & -(x|\delta)K_{00}^{-1} & 0 \\
-(a|a)p_{00}^{-1} & (a|x)p_{00}^{-1} & 0 & 0 & -(a|\delta)K_{00}^{-1} & 0 \\
0 & 0 & -(y|b)p_{00}^{-1} & (y|y)p_{00}^{-1} & 0 & 0 \\
0 & 0 & -(b|b)p_{00}^{-1} & (b|y)p_{00}^{-1} & 0 & 0 \\
-(t|a)p_{00}^{-1} & (t|x)p_{00}^{-1} & 0 & 0 & -(t|\delta)K_{00}^{-1} & K_{00}^{-1} \\
0 & 0 & 0 & 0 & -K_{00}^{-1}(1+\kappa)^{-1} & 0
\end{pmatrix}
$$

$$
\cdot
\begin{pmatrix}
(x|x) & (a|x) & 0 & 0 & (t|x) & 0 \\
(x|a) & (a|a) & 0 & 0 & (t|a) & 0 \\
0 & 0 & (y|y) & (b|y) & 0 & 0 \\
0 & 0 & (y|b) & (b|b) & 0 & 0 \\
0 & 0 & 0 & 0 & 1 & 0 \\
(x|\delta) & (a|\delta) & 0 & 0 & (t|\delta) & (1+\kappa)^{-1}
\end{pmatrix}
. \tag{8.26b}
$$

From this relation, we find directly,

$$
(x|x)(a|a) - (x|a)(a|x) = p_{00}/p_{0e}, \tag{8.27a}
$$

$$
(y|y)(b|b) - (y|b)(b|y) = p_{00}/p_{0e}, \tag{8.27b}
$$

i.e., well-known fact that the determinants of transfer matrices are equal to the ratio of the corresponding momenta p_{00} and p_{0e} as in Eqs. (2.23a) and (2.23b). In profile planes with vanishing vector potential A, this ratio of p_{00}/p_{0e} can also be replaced by the ratio of the refractive indices n_0/n_e. From Eq. (8.26b), we find, furthermore,

$$
(x|\delta)/K_{00} = [(x|x)(t|a) - (x|a)(t|x)]/p_{00}, \tag{8.28a}
$$

$$
(a|\delta)/K_{00} = [(a|x)(t|a) - (a|a)(t|x)]/p_{00}. \tag{8.28b}
$$

With the help of Eq. (8.27a), these relations can be transformed to

$$
(t|x)v_e = (p_{0e}v_e/-K_{00})[(x|x)(a|\delta) - (a|x)(x|\delta)], \tag{8.29a}
$$

$$
(t|a)v_e = (p_{0e}v_e/-K_{00})[(x|a)(a|\delta) - (a|a)(x|\delta)], \tag{8.29b}
$$

and

$$p_{0e}v_e/-K_{00} = [-2(1 + \eta_e)/(1 + 2\eta_e)](1 + \kappa),$$

where use is made of $p_{0e}v_e = m_{00}(1 + 2\eta_e)v_e^2$, with v_e given in Eq. (2.2b),
as well as of $\eta_e = K_{00}(1 + \kappa)/2m_{00}c^2$. Note that Eqs. (8.29a) and (8.29b)
are already known from Eqs. (4.80a) and (4.80b) for the case $\kappa = 0$ [see
also Wollnik and Matsuo (1981) and Wollnik and Matsuda (1981)].

8.3.3.2 Relations between Coefficients of Second Order

Using the first- as well as the second-order terms in Eq. (8.2), we can
again determine the Jacobi matrix of Eq. (8.22a). Introducing this matrix
into Eq. (8.24), we find by comparing coefficients of equal powers of r_k (see
also Wollnik and Berz, 1985):

$$\sum_{\mu=1}^{3} [(r_{2\mu-1}|r_i)(r_{2\mu}|r_k r_j) - (r_{2\mu}|r_i)(r_{2\mu-1}|r_k r_j)$$

$$+ (r_{2\mu-1}|r_i r_k)(r_{2\mu}|r_j) - (r_{2\mu}|r_i r_k)(r_{2\mu-1}|r_j)] = 0 \qquad (8.30a)$$

where $i, j, k = 1, 2, 3, 4, 5, 6$. Assuming that Eqs. (8.3a)–(8.3d) are valid
here, only those coefficients are nonzero that are listed in Tables 8.1 and
8.2 or in the transfer matrices of Tables 8.3 and 8.4. Rewriting Eq. (8.30a)
for all possible $i, j, k,$ and μ, we find by using Eqs. (8.27) and (8.29),

$$(t|xx)v_e = -[(x|x)(a|x\delta) - (a|x)(x|x\delta)$$
$$+ (x|xx)(a|\delta) - (a|xx)(x|\delta)]p_{0e}v_e/K_{00}, \qquad (8.31a)$$

$$(t|xa)v_e = -[(x|x)(a|a\delta) - (a|x)(x|a\delta)$$
$$+ (x|xa)(a|\delta) - (a|xa)(x|\delta)]p_{0e}v_e/K_{00}, \qquad (8.31b)$$

$$(t|aa)v_e = -[(x|a)(a|a\delta) - (a|a)(x|a\delta)$$
$$+ (x|aa)(a|\delta) - (a|aa)(x|\delta)]p_{0e}v_e/K_{00}, \qquad (8.31c)$$

$$(t|yy)v_e = -[(y|y)(b|y\delta) - (b|y)(y|y\delta)$$
$$+ (x|yy)(a|\delta) - (a|yy)(x|\delta)]p_{0e}v_e/K_{00}, \qquad (8.31d)$$

$$(t|yb)v_e = -[(y|y)(b|b\delta) - (b|y)(y|b\delta)$$
$$+ (x|yb)(a|\delta) - (a|yb)(x|\delta)]p_{0e}v_e/K_{00}, \qquad (8.31e)$$

$$(t|bb)v_e = -[(y|b)(b|b\delta) - (b|b)(y|b\delta)$$
$$+ (x|bb)(a|\delta) - (a|bb)(x|\delta)]p_{0e}v_e/K_{00}, \qquad (8.31f)$$

$$(t|x\delta)v_e = -[(x|x)(a|\delta\delta) - (a|x)(x|\delta\delta)$$
$$+ (x|x\delta)(a|\delta) - (a|x\delta)(x|\delta)]p_{0e}v_e/K_{00}, \qquad (8.31g)$$

$$(t|a\delta)v_e = -[(x|a)(a|\delta\delta) - (a|a)(x|\delta\delta)$$
$$+ (x|a\delta)(a|\delta) - (a|a\delta)(x|\delta)]p_{0e}v_e/K_{00}. \qquad (8.31h)$$

Furthermore, we find relations without $(t| \cdots)$ matrix elements that can be combined to:

$$(x|xa) = (x|a)[(x|a)(a|xx) - (a|a)(x|xx)]$$
$$- (x|x)[(a|x)(x|aa) - (x|x)(a|aa)], \qquad (8.32a)$$

$$(a|xa) = (a|a)[(x|a)(a|xx) - (a|a)(x|xx)]$$
$$- (a|x)[(x|x)(a|aa) - (a|x)(x|aa)], \qquad (8.32b)$$

$$(y|yx) = (y|b)[(x|x)(a|yy) - (a|x)(x|yy)]$$
$$- (y|y)[(x|x)(a|yb) - (a|x)(x|yb)], \qquad (8.32c)$$

$$(b|yx) = (b|b)[(x|x)(a|yy) - (a|x)(x|yy)]$$
$$- (b|y)[(x|x)(a|yb) - (a|x)(x|yb)], \qquad (8.32d)$$

$$(y|ya) = (y|b)[(x|a)(a|yy) - (a|a)(x|yy)]$$
$$- (y|y)[(x|a)(a|yb) - (a|a)(x|yb)], \qquad (8.32e)$$

$$(b|ya) = (b|b)[(x|a)(a|yy) - (a|a)(x|yy)]$$
$$- (b|y)[x|a)(a|yb) - (a|a)(x|yb)], \qquad (8.32f)$$

$$(y|bx) = (y|b)[(x|x)(a|by) - (a|x)(x|by)]$$
$$- (y|y)[(x|x)(a|bb) - (a|x)(x|bb)], \qquad (8.32g)$$

$$(b|bx) = (b|b)[(x|x)(a|by) - (a|x)(x|by)]$$
$$- (b|y)[(x|x)(a|bb) - (a|x)(x|bb)], \qquad (8.32h)$$

$$(y|ba) = (y|b)[(x|a)(a|by) - (a|a)(x|by)]$$
$$- (y|y)[(x|a)(a|bb) - (a|a)(x|bb)], \qquad (8.32i)$$

$$(b|ba) = (b|b)[(x|a)(a|by) - (a|a)(x|by)]$$
$$- (b|y)[(x|a)(a|bb) - (a|a)(x|bb)]. \qquad (8.32j)$$

Finally, there are two relations that can be written as

$$(a|a\delta) = [(a|x)(x|a\delta) + (a|x\delta)(x|a) - (x|x\delta)(a|a)]/(x|x), \qquad (8.33a)$$

or

$$(x|a\delta) = [(x|x)(a|a\delta) - (a|x\delta)(x|a) + (x|x\delta)(a|a)]/(a|x), \qquad (8.33b)$$

$$(b|b\delta) = [(b|y)(y|b\delta) + (b|y\delta)(y|b) - (y|y\delta)(b|b)]/(y|y), \qquad (8.34a)$$

or

$$(y|b\delta) = [(y|y)(b|b\delta) - (b|y\delta)(y|b) + (y|y\delta)(b|b)]/(b|y), \qquad (8.34b)$$

where for numerical calculations, only one part of Eq. (8.33) and one part of Eq. (8.34) can be used. Here we normally choose those relations for which the absolute value of the denominator is maximal.

By Eqs. (8.31) and (8.32), as well as by Eqs. (8.27), (8.29), (8.33), and (8.34), several of the matrix elements listed in Tables 8.1 and 8.2 are fixed and can thus be determined quite easily. To give some idea of the simplification so achieved, the first two and three lines of the matrices of Tables 8.3 and Table 8.4 plus the corresponding third-order elements are again listed in Table 8.5, where all elements marked by an asterisk are those that can be determined from the condition of symplecticity and other "asterisk free" elements. Note that all time coefficients can be so determined, except

TABLE 8.5

Elements of a Third-Order Transfer Matrix for a Sector Field[a]

$(x\|x)$	$(a\|x)$	$*(t\|x)$	$(y\|y)$	$(b\|y)$
$(x\|a)$	$(a\|a)$	$*(t\|a)$	$(y\|b)$	$(b\|b)$
$(x\|\delta)$	$(a\|\delta)$	$(t\|\delta)$		
$(x\|xx)$	$(a\|xx)$	$*(t\|xx)$	$*(y\|yx)$	$*(b\|yx)$
$*(x\|xa)$	$*(a\|xa)$	$*(t\|xa)$	$*(y\|bx)$	$*(b\|bx)$
$(x\|aa)$	$(a\|aa)$	$*(t\|aa)$	$*(y\|ya)$	$*(b\|ya)$
$(x\|yy)$	$(a\|yy)$	$*(t\|yy)$	$*(y\|ba)$	$*(b\|ba)$
$(x\|yb)$	$(a\|yb)$	$*(t\|yb)$	$(y\|y\delta)$	$(b\|y\delta)$
$(x\|bb)$	$(a\|bb)$	$*(t\|bb)$	$(y\|b\delta)$	$(b\|b\delta)$
$(x\|x\delta)$	$(a\|x\delta)$	$*(t\|x\delta)$		
$(x\|a\delta)$	$(a\|a\delta)$	$*(t\|a\delta)$		
$(x\|\delta\delta)$	$(a\|\delta\delta)$	$(t\|\delta\delta)$		
$(x\|xxx)$	$(a\|xxx)$	$*(t\|xxx)$	$(y\|yyy)$	$(b\|yyy)$
$*(x\|xxa)$	$*(a\|xxa)$	$*(t\|xxa)$	$(y\|yyb)$	$(b\|yyb)$
$*(x\|xaa)$	$*(a\|xaa)$	$*(t\|xaa)$	$(y\|ybb)$	$(b\|ybb)$
$(x\|aaa)$	$(a\|aaa)$	$*(t\|aaa)$	$(y\|bbb)$	$(b\|bbb)$
$(x\|xyy)$	$(a\|xyy)$	$*(t\|xyy)$	$*(y\|yxx)$	$*(b\|yxx)$
$(x\|xyb)$	$(a\|xyb)$	$*(t\|xyb)$	$*(y\|yxa)$	$*(b\|yxa)$
$(x\|xbb)$	$(a\|xbb)$	$*(t\|xbb)$	$*(y\|yaa)$	$*(b\|yaa)$
$(x\|ayy)$	$(a\|ayy)$	$*(t\|ayy)$	$*(y\|bxx)$	$*(b\|bxx)$
$(x\|ayb)$	$(a\|ayb)$	$*(t\|ayb)$	$*(y\|bxa)$	$*(b\|bxa)$
$(x\|abb)$	$(a\|abb)$	$*(t\|abb)$	$*(y\|baa)$	$*(b\|baa)$
$(x\|xx\delta)$	$(a\|xx\delta)$	$*(t\|xx\delta)$	$*(y\|yx\delta)$	$*(b\|yx\delta)$
$*(x\|xa\delta)$	$*(a\|xa\delta)$	$*(t\|xa\delta)$	$*(y\|ya\delta)$	$*(b\|ya\delta)$
$(x\|aa\delta)$	$(a\|aa\delta)$	$*(t\|aa\delta)$	$*(y\|bx\delta)$	$*(b\|bx\delta)$
$(x\|yy\delta)$	$(a\|yy\delta)$	$*(t\|yy\delta)$	$*(y\|ba\delta)$	$*(b\|ba\delta)$
$(x\|yb\delta)$	$(a\|yb\delta)$	$*(t\|yb\delta)$	$(y\|y\delta\delta)$	$(b\|y\delta\delta)$
$(x\|bb\delta)$	$(a\|bb\delta)$	$*(t\|bb\delta)$	$(y\|b\delta\delta)$	$(b\|b\delta\delta)$
$(x\|x\delta\delta)$	$(a\|x\delta\delta)$	$*(t\|x\delta\delta)$		
$(x\|a\delta\delta)$	$(a\|a\delta\delta)$	$*(t\|a\delta\delta)$		
$(x\|\delta\delta\delta)$	$(a\|\delta\delta\delta)$	$(t\|\delta\delta\delta)$		

[a] Asterisk indicates elements that can be determined as function of the other (no asterisk) matrix elements due to the symplectic condition.

$(t|\delta)$ and $(t|\delta\delta)$, and that all geometric second-order coefficients in the y direction can be so determined as well. Using the same procedures for nth-order coefficients as well, we find that all time coefficients are given, except $(t|\delta^n)$, and that all geometric coefficients $(y|\cdots x^k a^l \cdots)$ can be determined for $k > 0$, $l > 0$.

8.4   IMAGE ABERRATIONS OF nth ORDER

A detailed derivation of coefficients of image aberrations is rather involved and therefore is only outlined in the Appendix to this chapter. Very generally, these coefficients are derived separately for regions in which the k of Hill's equation [Eq. (1.16)] is constant, as in field-free regions, sector fields, and quadrupole lenses, or in which k varies with z, as in all fringing fields.

In regions with constant k, it is relatively simple to derive the explicit equations of motion if the detailed field distribution in an optical element is known in a power series in x and y. However, a solution of this differential equation is involved, since we can solve only the homogeneous equation directly and must then solve the full equation by some perturbation method. In Section 8.A1, in the appendix to this chapter, such an iterative method is outlined which yields a solution that improves by one order for each iteration step. For this derivation, z independent multipole field distributions are assumed,

$$B_y(x, y = 0) = B_y(0, 0) + x\frac{\partial B_y}{\partial x} + \frac{x^2}{2}\frac{\partial^2 B_y}{\partial x^2} + \frac{x^3}{2\cdot 3}\frac{\partial^3 B_y}{\partial x^3} + \cdots = \sum \frac{x^n}{n\pm}K_{Bn},$$
$$(8.35a)$$

$$E_x(x, y = 0) = E_y(0, 0) + x\frac{\partial E_x}{\partial x} + \frac{x^2}{2}\frac{\partial^2 E}{\partial x^2} + \frac{x^3}{2\cdot 3}\frac{\partial^3 E_x}{\partial x^3} + \cdots = \sum \frac{x^n}{n\pm}K_{En},$$
$$(8.35b)$$

where the coefficients to the nth power of x are determined by a magnetic or electrostatic $2(n + 1)$ multipole.

Because the ith-order solution of the particle trajectories (Section 8.A1) is determined from all $i - 1$, $i - 2, \ldots$, solutions, it depends on all optical elements with $n = 0, 1, \ldots, i$ but not on elements with $n = i + 1, i + 2, \ldots$, as is illustrated in Table 8.6. For $i = 2$, such calculations were performed by Hintenberger and Koenig (1957); Ewald and Liebl (1955, 1957); Boerboom et al. (1959); and Wollnik (1964, 1965). For magnetic sector fields, these results were expanded by Ludwig (1967) to $i = 3$, whereas for general

TABLE 8.6

Effects of Multipoles of Aberrations of Zeroth to Fourth Order[a]

Order	Dipole $(n = 0)$	Quadrupole $(n = 1)$	Hexapole $(n = 2)$	Octupole $(n = 3)$	Decapole $(n = 4)$
Zeroth (deflection)	*	0	0	0	0
First (focusing)	*	*	0	0	0
Second	*	*	*	0	0
Third	*	*	*	*	0
Fourth	*	*	*	*	*

[a] The ith-order aberration is found as the sum of all terms of the ith row. A term abbreviated by an asterisk indicates that the corresponding $2(n + 1)$-pole element has influence on the aberrations of ith order and that the corresponding term is nonzero; zero implies that this multipole element cannot change the aberration of the indicated order. Note that in the case of an ith-order approximation, only the first i rows are of interest.

magnetic and electrostatic sector fields, this was done by Matsuo and Matsuda (1971) and by Matsuo *et al.* (1972), and for magnetic and electrostatic quadrupole lenses by Smith (1969) and Lee-Whiting (1970). The results of such similar calculations are presented in Section 8.A2 in the appendix to this chapter.

In regions with variable k the approach used in Section 8.A1 can also be used to solve the resulting equations of motion; however, in this case, to obtain the corresponding differential equations is very involved, since here B and E are functions of z. In the case of a magnetic sector field, terms such as $\int B_y \, dz$ of $\iint B_y \, dz \, dz$ appear in the final solution for the particle trajectories. If the magnetic rigidity $B_{y0}\rho_0$ is large compared to these terms, i.e., $B_{y0}\rho_0 \gg \int B_y \, dz \gg \iint B_y \, dz \, dz \gg \cdots$ because of the limited extension of the fringing fields, we can neglect all but the first few terms for the solution of the final particle trajectory. In an ith-order approximation for the particle trajectory, only the elements of the first $(i + 1)$ columns of Table 8.7 are important. In the case of a second-order calculation for a magnetic sector field, the (normally easy to adjust) beam deflection depends on the distribution of B_y in detail; the (often easy to adjust) focusing properties depend only slightly on the details of this distribution, whereas the (normally more difficult to adjust) second-order aberrations are independent of the detailed fringing field distribution.

Very similar relations are found for an electrostatic sector field where only B_y must be replaced by E_x. Note that Table 8.7 is also meaningful for a magnetic or electrostatic $2(n + 1)$ pole with a straight optic axis if in this case all terms in the first n rows and n columns are put to zero, and B_y or E_x is replaced by $\partial^n B_y / \partial z^n$ or $\partial^n E_x / \partial z^n$.

TABLE 8.7

Effects of Fringing-Field Integrals on Aberrations of Zeroth to Fourth Order[a]

Order	Dipole	Quadrupole	Hexapole	Octupole	Decapole
Zeroth (deflection)	$\cdots B_0\rho_0$	$\cdots \int B_0\,dz$	$\cdots \iint B_y\,dz^2$	$\cdots \iiint B_y\,dz^3$	$\cdots \iiiint B_y\,dz^4$
First (focusing)	0	$\cdots B_0\rho_0$	$\cdots \int B_y\,dz$	$\cdots \iint B_y\,dz^2$	$\cdots \iiint B_y\,dz^3$
Second	0	0	$\cdots B_0\rho_0$	$\cdots \int B_y\,dz$	$\cdots \iint B_y\,dz^2$
Third	0	0	0	$\cdots B_0\rho_0$	$\cdots \int B_y\,dz$
Fourth	0	0	0	0	$\cdots B_0\rho_0$

[a] For a magnetic dipole the ith-order aberration is the sum of the terms in the ith row. For $B_0\rho_0 \gg \int B\,dz \gg \iint B\,dz^2 \gg \iiint B\,dz^3 \gg \iiiint B\,dz^4 \cdots$, we may accept that in an ith-order approximation of the particle trajectory, only the elements of the first $(i+1)$ columns are of importance. Note that in this case, the ith-order aberration depends only on $B_0\rho_0$ and, thus is independent of the detailed distribution of the fringing field similarly as the energy gain for a charged particle between two points does not depend on the potential distribution between them. For a magnetic $2(n+1)$ pole all terms in the first n rows and n columns vanish, and in all other terms B_y is replaced by $\partial^n B_y/\partial x_x^n$, i.e., for a quadrupole ($n=1$) B_y is replaced by the flux density gradient $\partial B_y/\partial x$.

8.4.1 Overall Image Aberrations

An optical system that consists of two parts is described by $T_{31} = T_{32}T_{21}$ or, explicitly, to first order by

$$\begin{pmatrix} (x_3|x)(x_3|a) \\ (a_3|x)(a_2|a) \end{pmatrix} = \begin{pmatrix} (x_3|x_2)(x_3|a_2) \\ (a_3|x_2)(a_3|a_2) \end{pmatrix}\begin{pmatrix} (x_2|x)(x_2|x) \\ (a_2|x)(a_2|x) \end{pmatrix}, \qquad (8.36)$$

where the index 1 has been omitted as a shorthand notation. Normally, we require T_{21} and T_{32} to be known and determine T_{31} by simple matrix multiplication. Alternatively, we could also require T_{21} and T_{31} to be known, in which case we would find $T_{32} = T_{31}T_{21}^{-1}$ with

$$T_{32} = \begin{pmatrix} (x_3|x)(a_2|a) - (x_3|a)(a_2|x) & (x_3|a)(x_2|x) - (x_3|x)(x_2|a) \\ (a_3|x)(a_2|a) - (a_3|a)(a_2|x) & (a_3|a)(x_2|x) - (a_3|x)(x_2|a) \end{pmatrix}.$$

If at $z = z_2$, the particle trajectory under investigation experiences a small bend ε_2, we find the particle trajectory at $z = z_3$ [characterized by $\Delta X_3 = (\Delta x_3, \Delta a_3)$] by applying to T_{32} the position vector $\Delta X_2 = (0, \varepsilon_2)$:

$$\Delta x_3 = \varepsilon_2[(x_3|a)(x_2|x) - (x_3|x)(x_2|a)], \qquad (8.37a)$$

$$\Delta a_3 = \varepsilon_2[(a_3|a)(x_2|x) - (a_3|x)(x_2|a)]. \qquad (8.37b)$$

If the particle trajectory experiences such small bends ε_2 at many z, we find
from Eqs. (8.37a) and (8.37b) the overall Δx_3 and Δa_3:

$$\Delta x_3 = (x_3|a) \int \varepsilon_2(x_2|x) \, dz - (x_3|x) \int \varepsilon_2(x_2|a) \, dz, \tag{8.38a}$$

$$\Delta a_3 = (a_3|a) \int \varepsilon_2(x_2|x) \, dz - (a_3|x) \int \varepsilon_2(x_2|a) \, dz. \tag{8.38b}$$

At a focus point, i.e., at a position z_3 at which $(x_3|a)$ vanishes and $(x_3|x) = M$
characterizes the corresponding x magnification, Eq. (8.38a) becomes

$$-\frac{\Delta x_3}{M} = \int_0^{z_3} \varepsilon_2(x_2|a) \, dz_2. \tag{8.38c}$$

Similar relations have been derived by Brown (1970).

8.4.1.1 The Rigidity Dispersion

If the ε_2 is created by a short dipole field of length dz, ε_2 is found as
the difference between trajectories of particles of different rigidities $\chi = \chi_0(1 + \Delta)$. Since these particles move along radii $\rho_0(1 + \Delta)$, their angles of
deflection are $dz/[\rho_0(1 + \Delta)] \approx (dz/\rho_0)(1 - \Delta + \cdots)$, and ε_2 becomes
$-(dz/\rho_0)(\Delta + \cdots)$. At a focus point, i.e., at a point with $(x_3|a) = 0$, we thus
find from Eq. (8.38c),

$$-\frac{\Delta x_3}{M} = -\frac{(x_3|\Delta)\Delta}{(x_3|x)} = -\Delta \int_0^{z_3} (x_2|a) \frac{dz}{\rho_0}. \tag{8.39}$$

As an example of what Eq. (8.39) implies, we may look at Fig. 4.5 and find
$(x_3|\Delta)/(x_3|x)$ proportional to the banana-shaped area enclosed by the
trajectories characterized by $\pm a_{10}$. Note that for two such $180°$ deflectors
in series, i.e., for a $360°$ deflection, $(x_3|\Delta)$ vanishes, since the second banana-
shaped region has the opposite sign compared to the first.

8.4.1.2 Aberrations of Second Order

If ε_2 is created by a short multipole field of second order, i.e., by a
hexapole field of length dz and strength K_2, we find ε_2 from Eq. (9.28a)
for $n = 2$ as

$$\varepsilon_2(n = 2) = K_2[(x^2 - y^2)/2\rho_0], \tag{8.40}$$

with $\rho = \rho_0(1 + \Delta)$. Independently, the right-hand transfer matrix of Eq.
(8.36) yields

$$x_2 = (x_2|x)x_1 + (x_2|a)a_1 + (x_2|\Delta)\Delta, \tag{8.41a}$$

where the Δ term was omitted in Eq. (8.36) and is added here as, for instance, in Eq. (4.8a). Analogously, Eq. (4.8b) or Eq. (4.61b) yields

$$y_2 = (y_2|y)y_1 + (y_2|b)b_1. \tag{8.41b}$$

Introducing the x_2 and y_2 of Eqs. (8.41a) and (8.41b) into Eq. (8.40) and then ε_2 into Eq. (8.38c), we find for optical systems with a twofold symmetry, i.e., for sector fields,

$$\Delta x_3 = (x_3|xx)x_1^2 + 2(x_3|xa)x_1a_1 + (x_3|aa)a_1^2$$
$$+ (x_3|yy)y_1^2 + 2(x_3|yb)y_1b_1 + (x_3|bb)b_1^2$$
$$+ 2(x_3|x\Delta)x_1\Delta + 2(x_3|a\Delta)a\Delta + (x_3|\Delta\Delta)\Delta^2 + \cdots, \tag{8.42a}$$

$$-\frac{(x_3|xx)}{(x_3|x)} = \int (x_2|a)(x_2|x)^2 K_2 \frac{dz}{2\rho_0^2}, \tag{8.42b}$$

$$-\frac{(x_3|xa)}{(x_3|x)} = \int (x_2|a)^2(x_2|x) K_2 \frac{dz}{2\rho_0^2}, \tag{8.42c}$$

$$-\frac{(x_3|aa)}{(x_3|x)} = \int (x_2|a)^3 K_2 \frac{dz}{2\rho_0^2}, \tag{8.42d}$$

$$\frac{(x_3|yy)}{(x_3|x)} = \int (x_2|a)(y_2|y)^2 K_2 \frac{dz}{2\rho_0^2}, \tag{8.42e}$$

$$\frac{(x_3|yb)}{(x_3|x)} = \int (x_2|a)(y_2|y)(y_2|b) K_2 \frac{dz}{2\rho_0^2}, \tag{8.42f}$$

$$\frac{(x_3|bb)}{(x_3|x)} = \int (x_2|a)(y_2|b)^2 K_2 \frac{dz}{2\rho_0^2}, \tag{8.42g}$$

$$-\frac{(x_3|x\Delta)}{(x_3|x)} = \int (x_2|a)(x_2|x)(x_2|\Delta) K_2 \frac{dz}{2\rho_0^2} - \Delta, \tag{8.42h}$$

$$-\frac{(x_3|a\Delta)}{(x_3|x)} = \int (x_2|a)(x_2|a)(x_2|\Delta) K_2 \frac{dz}{2\rho_0^2} - \frac{(x_3|a)}{(x_3|x)}\Delta, \tag{8.42i}$$

$$-\frac{(x_3|\Delta\Delta)}{(x_3|x)} = \int (x_2|a)(x_2|\Delta)^2 K_2 \frac{dz}{2\rho_0^2} - \frac{(x_3|\Delta)}{(x_3|x)}\Delta. \tag{8.42j}$$

Replacing all x, a in Eqs. (8.36), (8.37), and (8.38c) by y, b, the same reasoning would hold, and we would find for optical systems with a twofold symmetry,

$$\Delta y_3 = (y_3|yx)y_1x_1 + (y_3|ya)y_1a_1 + (y_3|bx)b_1x_1 + (y_3|ba)b_1a_1$$
$$+ (y_3|y\Delta)y_1\Delta + (y_3|b\Delta)b_1\Delta, \tag{8.43a}$$

$$-\frac{(y_3|yx)}{(y_3|y)} = \int (y_2|b)(y_2|y)(x_2|x)K_2\frac{dz}{2\rho_0^2}, \tag{8.43b}$$

$$-\frac{(y_3|bx)}{(y_3|y)} = \int (y_2|b)^2(x_2|x)K_2\frac{dz}{2\rho_0^2}, \tag{8.43c}$$

$$-\frac{(y_3|ya)}{(y_3|y)} = \int (y_2|b)(y_2|y)(x_2|a)K_2\frac{dz}{2\rho_0^2}, \tag{8.43d}$$

$$-\frac{(y_3|ba)}{(y_3|y)} = \int (y_2|b)^2(x_2|a)K_2\frac{dz}{2\rho_0^2}, \tag{8.43e}$$

$$-\frac{(y_3|y\Delta)}{(y_3|y)} = \int (y_2|b)(y_2|y)(x_2|\Delta)K_2\frac{dz}{2\rho_0^2} - \Delta, \tag{8.43f}$$

$$-\frac{(y_3|b\Delta)}{(y_3|y)} = \int (y_2|b)^2(x_2|\Delta)K_2\frac{dz}{2\rho_0^2} - \frac{(y_3|y)}{(y_2|y)}\Delta. \tag{8.43g}$$

The additional terms in Eqs. (8.42h)–(8.42j) and (8.43f)–(8.43g) are not caused by the assumed hexapole but are due to the fact that the transformation of Eq. (8.36), if written for particles of rigidity $\chi = \chi_0(1 + \Delta)$, reads [see also Eqs. (2.22a) and (2.22b)],

$$\begin{pmatrix} x_3 \\ a_3/(1+\Delta) \end{pmatrix} = \begin{pmatrix} (x_3|x) & (x_3|a) \\ (a_3|x) & (a_3|a) \end{pmatrix}\begin{pmatrix} x_1 \\ a_1/(1+\Delta) \end{pmatrix}, \tag{8.44a}$$

or

$$\begin{pmatrix} x_3 \\ a_3 \end{pmatrix} = \begin{pmatrix} (x_3|x)_\Delta(x_3|a)_\Delta \\ (a_3|x)_\Delta(a_3|a)_\Delta \end{pmatrix}\begin{pmatrix} x_1 \\ a_1 \end{pmatrix} = \begin{pmatrix} \dfrac{(x_3|x)}{1+\Delta} & (x_3|a)(1+\Delta) \\ \dfrac{(a_3|x)}{1+\Delta} & (a_3|a)(1+\Delta) \end{pmatrix}\begin{pmatrix} x_1 \\ a_1 \end{pmatrix}. \tag{8.44b}$$

Given a first-order design of the optical system under consideration, we find all $(x_2|\cdots)$, $(y_2|\cdots)$ as functions of z by drawing into the designed system trajectories with initial position vectors:

$$\mathbf{X}_{x1} = (1, 0, 0) \rightarrow (x_2|x), \tag{8.45a}$$

$$\mathbf{X}_{a1} = (0, 1, 0) \rightarrow (x_2|a), \tag{8.45b}$$

$$\mathbf{X}_{\Delta1} = (0, 0, 1) \rightarrow (x_2|\Delta), \tag{8.45c}$$

$$\mathbf{Y}_{y1} = (1, 0) \quad\;\; \rightarrow (y_2|y), \tag{8.45d}$$

$$\mathbf{Y}_{b1} = (0, 1) \quad\;\; \rightarrow (y_2|x). \tag{8.45e}$$

By simple inspection of the geometry of a given optical design, the image aberrations of the system are known from Eqs. (8.42) and (8.43). From the same inspection, we also find at which positions z it is useful to add a K_2, i.e., a multipole element of second order—a hexapole. The driving terms, namely, the other coefficients under the different integrals of Eqs. (8.42) and (8.43), should be reasonably large, so as to correct a certain aberration, and others reasonably small, so as not to aggravate other corresponding aberrations.

8.4.2 The Aberration Driving Terms of all Orders

In the case of third-, fourth-, or higher-order multipoles, we find similar to Eq. (8.40),

$$\varepsilon_2(n = 3) = K_3[(x^3 - 3xy^2)/3!\rho_0^3],$$

$$\varepsilon_2(n = 4) = K_4[(x^4 - 6x^2y^2 + y^4)/4!\rho_0^4] \tag{8.46}$$

[see also Eq. (9.28a)]. Introducing into these relations the x_2, y_2 of Eqs. (8.41a) and (8.41b), we find similar relations to those in Eqs. (8.42) and (8.43). These relations can be combined (see also Brown, 1970) to read for a nonvanishing nth-order aberration, with $n = i + j + k + l + m$ and $r \in \{x, y\}$,

$$\frac{(r_3|x^i a^j y^k b^l \Delta^m)}{(r_3|r)} = \frac{n!(-1)^{(k+l)}}{i!j!k!l!m!}$$

$$\times \int_0^{z_3} [(x_2|x)^i (x_2|a)^{j+1} (y_2|y)^k (y_2|b)^l (x_2|\Delta)^m K_n] \, dz$$

$$+ \cdots K_{n+1} + \cdots K_{n-2} + \cdots. \tag{8.47}$$

For an nth-order image aberration, terms with K_{n-1}, K_{n-2}, ..., K_3, K_2 are listed which are caused by the fact that the transfer matrices of Eq. (8.36) must contain all $n - 1$, $n - 2$, ..., order effects and not only those of first order, as was sufficient in the case $n = 2$ of Section 8.4.1.2. Thus, Eq. (8.47) is not directly suited to calculate image aberrations of nth order. However, assuming that an optical system is fixed to $(n - 1)$th order, the driving terms, i.e., the coefficients of K_n, show where the nth-order multipole should be placed in order to correct a certain aberration of nth order and not to enlarge the others too much.

In optical systems with a straight optic axis, K_2 terms appear only if an explicit hexapole element has been added. As long as there is no hexapole element, Eq. (8.47) thus yields the correct values for all image aberrations of third order. However, the chromatic aberrations must be calculated with the matrix elements of Eq. (8.44b).

8.4.3 Image Aberrations due to Fringing Fields

The action of the fringing fields to zeroth and first order was derived in Chapter 7. We shall now look for the higher-order effects, which were derived in second- and third-order theories by Wollnik (1964), Wollnik and Ewald (1965), Enge (1967), Matsuda and Wollnik (1970), and Matsuda (1971).

8.4.3.1 Fringing Field Effects in Magnetic Sector Fields

The general equations of motion in the fringing fields of dipole magnets are given in Eqs. (7.1) and (7.2). A detailed description of the distribution of the magnetic flux density would be found to second order,

$$B_\xi(\xi, \eta, \zeta) = B_\xi\left(0, \eta, \zeta - \frac{\xi^2}{2R}\right)\left(\frac{-\xi}{\xi^2 + (R - \zeta)^2}\right)$$

$$= -\frac{\xi \eta}{R}\left(\frac{\partial B_\xi}{\partial \eta}\right) + \cdots, \tag{8.48a}$$

$$B_\eta(\xi, \eta, \zeta) = B_\eta\left(0, \eta, \zeta - \frac{\xi^2}{2R}\right)$$

$$= B_\eta(0, 0, \zeta) - \frac{\xi^2}{2R}\left(\frac{\partial B_\eta}{\partial \zeta}\right) + \frac{\eta^2}{2}\left(\frac{\partial^2 B_\eta}{\partial \eta^2}\right)$$

$$- \frac{\eta^2 \xi^2}{4R}\left(\frac{\partial^3 B_\eta}{\partial \zeta \, \partial \eta^2}\right) + \cdots, \tag{8.48b}$$

$$B_\zeta(\xi, \eta, \zeta) = B_\zeta\left(0, \eta, \zeta - \frac{\xi^2}{2R}\right)\left(\frac{R - \zeta}{\xi^2 + (R - \zeta)^2}\right)$$

$$= \eta\left(\frac{\partial B_\xi}{\partial \eta}\right) - \frac{\eta \xi^2}{2R}\left(\frac{2B_\zeta}{\partial \zeta \, \partial \eta}\right) + \cdots. \tag{8.48c}$$

Here, all derivatives are to be taken at $\xi = \eta = 0$ and at an arbitrary ζ. Furthermore, here R is the radius of curvature of the effective field boundary. In the case of a magnet with good fringing field shunts, we find (Fig. 8.6)

$$1/R \approx 1/(R_{\text{mech}} + \zeta^*), \tag{8.49a}$$

where R_{mech} is the radius of curvature of the pole-shoe boundary and ζ^* the distance between the pole shoe and the effective field boundary; however, for a straight pole-shoe boundary the effective field boundary is slightly curved for large $\pm\xi$. For unshunted magnets, this effect is almost never negligible. Enge (1975) describes this radius of curvature as

$$1/R = 1/R_{\text{mech}} + 0.7G_0/w^2 \tag{8.49b}$$

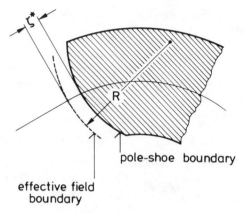

effective field
boundary

Fig. 8.6. A pole-shoe boundary curved by a radius R_{mech} and the corresponding effective field boundary curved by a radius $R = R_{mech} + \zeta^*$. For optical calculations, we postulate the effective field boundary to be curved by a radius R and design and curvature $1/R_{mech}$ of the pole-shoe boundary later.

with $2w$ the pole-shoe width and $2G_0$ the magnet air gap. Unfortunately, the factor 0.7 depends on the specific magnet design and thus can vary considerably (see Fig. 8.7).

Using the relations curl $\mathbf{B}$ = div $\mathbf{B}$ = 0, we find generally,

$$\partial B_\zeta/\partial \eta = \partial B_\eta/\partial \zeta, \qquad (\partial^2 B_\eta/\partial \xi^2) + (\partial^2 B_\eta/\partial \eta^2) + (\partial^2 B_\eta/\partial \zeta^2) = 0,$$

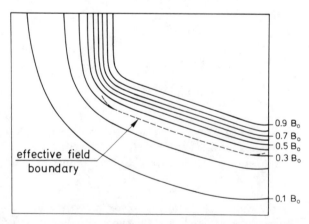

Fig. 8.7. Measured field lines of equal flux density for a magnetic dipole without fringing field shunts. In the investigated magnet, the magnet air gap was $2G_0 = 8$ mm and the pole-shoe width $2w = 60$ mm. Note the slightly curved effective field boundary shown as a dashed line, which more or less coincides with the 30% contour line, a typical situation. Note also that the measured curvature of the effective field boundary only roughly agrees with the result of Eq. (8.49b).

and from geometry, $\partial B_\eta / \partial \xi = [-\xi/(R - \eta)](\partial B_\eta / \partial \zeta)$, or

$$\frac{\partial^2 B_\eta}{\partial \xi^2} + \frac{\partial B_\eta}{\partial \zeta} \frac{1}{R - \zeta} = 0.$$

Thus, Eqs. (8.48a)–(8.48c) become

$$B_\xi(\xi, \eta, \zeta) = -\frac{\xi \eta}{R}\left(\frac{\partial^2 B_\eta}{\partial \zeta^2}\right) + \cdots, \tag{8.50a}$$

$$B_\eta(\xi, \eta, \zeta) = B_\eta(0, 0, \zeta) + \frac{\eta^2 - \xi^2}{2R}\left(\frac{\partial B_\eta}{\partial \zeta}\right) - \frac{\eta^2}{2}\left(\frac{\partial^2 B_\eta}{\partial \zeta^2}\right)$$

$$+ \frac{\eta^2 \xi^2}{4R}\left(\frac{\partial^3 B_\eta}{\partial \zeta^3}\right), \tag{8.50b}$$

$$B_\zeta(\xi, \eta, \zeta) = \eta\left(\frac{\partial B_\eta}{\partial \zeta}\right) - \frac{\eta \xi^2}{2R}\left(\frac{\partial^2 B_\eta}{\partial \zeta^2}\right). \tag{8.50c}$$

Introducing Eqs. (8.50a), (8.50b), and (8.50c) into Eqs. (7.2a) and (7.2b), we find the detailed particle trajectories in the real or the effective fringing field of magnetic dipoles. Postulating that the particle trajectories must coincide in the main field whether the particles have passed through the real or the effective fringing field, all trajectories must experience shifts $\Delta \xi$, $\Delta \eta$ and bends $\Delta \varepsilon$, $\Delta \beta$ when passing through the field boundary.

By subtracting the two solutions obtained Wollnik (1964) and Wollnik and Ewald (1965) found with $\xi_1 = x_1/\cos \varepsilon_a$, $\eta_1 = y_1$,

$$\rho_0 \Delta \varepsilon = \frac{x_1^2 - y_1^2}{2R \cos^3 \varepsilon_a} + \frac{y_1^2 \tan^3 \varepsilon_a}{2\rho_0} - y_1 \beta_1 \tan^2 \varepsilon_a$$

$$- \frac{\sin^2 \varepsilon_a}{R \cos^3 \varepsilon_a} \frac{G_0^2}{\rho_0^2} I_2 + \cdots, \tag{8.51a}$$

$$\Delta \xi = \frac{1}{\cos^3 \varepsilon_a}\left(I_2 \frac{G_0^2}{\rho_0} + \frac{y_1^2}{2}\right) + \cdots, \tag{8.51b}$$

$$\rho_0 \Delta \beta = -y_1 \tan \varepsilon_a + \frac{y_1(\delta_K + \delta_m)}{2} \tan \varepsilon_a - \frac{y_1 x_1}{R \cos^3 \varepsilon_a}$$

$$- \frac{y_1}{\cos \varepsilon_a}(1 + 2\tan^2 \varepsilon_a)I_3 \frac{G_0}{\rho_0} + \cdots, \tag{8.51c}$$

$$\Delta \eta = 0 + \cdots. \tag{8.51d}$$

The fringing field integrals I_2, I_3 necessary here were given earlier in Eqs. (7.5) and (7.10).

The expressions of Eqs. (8.51a)–(8.51d) can be divided into three groups:

(1)   Terms that deflect or shift all particle trajectories of a beam equally, as do the terms with I_2 in Eqs. (8.51a) and (8.51b).

(2)   Terms, that change the focusing power, as do the first and the last terms in Eq. (8.51c). Note here that $I_3 G_0$ is always small compared to ρ_0.

(3)   Terms that modify the aberrations of second order are all the other terms of Eqs. (8.51a)–(8.51d).

Terms of the first group normally must not be taken into account during the design of an optical system. However, they must be taken into account during the design of the optical elements and during their exact positioning. Small errors in the terms of the first group, for instance, due to only an approximate knowledge of the fringing field distribution, can normally—at the very end—be counterbalanced by slightly increased deflecting field strengths and slightly shifted optical elements.

The terms of the second group are widely independent of the detailed fringing field distribution. Deviations in the initially assumed and finally existing distributions thus have only small effects. Furthermore, these effects normally can be counterbalanced by a slightly increased focusing strength of one of the optical elements or by a small shift of the position of the final particle detector.

The terms of the third group are independent of the detailed fringing field distribution, similar to the energy gain of a charged particle that does not depend on the potential distribution between the initial and the final position. Thus, if the finally existing fringing field distribution deviates a little from the one that originally was assumed, no drastic deteriorations of the optical performance of the optical system must be expected.

8.4.3.2 Fringing Field Effects in Electrostatic Sector Fields

The systematics of fringing field effects of magnetic sector fields persists also for electrostatic sector fields. The main difference is the fact that there are fringing field focusing effects in the y direction for a magnetic deflector and in the x direction for an electrostatic deflector. However, matters are somewhat simplified, since the particle beam is always assumed to enter an electrostatic sector field perpendicular, i.e., ε_a is always zero.

APPENDIX.   COEFFICIENTS OF IMAGE ABERRATIONS OF nth ORDER

To derive the coefficients of image aberrations, we must perform perturbation theories (see for instance, Berz, 1986) with the first-order solutions treated in Chapters 3 and 4. Particle trajectories are described in detail by the canonical equations [Eq. (8.17a)],

$$\partial H/\partial p_x = \dot{x}, \qquad p_x = -\partial H/\partial x = F_x, \qquad (8.52a)$$

$$\partial H/\partial p_y = \dot{y}, \qquad p_y = -\partial H/\partial y = F_y, \qquad (8.52b)$$

$$\partial H/\partial p_z = \dot{z}, \qquad p_z = -\partial H/\partial z = F_z, \qquad (8.52c)$$

with F_x, F_y, F_z denoting the x, y, z forces on the particle under consideration. With $\mathbf{F} = (F_x, F_y, F_z)$, as well as $\mathbf{p}_0 = (p_{x0}, p_{y0}, p_{z0})$ and $\mathbf{p}_e = (p_{xe}, p_{ye}, p_{ze})$ at times t_0 and t_e, we can transform Eqs. (8.52a)–(8.52c) to

$$\mathbf{p}_e = \mathbf{p}_0 + \int_{t_0}^{t_e} F(t)\, dt = \mathbf{p}_0 + \int_{s_0}^{s_e} \mathbf{F}(s)\, \frac{dt}{ds}\, ds. \qquad (8.53)$$

We now assume that the coordinate system moves with a reference particle of velocity v_0. If this reference particle moves on a radius of curvature ρ_0, the final coordinate system is rotated with respect to the initial one by the angle

$$\phi = v_0 t/\rho_0.$$

We may now describe an arbitrary trajectory at the position s_e relative to the trajectory of the reference particle, i.e., in a Cartesian coordinate system rotated by the angle ϕ. From Eq. (8.53) we find the particle momentum $\bar{p}_e = (\bar{p}_x, \bar{p}_y, \bar{p}_z)$, at s_e as

$$\bar{p}_x = \left[p_{x0} + \int F_x \frac{dt}{d(\rho_0\phi)}\, d(\rho_0\phi) \right] \cos\phi + \left[p_z + \int F_z \frac{dt}{d(\rho_0\phi)}\, d(\rho_0\phi) \right] \sin\phi,$$

$$\bar{p}_y = p_y + \int F_y \frac{dt}{d(\rho_0\phi)}\, d(\rho_0\phi),$$

$$\bar{p}_z = \left[p_z + \int F_z \frac{dt}{d(\rho_0\phi)}\, d(\rho_0\phi) \right] \cos\phi + \left[p_x + \int F_x \frac{dt}{d(\rho_0\phi)}\, d(\rho_0\phi) \right] \sin\phi.$$

Differentiating these relations with respect to $\rho_0\phi$ and evaluating them at $\rho_0\phi = 0$ yields

$$\frac{d\bar{p}_x}{d(\rho_0\phi)} = F_x \frac{dt}{d(\rho_0\phi)} + \frac{p_z}{\rho_0} = (E_x - v_z B_y)\frac{(ze)\, dt}{d(\rho_0\phi)} + \frac{p_z}{\rho_0}, \qquad (8.54a)$$

$$\frac{d\bar{p}_y}{d(\rho_0\phi)} = F_y \frac{dt}{d(\rho_0\phi)} = (E_y + v_z B_x)\frac{(ze)\, dt}{d(\rho_0\phi)}, \qquad (8.54b)$$

$$\frac{d\bar{p}_z}{d(\rho_0\phi)} = F_z \frac{dt}{d(\rho_0\phi)} - \frac{p_x}{\rho_0} = (E_z + v_x B_y - v_y B_x)\frac{(ze)\, dt}{d(\rho_0\phi)} - \frac{p_x}{\rho_0}. \qquad (8.54c)$$

For the reference particle moving along the optic axis, i.e., for $p_x = p_y = 0$ and $p_z = p_0$, we find here, with $dt/d(\rho_0\phi) = 1/v_0$,

$$F_x = -p_0 v_0/\rho_0.$$

For the magnetic or the electrostatic case, F_x equals $-(z_0 e)v_0 B_{y0}$, or $-(z_0 e)E_{x0}$, where B_{y0} and E_{x0} denote the magnetic flux density and the electrostatic field strength at the optic axis. Thus, the relation $F_x = -(p_0 v_0)/\rho_0$ becomes

$$B_{y0}\rho_0 = p_0, \qquad -E_{x0}\rho_0 = p_0 v_0,$$

which suggests the definition of the magnetic and electrostatic rigidities as given in Eqs. (2.10) and (2.13):

$$\chi_B = B_{y0}\rho_0/(z_0 e) = p_0/(z_0 e), \qquad \chi_E = -E_{x0}\rho_0/(z_0 e) = p_0 v_0/(z_0 e).$$

8.A1 Equations of Motion in Time-Independent Fields

Using the definitions $a = p_x/p_0 = v_x/v_0$, $b = p_y/p_0 = v_y/v_0$ of Eqs. (2.7a) and (2.7b) for $m_0 = m_{00}$, we find from Eqs. (8.54a)–(8.54c), with $p^2 = p_x^2 + p_y^2 + p_z^2$ and $v^2 = v_x^2 + v_y^2 + v_z^2$ as taken from Fig. 2.6,

$$\frac{da}{d(\rho_0\phi)} = \frac{F_x\, dt}{p_0\, d(\rho_0\phi)} + \frac{p_z}{\rho_0 p_0} = \left[\frac{E_x}{\chi_{E0}} - \left(\frac{v_z}{v_0}\right)\frac{B_y}{\chi_{B0}}\right]v_0\frac{dt}{d(\rho_0\phi)} + \left(\frac{p_z}{p_0}\right)\frac{1}{\rho_0}$$

$$= \left[\frac{E_x}{\chi_{E0}} - \frac{B_y}{\chi_{B0}}\sqrt{\left(\frac{v}{v_0}\right)^2 - a^2 - b^2}\right]\frac{(ze)}{v}$$

$$\times \sqrt{\left[1+\frac{x}{\rho_0}\right]^2 + \left[\frac{dx}{d(\rho_0\phi)}\right]^2 + \left[\frac{dy}{d(\rho_0\phi)}\right]^2}$$

$$+ \frac{1}{\rho_0}\sqrt{\left(\frac{p}{p_0}\right)^2 - a^2 - b^2}, \tag{8.55a}$$

$$\frac{db}{d(\rho_0\phi)} = \frac{F_y\, dt}{p_0\, dz} = \left[\frac{E_y}{\chi_{E0}} - \left(\frac{v_z}{v_0}\right)\frac{B_x}{\chi_{B0}}\right]v_0\frac{dt}{d(\rho_0\phi)},$$

$$= \left[\frac{E_y}{\chi_{E0}} - \frac{B_x}{\chi_{B0}}\sqrt{\left(\frac{v}{v_0}\right)^2 - a^2 - b^2}\right]\frac{(ze)}{v}$$

$$\times \sqrt{\left[1+\frac{x}{\rho_0}\right]^2 + \left[\frac{dx}{d(\rho_0\phi)}\right]^2 + \left[\frac{dy}{d(\rho_0\phi)}\right]^2}. \tag{8.55b}$$

In the curvilinear coordinates of Fig. 2.6, we find further,

$$\frac{dx}{d(\rho_0\phi)} = \left(1+\frac{x}{\rho_0}\right)\frac{p_x}{p_z} = \left(1+\frac{x}{\rho_0}\right)\frac{a}{\sqrt{(p/p_0)^2 - a^2 - b^2}}, \tag{8.55c}$$

$$\frac{dy}{d(\rho_0\phi)} = \left(1+\frac{x}{\rho_0}\right)\frac{p_y}{p_z} = \left(1+\frac{x}{\rho_0}\right)\frac{b}{\sqrt{(p/p_0)^2 - a^2 - b^2}}. \tag{8.55d}$$

8.A2 Explicit Particle Trajectories in Radially Inhomogeneous Electromagnetic Sector Fields

Integrating the differential equations [Eq. (8.55)] in ϕ-independent fields would yield the explicit particle trajectories. However, due to the complex structure of these relations, a direct integration in general is impossible. Thus, we shall expand Eqs. (8.55a)–(8.55d) in a power series in the small quantities x, a, y, b, as well as $\delta_K = (K/z - K_0/z_0)/(K_0/z_0)$ and $\delta_m = (m/z - m_0/z_0)/(m_0/z_0)$ taken from Eqs. (2.15) and (2.16). Here, v and $p = \chi_B/(ze)$ are expressed as functions of δ_K, $\hat{\delta}_K$, and δ_m in Eqs. (2.18a) and (2.18b) [see also Eq. (4.52b) for the magnitude of $\hat{\delta}_K$]. Also, the distribution of the magnetic flux density and the electrostatic field must be expanded in a power series in x and y. Assuming general field strength distributions in the x, z plane as in Eq. (8.35),

$$B_y(x, y = 0) = B_0 \sum_{n=1}^{N} K_{nB} \frac{(x/\rho_0)^n}{n!}, \qquad E_x(x, y = 0) = E_0 \sum_{n=1}^{N} K_{nE} \frac{(x/\rho_0)^n}{n!},$$

we find B_x as well as E_y from B_y and E_x. For this purpose, we need the relations curl $B = \text{div } B = 0$ and curl $E = \text{div } E = 0$ for $B_z = E_z = 0$, i.e., for ϕ independent fields:

$$\partial B_x/\partial x = -\partial B_y/\partial y, \qquad \partial B_y/\partial x = \partial B_x/\partial y,$$

$$\partial E_x/\partial x = -\partial E_y/\partial y, \qquad \partial E_y/\partial x = \partial E_x/\partial y.$$

The potential V_E, also necessary, is found from grad $V_E = E$.

Berz (1986) expanded Eqs. (8.55a)–(8.55d) in a power series in x, a, y, b, δ_K, δ_m,

$$[dx/d(\rho_0\phi)] - a = f_x^{(2)} + f_x^{(3)} + f_x^{(4)} + \cdots, \tag{8.56a}$$

$$[da/d(\rho_0\phi)] + xk_x^2 = \rho_0(\delta_K N_K + \delta_m N_m)$$
$$+ f_a^{(2)} + f_a^{(3)} + f_a^{(4)} + \cdots, \tag{8.56b}$$

$$[dy/d(\rho_0\phi)] - b = f_y^{(2)} + f_y^{(3)} + f_y^{(4)} + \cdots, \tag{8.56c}$$

$$[db/d(\rho_0\phi)] + yk_y^2 = f_b^{(2)} + f_b^{(3)} + f_b^{(4)} + \cdots. \tag{8.56d}$$

Here, the $f^{(i)}$ are polynomials in the phase-space coordinates x, a, y, b, δ_K, δ_m, where each polynomial consists of monomials of order i, and we have used again the abbreviations

$$\rho_0 k_x^2 = 1 + h(1 + 2\eta_0)^2 - \rho_0 k_y^2 = n_{B1}(1 - h) + hn_{E1}$$

$$N_K = \frac{(1 + 2\eta_0)^2 + h}{2(1 + \eta_0)(1 + 2\eta_0)}, \qquad N_m = \frac{1 + 2\eta_0 - h}{2(1 + \eta_0)(1 + 2\eta_0)}.$$

With $h = 0$ for magnetic and $h = 1$ for electrostatic fields. The final particle trajectories are described as:

$$x = x^{(1)} + x^{(2)} + x^{(3)} + x^{(4)} + \cdots ,$$

$$a = a^{(1)} + a^{(2)} + a^{(3)} + a^{(4)} + \cdots , \tag{8.57a}$$

$$y = y^{(1)} + y^{(2)} + y^{(3)} + y^{(4)} + \cdots ,$$

$$b = b^{(1)} + b^{(2)} + b^{(3)} + b^{(4)} + \cdots . \tag{8.57b}$$

where the $x^{(i)}$, $a^{(i)}$, $y^{(i)}$, $b^{(i)}$ are also polynomials in the phase-space coordinates x, a, y, b, δ_K, δ_m, and each of the polynomials consists of monomials of order i.

Introducing the x, a, y, b of Eqs. (8.57a) and (8.57b) into Eqs. (8.56), we find that all $f^{(i)}$ depend only on δ_K and δ_m as well as on $x^{(\nu)}$, $a^{(\nu)}$, $y^{(\nu)}$, $b^{(\nu)}$, with $\nu = 1, 2, \ldots, i - 1$. Comparing coefficients of equal power thus yields:

$$dx^{(i)}/d(\rho_0\phi) = a^{(i)} + f_x^{(i)}, \qquad da^{(i)}/d(\rho_0\phi) = -k_x^2 x^{(i)} + f_a^{(i)}, \tag{8.58a}$$

$$dy^{(i)}/d(\rho_0\phi) = b^{(i)} + f_y^{(i)}, \qquad db^{(i)}/d(\rho_0\phi) = -k_y^2 y^{(i)} + f_b^{(i)}. \tag{8.58b}$$

The solutions of these differential equations are found by solving first the homogeneous equations $[f_x^{(i)} = f_a^{(i)} = f_y^{(i)} = f_b^{(i)} = 0]$ with the result

$$x^{(i)} = A_x^{(i)} c_x + B_x^{(i)} s_x, \qquad a^{(i)} = B_x^{(i)} c_x - k_x^2 A_x^{(i)} s_x, \tag{8.59a}$$

$$y^{(i)} = A_y^{(i)} c_y + B_y^{(i)} s_y, \qquad b^{(i)} = B_y^{(i)} c_y - k_y^2 A_y^{(i)} s_y, \tag{8.59b}$$

where

$$c_x = \cos(k_x\rho_0\phi), \qquad s_x = k_x^{-1} \sin(k_x\rho_0\phi),$$

$$c_y = \cos(k_y\rho_0\phi), \qquad s_y = k_y^{-1} \sin(k_y\rho_0\phi).$$

Note here that as outlined in Sections 3.1.1 and 4.3.2, the sin and cos transform into sinh and cosh for $k_x^2 < 0$ or $k_y^2 < 0$.

To find a solution for the inhomogeneous equations $[f_x^{(i)} \neq 0, f_a^{(i)} \neq 0, f_y^{(i)} \neq 0, f_b^{(i)} \neq 0]$ we must assume that the A_x, B_x, A_y, B_y are functions of $(\rho_0\phi)$. These coefficients are obtained by differentiating Eqs. (8.59a) and (8.59b) with respect to $(\rho_0\phi)$:

$$\frac{dx^{(i)}}{d(\rho_0\phi)} = \frac{dA_x^{(i)}}{d(\rho_0\phi)} c_x + \frac{dB_x^{(i)}}{d(\rho_0\phi)} s_x + a^{(i)}, \tag{8.60a}$$

$$\frac{da^{(i)}}{d(\rho_0\phi)} = \frac{dB_x^{(i)}}{d(\rho_0\phi)}c_x + k_x^2\frac{dA_x^{(i)}}{d(\rho_0\phi)}s_x - k_x^2 x^{(i)}, \tag{8.60b}$$

$$\frac{dy^{(i)}}{d(\rho_0\phi)} = \frac{dA_y^{(i)}}{d(\rho_0\phi)}c_y + \frac{dB_y^{(i)}}{d(\rho_0\phi)}s_y + a^{(i)}, \tag{8.60c}$$

$$\frac{db^{(i)}}{d(\rho_0\phi)} = \frac{dB_y^{(i)}}{d(\rho_0\phi)}c_y + k_y^2\frac{dA_y^{(i)}}{d(\rho_0\phi)}s_y - k_y^2 y^{(i)}. \tag{8.60d}$$

Comparing Eqs. (8.60a)–(8.60d) with Eqs. (8.58a) and (8.58b) yields

$$\frac{dA_x^{(i)}}{d(\rho_0\phi)}c_x + \frac{dB_x^{(i)}}{d(\rho_0\phi)}s_x = f_x^{(i)}, \qquad \frac{dA_y^{(i)}}{d(\rho_0\phi)}c_y + \frac{dB_y^{(i)}}{d(\rho_0\phi)}s_y = f_y^{(i)},$$

$$\frac{dB_x^{(i)}}{d(\rho_0\phi)}c_x + \frac{dA_x^{(i)}}{d(\rho_0\phi)}s_x k_x^2 = f_a^{(i)}, \qquad \frac{dB_y^{(i)}}{d(\rho_0\phi)}c_y + \frac{dA_y^{(i)}}{d(\rho_0\phi)}s_y k_y^2 = f_b^{(i)}.$$

Solving these linear equations and integrating over $\rho_0\phi$, we find

$$A_x^{(i)} = \int_0^{\rho_0\phi} [c_x f_x^{(i)} - s_x f_a^{(i)}] \, d(\rho_0\phi),$$

$$B_x^{(i)} = \int_0^{\rho_0\phi} [-k_x^2 s_x f_x^{(i)} + c_x f_a^{(i)}] \, d(\rho_0\phi), \tag{8.61a}$$

$$A_y^{(i)} = \int_0^{\rho_0\phi} [c_y f_y^{(i)} - s_y f_b^{(i)}] \, d(\rho_0\phi),$$

$$B_y^{(i)} = \int_0^{\rho_0\phi} [-k_y^2 s_y f_y^{(i)} + c_y f_b^{(i)}] \, d(\rho_0\phi), \tag{8.61b}$$

which must be introduced into Eqs. (8.59a) and (8.59b) to obtain the $x^{(i)}$, $a^{(i)}$, $y^{(i)}$, and $b^{(i)}$ of any order i. The $f_x^{(i)}$, $f_a^{(i)}$, $f_y^{(i)}$, $f_b^{(i)}$ in Eq. (8.61) are known from Eqs. (8.56a)–(8.56d) and contain power series expansions in x, a, y, b, δ_K, δ_m of field distributions B_x, B_y, E_x, E_y, and V_E as well as of the square root expressions in Eqs. (8.55a)–(8.55b). To obtain the $f_x^{(i)}$, $f_a^{(i)}$, $f_y^{(i)}$, $f_b^{(i)}$ sufficiently accurate, it is necessary to introduce for all x, a, y, b, δ_K, δ_m the solution of νth order determined earlier, i.e., $x^{(\nu)}$, $a^{(\nu)}$, $y^{(\nu)}$, $b^{(\nu)}$, $\delta_K^{(\nu)}$, $\delta_m^{(\nu)}$, with $\nu = 1, 2, \ldots, i-1$.

REFERENCES

Berz, M. (1986). Thesis, Univ. Giessen, West Germany, *unpublished.*

Berz, M., Hoffmann, H. C., and Wollnik, H. (1986). *Nucl. Instrum. Meth.*, in print.

Boerboom, A. J. H., Tasman, H. A., and Wachsmuth, A. (1959). *Z. Naturforsch.* **14a**, 121, 816, 818.

Brown, K. L. (1970). *Proc. Int. Conf. Magnet. Technol.*, DESY Hamburg, W. Germany.

Brown, K. L., Belbeoch, R., and Bounin, P. (1964). *Rev. Sci. Instrum.* **35**, 481.

Dragt, A. J. (1982). "Lectures on nonlinear orbit dynamics," *AIP Conf. Proc.* **87**.

Enge, H. (1967). *In* "Focusing of Charged Particles," A. Septier (ed.) p. 203. Academic, New York.

Enge, H. (1975). ICI-3038-2/75 Industrial Coils Inc., Boston, Massachusetts.

Ewald, H., and Liebl, H. (1955). *Z. Naturforsch.* **10a**, 892.

Ewald, H., and Liebl, H. (1957). *Z. Naturforsch.* **12a**, 28.

Glaser, W. (1956). "Handbuch der Physik," vol. XXXIII, Springer-Verlag, Berlin.

Goldstein, H. (1980). "Classical Mechanics." Addison-Wesley, Reading, Massachusetts.

Halbach, K. (1969). *Nucl. Instrum. Meth.* **74**, 147.

Hintenberger, H., and Koenig, L. A. (1957). *Z. Naturforsch.* **12a**, 773.

Lee-Whiting, G. E. (1970). *Nucl. Instrum. Methods* **83**, 232.

Lee-Whiting, G. E. (1972). *Nucl. Instrum. Methods* **99**, 609.

Ludwig, R. (1967). *Z. Naturforsch.* **22a**, 553.

Matsuda, H. (1971). *Nucl. Instrum. Methods* **91**, 127.

Matsuda, H., and Wollnik, H. (1970). *Nucl. Instrum. Methods* **77**, 40, 283.

Matsuda, H., and Wollnik, H. (1972). *Nucl. Instrum. Methods* **103**, 117.

Matsuo, T., and Matsuda, H. (1971). *Int. J. Mass Spectr. Ion Phys.* **6**, 361.

Matsuo, T., Matsuda, H., and Wollnik, H. (1972). *Nucl. Instrum. Methods* **103**, 515.

Plies, E., and Rose, H. (1971). *Optik* **34**, 71.

Smith, D. L. (1970). *Nucl. Instrum. Methods* **79**, 144.

Stromberg, K. (1981). "An Introduction to Classical Real Analysis." Wadsworth, New York.

Takeshita, T. (1966). *Z. Naturforsch.* **21a**, 9.

Thirring, W. (1977). "Lehrbuch der mathematischen Physik," Vol. 1. Springer-Verlag, Berlin.

Wollnik, H. (1964). Thesis, Techn. Hochschule Muenchen, West Germany, *unpublished*.

Wollnik, H. (1965). *Nucl. Instrum. Methods* **34**, 213.

Wollnik, H. (1967a). *Nucl. Instrum. Methods* **52**, 250.

Wollnik, H. (1967b). *In* "Focusing of Charged Particles," A. Septier (ed.) p. 163. Academic, New York.

Wollnik, H., and Berz, M. (1985). *Nucl. Instrum. Methods* **238**, 127.

Wollnik, H., and Ewald, H. (1965). *Nucl. Instrum. Methods* **36**, 93.

Wollnik, H., and Matsuda, H. (1981). *Nucl. Instrum. Meth.* **189**, 361.

Wollnik, H., and Matsuo, T. (1981). *Int. J. Mass. Spectr. Ion Phys.* **37**, 209.

Wollnik, H., Matsuo, T., and Matsuda, H. (1972). *Nucl. Instrum. Methods* **102**, 13.

9

Design of Particle Spectrometers and Beam Guide Lines

Using the ideas presented in the first seven chapters of this book, we can calculate the first-order properties of particle spectrometers and of beam lines. The corresponding image aberrations can be determined from the relations outlined in Chapter 8. These eight chapters thus provide the tools to calculate the properties of a given optical system and to judge its performance. In this chapter, we shall discuss how such a system shall be designed and optimized.

At the beginning of the design of an optical system the dimensions of all sector fields and quadrupole lenses can usually be chosen from fairly wide limits. In most cases not even the number of fields is fixed. To obtain a system that performs optimally we must choose a suitable arrangement of sector fields, quadrupole lenses, and field free regions and then theoretically vary some of the dimensions of the system appropriately. The second step requires an iterative approximation procedure, which is normally carried through with the help of some complex program such as TRANS-

PORT, TRIO, or GIOS on a large computer (Brown *et al.*, 1980; Matsuo *et al.*, 1976; Wollnik *et al.*, 1984). The first step, normally the most important, requires the intuition of an experienced designer. Only if this first choice of parameters is reasonable, can the iterative procedure converge to some acceptable solution. To improve this intuition for the interested reader is the goal of the ideas presented in this chapter.

9.1 DISPERSION AND RESOLVING POWER OF MULTIFIELD PARTICLE SPECTROMETERS

The magnification, dispersion, and resolving power of a single sector field are connected by simple relations found in Chapter 4. Corresponding relations for particle spectrometers consisting of several sector fields and quadrupole lenses are discussed here.

9.1.1 Particle Spectrometers Consisting of Only One Sector Field

For an x-focusing one-sector field system, the relation between position vectors $X_0 = (x_0, a_0, t_0, \Delta)$ and $X_1 = (x_1, a_1, t_1, \Delta)$ at object and image is given as $X_1 = T_{10}X_0$, with the transfer matrix T [see Eq. (4.62a)],

$$T_{10} = \begin{pmatrix} (x|x) & (x|a) & (x|\Delta) \\ (a|x) & (a|a) & (a|\Delta) \\ 0 & 0 & 0 \end{pmatrix} = \begin{pmatrix} M & 0 & (x|\Delta) \\ 1/f & 1/M & (a|\Delta) \\ 0 & 0 & 1 \end{pmatrix}. \quad (9.1)$$

Here, $(x|x) = M$ is the lateral x magnification, and

$$(x|\Delta) = (1 - M)/k_x^2\rho_0$$

the rigidity dispersion [Eq. (4.19a) and (4.70a)]. For an object of size $2x_{00}$ and $M < 0$, the lateral resolving power R_Δ can be defined [Eq. (4.19c)] as

$$R_\Delta = -(x|\Delta)/2x_{00}M = \overleftarrow{(x|\Delta)}/2x_{00}, \quad (9.2)$$

with $(x|\Delta) = [1 - (1/M)]/k_x^2\rho_0$ the dispersion of the reversed spectrometer for which object and image are exchanged. In both cases, ρ_0 describes the radius of curvature of the optic axis, and k_x is the inhomogeneity parameter of the sector field [Eq. (4.58a)].

9.1.2 Particle Spectrometers Consisting of Several Sector Fields and Quadrupoles Lenses

The simplest multifield particle spectrometer consists of two cascaded sector fields. With transfer matrices T_{21} and T_{10} characterizing the first and the second sector fields, respectively, we find the transfer matrix of the two-sector field system to be $T_{20} = T_{21}T_{10}$, which reads explicitly,

$$
\begin{pmatrix} (x|x) & (x|a) & (x|\Delta) \\ (a|x) & (a|a) & (a|\Delta) \\ 0 & 0 & 0 \end{pmatrix} = \begin{pmatrix} M_2 & 0 & (x_2|\Delta) \\ -f_2^{-1} & M_2^{-1} & (a_2|\Delta) \\ 0 & 0 & 1 \end{pmatrix} \begin{pmatrix} M_1 & 0 & (x_1|\Delta) \\ -f_1^{-1} & M_1^{-1} & (a_1|\Delta) \\ 0 & 0 & 1 \end{pmatrix}.
$$

(9.3)

The rules of matrix multiplications here yield

$$(x|x) = M_2 M_1,$$

$$(x|a) = 0,$$

$$(x|\Delta) = (x_2|\Delta) - M_2(x_1|\Delta).$$

For N cascaded spectrometers, we find analogously,

$$(x|x) = M_N M_{N-1} \cdots M_2 M_1, \tag{9.4a}$$

$$(x|a) = 0, \tag{9.4b}$$

$$(x|\Delta) = (x_N|\Delta) + M_N(x_{N-1}|\Delta) + M_N M_{N-1}(x_{n-2}|\Delta) + \cdots$$

$$= \sum_{i=1}^{N} (x_i|\Delta) \prod_{k=i+1}^{N} M_k. \tag{9.4c}$$

The resolving power R_Δ of such N cascaded systems is found [see also Eq. (9.2)] to be

$$2x_{00} R_\Delta = (-1)^N \frac{(x|\Delta)}{(x|x)} = (-1)^N \left[\frac{(x_1|\Delta)}{M_1} + \frac{(x_2|\Delta)}{M_1 M_2} + \frac{(x_3|\Delta)}{M_1 M_2 M_3} + \cdots \right]$$

$$= (-1)^N \sum_{i=1}^{N} (x_i|\Delta) \prod_{k=1}^{i} M_k^{-1}. \tag{9.4d}$$

This is identical to the dispersion $\overleftarrow{(x|\Delta)}$ of the reversed system for which the ith stage has a magnification M_i^{-1} and a dispersion $(x_i|\Delta)/M_i$.

Equations (9.4a)-(9.4d) are also correct if some or all of the intermediate images are only virtual. They are still correct if the ith stage is not dispersive $[(x_i|\Delta) = 0]$, as would be the case for a quadrupole lens or the action of a fringing field.

9.1.3 A Particle Spectrometer Consisting of One Dispersive Element Preceded and Followed by a Nondispersive Element

For a particle spectrometer consisting of one sector field preceded and followed by a quadrupole (as shown in Fig. 9.1), Eqs. (9.4a)-(9.4d) become

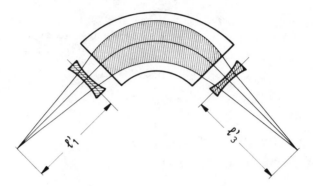

Fig. 9.1. A sector field spectrometer preceded and followed by quadrupoles that are defocusing in the x direction and focusing in the y direction.

with $(x_2|\Delta) \neq 0$; $(x_1|\Delta) = (x_3|\Delta) = 0$,

$$(x|x) = M_3 M_2 M_1, \tag{9.5a}$$

$$(x|a) = 0, \tag{9.5b}$$

$$(x|\Delta) = M_3(1 - M_2)/k_x^2 \rho_0. \tag{9.5c}$$

Here, ρ_0 and k_x denote the radius of curvature of the optic axis and the inhomogeneity parameter of the sector field [Eq. (4.58a)]. The resolving power of this spectrometer [Eq. (9.4d)] is found to be

$$R_\Delta = -\frac{(x|\Delta)}{2x_{00}(x|x)} = \frac{1 - (1/M_2)}{2x_{00}M_1 k_x^2 \rho_0}. \tag{9.5d}$$

From inspecting Eqs. (9.5c) and (9.5d), we find that R_Δ is independent of M_3, and $(x|\Delta)$ is independent of M_1. This can be expressed (Wollnik, 1971, 1980) as

$$(x|\Delta) \neq F(M_1), \tag{9.6a}$$

$$R_\Delta \neq G(M_3). \tag{9.6b}$$

Thus, the dispersion $(x|\Delta)$ can be varied by varying M_3 or the field strength in the second quadrupole, whereas R_Δ stays unchanged. Conversely, the resolving power R_Δ can be varied by varying M_1, or the field strength in the first quadrupole, whereas $(x|\Delta)$ remains unchanged. Since the action of inclined sector field boundaries can be understood as the action of infinitely thin quadrupole lenses, the Eqs. (9.6a) and (9.6b) apply also to sector fields with inclined boundaries for which case [see also Eqs. (4.25) and (4.31)], the correctness of Eq. (9.6b) has been described by Ewald and Hintenberger (1953) and Chavet *et al.* (1966).

9.1.4 The Optical Mode and the Resulting Dispersion and Resolving Power

For two thin quadrupole lenses (Fig. 9.1), corresponding lens equations [Eq. (1.11)] can be established:

$$(1/l_1') + (1/l_1'') = 1/f_1, \qquad (1/l_3') + (1/l_3'') = 1/f_3.$$

Here, l_1' and l_3' describe object lengths, l_1'' and l_3'' image lengths, and f_1 and f_3 the focal lengths of the two lenses. According to Eq. (1.12a), the magnifications of these elements are

$$M_1 = -l_1''/l_1' = [1 - (l_1'/f_1)]^{-1}, \tag{9.7a}$$

$$M_3 = -l_3''/l_3' = 1 - (l_3''/f_3). \tag{9.7b}$$

Introducing Eqs. (9.7a) and (9.7b) into Eqs. (9.5c) and (9.5d), we find with $M = M_1 M_2 M_3$,

$$(x|\Delta)k_x^2 \rho_0 = (1 - M_2)M_3 = \left(1 - \frac{l_3''}{f_3}\right) - M\left(1 - \frac{l_1'}{f_1}\right), \tag{9.8a}$$

$$R_\Delta 2 x_{00} k_x^2 \rho_0 = \frac{1 - (1/M_2)}{M_1} = -(x|\Delta)\frac{k_x^2 \rho_0}{M}. \tag{9.8b}$$

Assume now a sector field of fixed k_x and ρ_0 and quadrupole lenses that may be varied such that $M = M_1 M_2 M_3$ is a constant. In this case, the Eqs. (9.8a) and (9.8b) yield increased values of $(x|\Delta)$ and R_Δ in the case of negative f_1 and f_3. This corresponds to both quadrupole lenses being defocusing in the x direction (Fig. 9.1) and consequently focusing in the y direction (Fig. 9.2). To get a quantitative idea of what Eqs. (9.8a) and (9.8b) express, let us assume a homogeneous magnetic sector field $n_1 = 0$ or $\rho_0^2 k_x^2 = 1$, taken from Eq. (4.58a), and two equally strong quadrupole lenses, as indicated in Fig. 9.2, for three values of f_1 and f_3, respectively. We shall refer to these three cases as the optical modes 0, 1, and 2 of the particle spectrometer (Wollnik, 1970).

In *mode* 0, lengths l_1' and l_3'' are both zero, so that Eq. (9.8) yields for arbitrary f_1 and f_3,

$$(x|\Delta)/\rho_0 = 1 - M \approx 2, \tag{9.9a}$$

$$(R_\Delta 2 x_{00})/\rho_{00} = 1 - (1/M) \approx 2. \tag{9.9b}$$

The approximate signs here are valid for $M \approx -1$.

In *mode* 1, the focal lengths f_1 and f_3 are adjusted such that at the end a stigmatic image is achieved. If the beam is to be parallel in the y direction within the homogeneous sector magnet, we must choose $l_1' = f_1$ and $l_3'' = f_3$

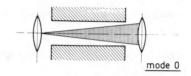

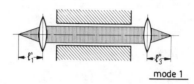

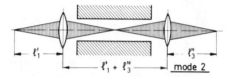

Fig. 9.2. The same field arrangement as in Fig. 9.1 where particle trajectories are shown with $y \neq 0$ in a projection on a surface $x = 0$. Note that both quadrupoles are focusing in the y direction since they were both assumed to be defocusing in the x direction (see, Fig. 9.1). Note also that there are three cases in which the object and image distances l'_1 and l''_3 are both assumed to be 0, f, or $2f$, and f the focal lengths of the quadrupoles. Note also that these three modes have resolving powers of about two, four, or six times $\rho_0/2x_{00}$, as shown in Eqs. (9.9b)–(9.11b).

such that Eqs. (9.8a) and (9.8b) yield

$$(x|\Delta)/\rho_0 = 2(1 - M) \approx 4, \tag{9.10a}$$

$$(R_\Delta 2x_{00})/\rho_0 = 2[1 - (1/M)] \approx 4. \tag{9.10b}$$

The approximate signs here again are valid for $M \approx -1$.

In *mode* 2, the focal lengths f_1 and f_2 are reduced relative to mode 1 such that at the end a stigmatic image is achieved, and additionally an intermediate y image occurs somewhere in the homogeneous sector magnet. In this case, we find $l'_1 = 2f_1$ and $l''_3 = 2f_3$, and the path length between the two quadrupoles is $2f_1 + 2f_2$. Under this side condition, Eqs. (9.8a) and (9.8b) yield

$$(x|\Delta)/\rho_0 = 3(1 - M) \approx 6, \tag{9.11a}$$

$$(R_\Delta 2x_{00})/\rho_0 = 3[1 - (1/M)] \approx 6, \tag{9.11b}$$

where again the approximate signs are valid for $M \approx -1$.

From Eqs. (9.9)–(9.11), it may be read that $(x|\Delta)$ and $R_\Delta x_{00}$ are both proportional to ρ_0 and thus to the size of the instrument. Furthermore, it

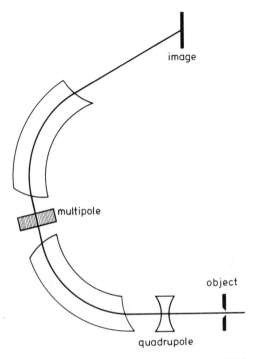

Fig. 9.3.   Principle of a particle spectrometer that uses two sector fields with an intermediate *y* focus at which a multipole correction element is placed (Enge, 1958; Enge and Kowalski, 1970; Drentje *et al.*, 1974; Ikegami, 1981).

may be concluded that the dispersion $(x|\Delta)$ and the resolving power R_Δ are mainly functions of the optical mode of the separator and not so much of its special design. This picture is valid approximately also for a magnification *M*, which deviates from −1 considerably.

Today, most existing particle spectrometers operate in mode 0 (Fig. 4.1) and only a few in mode 1 (Fig. 4.8) or mode 2 (Fig. 9.3).

9.2   AN OPTICAL Q_x VALUE FOR PARTICLE SPECTROMETERS

From experimental results it is known that a certain particle spectrometer can either achieve a high resolving power at low transmission or a low resolving power at high transmission. Thus, it seems reasonable to define a quality factor, or optical Q value, which characterizes the product of resolving power and transmission. In order to do this, we may look at an optical system in which some single sector field is preceded by an arbitrary part consisting of $i - 1$ elements. Such a system is described by the product of two transfer matrices:

$$
\begin{pmatrix} x_i \\ a_1 i \\ t_i \\ \delta_K \\ \delta_m \end{pmatrix} =
\begin{pmatrix}
c_{xi} & s_{xi} & 0 & N_{Ki}d_{xi} & N_{mi}d_{xi} \\
k_{xi}^2 s_{xi} & c_{xi} & 0 & N_{Ki}\dfrac{s_{xi}}{\rho_{0i}} & N_{mi}\dfrac{s_{xi}}{\rho_{0i}} \\
N_t \dfrac{s_{xi}}{v_0} & N_t \dfrac{d_{xi}}{v_0} & 1 & (t_i|\delta_{Ki}) & (t_i|\delta_{mi}) \\
0 & 0 & 0 & 1 & 0 \\
0 & 0 & 0 & 0 & 0
\end{pmatrix}
$$
$$
\times
\begin{pmatrix}
(x_{i-1}|x) & (x_{i-1}|a) & 0 & (x_{i-1}|\delta_K) & (x_{i-1}|\delta_m) \\
(a_{i-1}|x) & (a_{i-1}|a) & 0 & (a_{i-1}|\delta_K) & (a_{i-1}|\delta_m) \\
(t_{i-1}|a) & (t_{i-1}|a) & 1 & (t_{i-1}|\delta_K) & (t_{i-1}|\delta_m) \\
0 & 0 & 0 & 1 & 0 \\
0 & 0 & 0 & 0 & 1
\end{pmatrix}
\begin{pmatrix} x_0 \\ a_0 \\ t_0 \\ \delta_K \\ \delta_m \end{pmatrix}, \quad (9.12)
$$

with

$$
N_{Ki} = \frac{(1 + 2\eta_0)^2 + h_i}{2(1 + \eta_0)(1 + 2\eta_0)}, \qquad N_{mi} = \frac{(1 + 2\eta_0) - h_i}{2(1 + \eta_0)(1 + 2\eta_0)},
$$

$$
N_{ti} = 1 + \frac{h_i}{(1 + 2\eta_0)^2},
$$

$$
v_0(t_i|\delta_{Ki}) = -\frac{w_i}{2(1 + \eta_0)(1 + 2\eta_0)} + N_{Ki}N_{ti}\frac{w_i - s_{xi}}{k_{xi}^2\rho_{0i}^2},
$$

$$
v_0(t_i|\delta_{mi}) = \frac{w_i}{2(1 + \eta_0)(1 + 2\eta_0)} + N_{mi}N_{ti}\frac{w_i - s_{xi}}{k_{xi}^2\rho_{0i}^2},
$$

according to Eqs. (4.61), (4.77) and (4.79). With $w = \rho_0\phi$ as well as $c_{xi} = \cos(k_{xi}w_i)$, $s_{xi} = k_{xi}^{-1}\sin(k_{xi}w_i)$ and $d_{xi} = [1 - \cos(k_{xi}w_i)]/(\rho_{0i}k_x^2)$. Here the left-hand matrix describes the single-sector field under consideration, where h_i characterizes this sector field as magnetic ($h_i = 0$) or electrostatic ($h_i = 1$). This matrix is identical to that of Eq. (4.77), where the relative rigidity deviation $\Delta = \chi/\chi_0 - 1$ is replaced by the relative energy/charge deviation $\delta_K = Kz_0/K_0z - 1$ and the relative mass/charge deviation $\delta_m = m_0z_0/m_{00}z - 1$, as in Eqs. (2.15) and (2.16).

9.2.1 An Optical Q_x Value for Single-Sector Field Spectrometers

Homogeneous or inhomogeneous single-sector-field spectrometer may be arranged as in Fig. 9.4. In both cases, objects of size $2x_{00}$ are assumed,

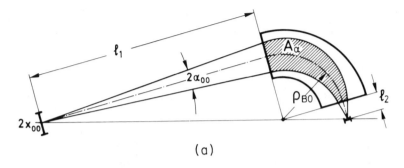

(a)

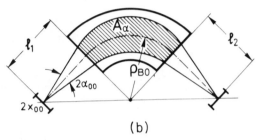

(b)

Fig. 9.4. Two single sector fields are shown with bundles of particles that started at the center of the objects under maximal angles $\pm\alpha_{00}$. Note that these two most inclined trajectories form in the sector field a region A_α, the magnitude of which was determined in Eqs. (9.15a) or (9.15b). Note also that the A_α ia larger in (b) than in (a), although the sector fields are equal. The only difference is that the x magnifications are -0.3 and -1 in the two cases, respectively.

from the center of which particles originate with angles $\pm\alpha_{00}$. The corresponding trajectories in the sector fields are found by applying a position vector $\mathbf{X}_0 = (0, a_{00}, 0, 0)$ to Eq. (9.12), the right-hand transfer matrix being that of a drift distance, i.e., a unity matrix with $(x_1|a) = l_1$, the length of the initial field-free region. Thus, $\mathbf{X}_0$ transforms with this matrix into $\mathbf{X}_1 = (\pm l_1 a_{00}, \pm a_{00}, 0, 0, 0)$, and we find

$$x_2 = \pm(c_x l_1 + s_x)a_{00}.$$

Consequently, the areas A_α of Fig. 9.4 are found to be $2\int x_2\,d(\rho_0\phi)$, or

$$A_\alpha = 2\int_0^{\rho_0\phi}(c_x l_1 + s_x)a_{00}\,d(\rho_0\phi) = 2(d_x\rho_0 + s_x l_1)a_{00}. \qquad (9.13a)$$

Using $(x|\delta_K)$ and $(a|\delta_K)$ of Eq. (4.61c), we can rewrite Eq. (9.13a) as

$$A_\alpha = 2[(x|\delta_K) + (a|\delta_K)l_1](\rho_0 a_{00}/N_K), \qquad (9.13b)$$

$$A_\alpha = 2[(x|\delta_m) + (a|\delta_m)l_1](\rho_0 a_{00}/N_m). \qquad (9.13c)$$

At this point, we may very generally describe the dispersions of the optical system of Eq. (9.12) followed by a field-free region of length l_2 as

$$D_K = (x|\delta_K) + l_2(a|\delta_K), \qquad D_m = (x|\delta_m) + l_2(a|\delta_m)$$

[see also Eq. (4.62a)]. Consequently, the dispersions of the reversed optical system are

$$\breve{D}_K = (x|\delta_K) + l_1(a|\delta_K), \qquad \breve{D}_m = (x|\delta_m) + l_1(a|\delta_m),$$

and Eqs. (9.13b) and (9.13c) can be rewritten as [see also Eq. (8.39)],

$$(A_\alpha/\rho_0)N_K = 2a_{00}\breve{D}_K, \tag{9.14a}$$

$$(A_\alpha/\rho_0)N_m = 2a_{00}\breve{D}_m. \tag{9.14b}$$

According to Eq. (4.19c), these $\breve{D}_K$ and $\breve{D}_m$ equal $-R_K 2x_{00}$ and $-R_m 2x_{00}$, so that Eqs. (9.14a) and (9.14b) become

$$Q_x(\delta_K) = (2x_{00}2a_{00})R_K = (A_\alpha/\rho_0)N_K, \tag{9.15a}$$

$$Q_x(\delta_m) = (2x_{00}2a_{00})R_m = (A_\alpha/\rho_0)N_m. \tag{9.15b}$$

This result shows (see also Wollnik, 1971) that the area A_α divided by ρ_0 is proportional to the total parallelogram-like radial phase space $2x_{00}2a_{00}$ and the energy-resolving power R_K or the mass-resolving power R_m. Note that both Q_α terms do not depend on the angle of deflection ϕ_0 of the sector field nor on k_x, the inhomogeneity parameter of the sector field. An inhomogeneous sector field thus offers no advantages over a homogeneous one in the case of equal areas A_a and equal radii of deflection ρ_0. Since A_a increases with the square of the size of an instrument, the ratio A_a/ρ_0 increases linearly. This implies that a spectrometer should always be built as large as possible, which speaks against superconducting magnets as long as their price per unit volume is higher than that for classical magnets.

At this point we should remember that the electrode distance (Fig. 9.5) does not vary with w and that the pole faces (Fig. 9.4) normally have a constant width in the x direction. For a given maximal beamwidth on the other hand, we read from Fig. 9.4 that for an approximately symmetric sector field of magnification $M_x \approx -1$, the Q_α value is always larger than that for a strongly magnifying or demagnifying system $|M_x| \ll 1$. In the second case, namely the area A_α has an almost triangular shape, whereas there is an almost rectangular one in the first case.

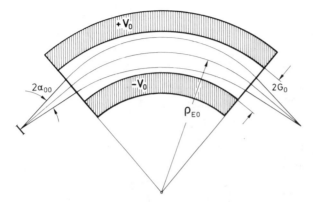

Fig. 9.5. A single electrostatic sector field is shown together with a bundle of particles that started at the center of the object with maximal angles $\pm\alpha_{00}$. Note the width $2G_0$ of the electrode separation and the symmetric potentials $\pm V_0$ at the electrodes.

For a magnetic ($h = 0$ and $\chi_{B0} = B_0\rho_{B0}$) or an electrostatic system ($h = 1$ and $\chi_{E0} = E_0\rho_{E0}$), Eqs. (9.15a) and (9.15b) can be rewritten with $\varepsilon_x = x_{00}a_{00}$ as

$$Q_{Bx}(\delta_m) = 4\varepsilon_x R_m = \frac{A_\alpha B_0}{\chi_{B0}}\frac{1}{2(1 + \eta_0)}, \qquad (9.15c)$$

$$Q_{Bx}(\delta_K) = 4\varepsilon_x R_K = \frac{A_\alpha B_0}{\chi_{B0}}\frac{1 + 2\eta_0}{2(1 + \eta_0)}, \qquad (9.15d)$$

$$Q_{Ex}(\delta_m) = 4\varepsilon_x R_m = \frac{A_\alpha E_0}{\chi_{E0}}\frac{\eta_0}{(1 + \eta_0)(1 + 2\eta_0)}, \qquad (9.15e)$$

$$Q_{Ex}(\delta_K) = 4\varepsilon_x R_K = \frac{A_\alpha E_0}{\chi_{E0}}\frac{1 + 2\eta_0(1 + \eta_0)}{(1 + \eta_0)(1 + 2\eta_0)}. \qquad (9.15f)$$

For particles of given rigidities χ_{B0} and χ_{E0}, the Q_α value is thus proportional to $A_\alpha B_0$ and $A_\alpha E_0$. In the electrostatic case (Fig. 9.5), we also may approximate A_α by $2\rho_0\phi_0 G_0$ (where $2G_0$ is the electrode separation) so that

$$A_\alpha E_0 \approx 2\rho_0\phi_0 V_0,$$

with V_0/G_0 the electrostatic field strength E_0.

All these considerations imply that it is advantageous to build a spectrometer as large as possible with limits set only by the available floor space or by the portability of the instrument. Fortunately, the costs normally do not increase linearly as well, since for particles of a given rigidity and for a fixed instrument geometry, the deflecting fields B_0 and E_0 decrease linearly.

Thus, the electrode potential is independent of instrument size, whereas the power consumption of a magnet actually decreases.[*]

9.2.2 An Optical Q_x Value for N Cascaded Spectrometers

The magnification and dispersion of N cascaded spectrometers has been determined in Eqs. (9.4a) and (9.4b). Let us now look at the ith stage of such an arrangement. The aperture angle $\alpha_i \approx a_i$ for this stage must be a_{00} multiplied by the overall angle magnification, i.e., $M_1^{-1} \cdot M_2^{-1} \cdot \cdots \cdot M_{i-1}^{-1}$ up to this stage, so that Eqs. (9.14a) and (9.14b) read

$$\frac{A_{\alpha i}}{\rho_{0i}} N_{ki} = \frac{2a_{00}}{M_1 M_2 \cdots M_{i-1}} \overleftarrow{(x_i|\delta_K)} = \frac{2a_{00}(x_i|\delta_K)}{M_1 M_2 \cdots M_{i-1} M_i},$$

$$\frac{A_{\alpha i}}{\rho_{0i}} N_{mi} = \frac{2a_{00}}{M_1 M_2 \cdots M_{i-1}} \overleftarrow{(x_i|\delta_m)} = \frac{2a_{00}(x_i \leftarrow \delta_m)}{M_1 M_2 \cdots M_{i-1} M_i},$$

where N_{Ki} and N_{mi} are defined in Eq. (9.12). Note that both coefficients depend on whether the ith field is magnetic ($h_i = 0$) or electrostatic ($h_i = 1$). Summing these expressions over $i = 1, 2, \ldots, j$, we find

$$\sum_{i=1}^{j} \left(\frac{A_{\alpha i}}{\rho_{0i}} N_{Ki}\right) = 2a_{00} \sum_{i=1}^{j} (x_i|\delta_K) \prod_{k=1}^{i} M_k^{-1} = 2a_{00} \frac{(x_f|\delta_K)}{(x_f|x)}$$

$$= (-1)^j 2a_{00} 2x_{00} R_K, \tag{9.16a}$$

$$\sum_{i=1}^{j} \left(\frac{A_{\alpha i}}{\rho_{0i}} N_{mi}\right) = 2a_{00} \sum_{i=1}^{j} (x_i|\delta_m) \prod_{k=1}^{i} M_k^{-1} = 2a_{00} \frac{(x_f|\delta_m)}{(x_f|x)}$$

$$= (-1)^j 2a_{00} 2x_{00} R_m, \tag{9.16b}$$

where the index f denotes the elements of the overall matrix. Thus, the resolving power of a complex multistage spectrometer can be judged by inspecting the geometry of the overall system, i.e., by looking at the individual ratios $A_{\alpha i}/\rho_{0i}$ of the different fields and estimating which percentage of the performance of the overall system is due to the ith field. This is a simple and most effective method of judging where and how the particle beam could be widened or should be reduced in the x direction by additional quadrupole lenses or oblique sector field boundaries to maximize R_K or R_m.

---

[*] To achieve a flux density B_0 in a magnet with an air gap $2G_0$ requires $NI = \int H\,ds = B_0 2G_0/\mu\mu_0$ ampere turns in the magnet coil. Since G_0 increases and B_0 decreases with size, the required number of ampere turns, stays constant. The power consumption of a magnet furthermore is $W = (NI)^2 R \sim (NI)^2 L_c/F$, with the coil resistance R proportional to the coil circumference L_c and inversely proportional to the coil cross section F. Thus, the necessary power decreases linearly with the instument size, whereas for not quadratically but only linearly increased coil cross section, the power consumption stays constant.

Note also that the ratio $A_{\alpha i}/\rho_{0i}$ vanishes in all cases in which ρ_{0i} becomes infinite. Note further that these ratios have signs that vary each time the deflection changes direction (ρ_0 and $-\rho_0$) or an intermediate x image occurs (A_α and $-A_\alpha$), because the two limiting outermost trajectories then exchange their positions. Note finally that h_i and thus N_{Ki} and N_{mi} have different values in magnetic ($h_i = 0$) and in electrostatic ($h_i = 1$) sector fields.

9.2.3 Achromatic Systems

For many applications we would like to deflect a particle beam so that the foci for particles of different energies coincide at least to first order. This type of achromaticity is identical to postulating that the sum of Eq. (9.16a), and thus R_K, vanishes.

The simplest such device is an optical system that contains field-free regions and quadrupole lenses but no deflecting fields, since in the case of all ρ_{0i} being infinite, the R_K and R_m of Eqs. (9.16a) and (9.16b) vanish. The next simple such device employs two identical sector fields that deflect the particle beam in opposite directions with a parallel beam inbetween (Fig. 9.6a) or that deflect the particle beam symmetrically in the same direction with an intermediate image in between (Fig. 9.6b). The systems of Fig. 9.6 are often used in beam guidance systems, where we may also include quadrupole lenses. In the example of Fig. 9.7, such a system is shown for which a sector field is preceded and followed by a quadrupole doublet, the sector field is split, and a single quadrupole is inserted in this region. If the focusing strength of this quadrupole is chosen such that in its middle, originally coinciding trajectories of different rigidities become parallel (Fig. 9.7b), these trajectories at the end will coincide again.

A very special system is an angle and energy focusing (Aston, 1919; Dempster, 1937; Mattauch and Herzog, 1934) mass spectrometer (Fig. 9.8), i.e., $(x|a) = (x|\delta_K) = 0$, which still preserves a mass dispersion. Such a system consists of x-focusing magnetic and electrostatic sector fields that are designed such that R_K vanishes. For $j = 2$, Eq. (9.16a) thus becomes $A_{B\alpha}N_{BK}/\rho_{B0} = A_{E\alpha}N_{EK}/\rho_{E0}$, or for low-energy ions ($\eta_0 = 0$):

$$A_{B\alpha}/2\rho_{B0} = A_{E\alpha}/\rho_{E0}. \tag{9.17}$$

This corresponds to field arrangements like those shown in Fig. 9.6, where one of the ratios A_α/ρ_0 is decreased to one-half before the corresponding sector field is exchanged for an electrostatic field (Fig. 9.9). Note that it makes no difference whether the electrostatic sector field precedes or follows the magnetic field (Maurer *et al.*, 1971).

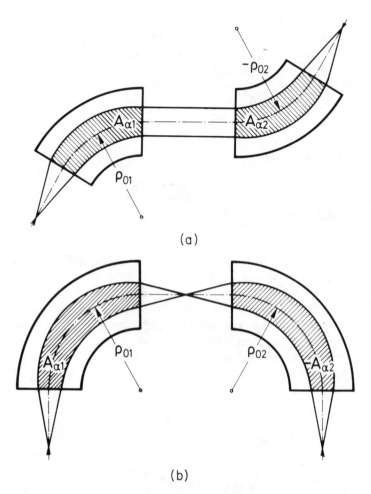

(a)

(b)

Fig. 9.6. Two double sector field arrangements that both cause achromatic beam deflections. Note that there are opposite signs for the radii of deflection in (a) and opposite signs for the area A_α in (b).

9.3.   A Q_t VALUE FOR TIME-OF-FLIGHT PARTICLE SPECTROMETERS

Analogously to Q_α, which is a quality value for laterally dispersive particle spectrometers, we can also define a Q_t value for longitudinally dispersive systems, i.e., for time-of-flight particle spectrometers. For this purpose, we assume a radially inhomogeneous sector field with an angle of deflection ϕ_i preceded by an arbitrary optical system [Eq. (9.12)]. From

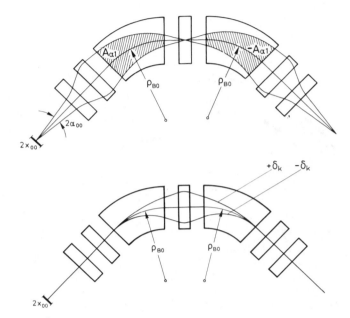

Fig. 9.7. An achromatic beam deflector consisting of two sector fields preceded and followed by quadrupole doublets. Note that the areas $A_{\alpha 1}$ and $-A_{\alpha 1}$ are of equal size. The quadrupole in the middle between the sector fields can be adjusted so that the overall system is not only dispersion free $(x|\delta_K) = (x|\delta_m) = 0$ but achromatic $(a|\delta_K) = (a|\delta_m) = 0$; i.e., particles of different rigidities $(\delta_K \neq 0)$ are parallel in the middle of the center quadrupole and coincide at the end of the systems if they coincided at the beginning.

a multiplication of the transfer matrices of Eq. (9.12), we find,

$$(x_i|\delta_K) = c_{xi}(x_{i-1}|\delta_K) + s_{xi}(a_{i-1}|\delta_K) + N_{Ki}\,d_{xi}, \qquad (9.18a)$$

$$(x_i|\delta_m) = c_{xi}(x_{i-1}|\delta_m) + s_{xi}(a_{i-1}|\delta_m) + N_{mi}\,d_{xi}, \qquad (9.18b)$$

$$[(t_i|\delta_K) - (t_{i-1}|\delta_K)]v_0$$
$$= N_{ti}\left[s_{xi}(x_{i-1}|\delta_K) + d_{xi}(a_{i-1}|\delta_K) + N_{Ki}\frac{w_i - s_{xi}}{k_{xi}^2\rho_{0i}^2} \right]$$
$$\qquad - \frac{w_i}{2(1 + \eta_0)(1 + 2\eta_0)}, \qquad (9.18c)$$

$$[(t_i|\delta_m) - (t_{i-1}|\delta_m)]v_0$$
$$= N_{ti}\left[s_{xi}(x_{i-1}|\delta_m) + d_{xi}(a_{i-1}|\delta_m) + N_{mi}\frac{w_i - s_{xi}}{k_{xi}^2\rho_{0i}^2} \right]$$
$$\qquad + \frac{w_i}{2(1 + \eta_0)(1 + 2\eta_0)}. \qquad (9.18d)$$

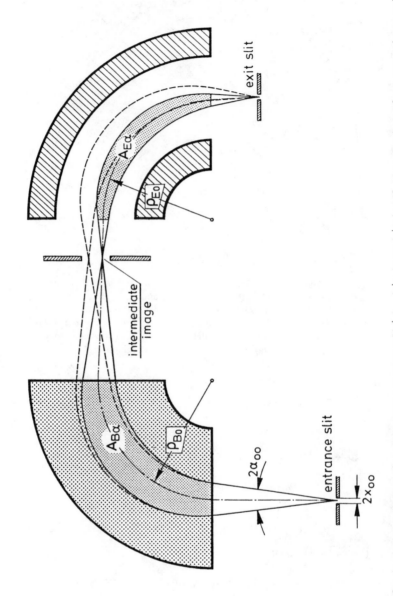

Fig. 9.8. An angle- and energy-focusing mass spectrometer $[(x|a) = (x|\delta_K) = 0, (x|\delta_m) \neq 0]$. Note that ions of one mass–charge ratio but two energy–charge ratios are focused to different positions in the middle and one at the end of the mass spectrometer.

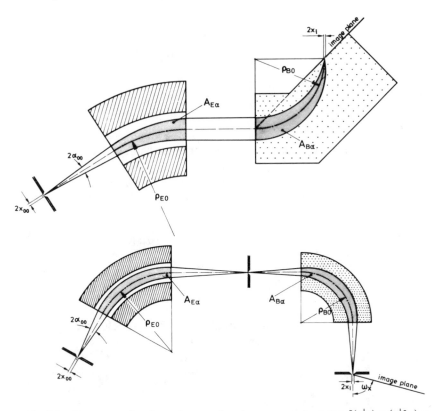

Fig. 9.9. Two types of angle- and energy-focusing mass spectrometers $[(x|a) = (x|\delta_K) = 0, (x|\delta_m) \neq 0]$ are shown; for both, of which the relation $A_{E\alpha}/\rho_{E0} = A_{B\alpha}\alpha_a/2\rho_{B0}$ holds.

For the sector field under consideration, we may now determine the areas enclosed by trajectories characterized by initial position vectors $\mathbf{X}_0(0, 0, 0, \pm\delta_K, 0)$ or $\mathbf{X}_0 = (0, 0, 0, 0, \pm\delta_m)$:

$$A_{\delta i}N_{Ki}\delta_K = 2\delta_K \int_0^{\pi_{0i}\phi_i} (x_i|\delta_K)\, d(\rho_0\phi)$$

$$= \left\{ [(t_i|\delta_K) - (t_{i-1}|\delta_K)]v_0 + \frac{w_i}{(1 + \eta_0)(1 + 2\eta_0)} \right\} \frac{2\delta_K\rho_{0i}}{N_{ti}}, \quad (9.19a)$$

$$A_{\delta i}N_{mi}\delta_m = 2\delta_m \int_0^{\rho_{0i}\phi_i} (x_i|\delta_m)\, d(\rho_0\phi)$$

$$= \left\{ [(t_i|\delta_m) - (t_{i-1}|\delta_m)]v_0 - \frac{w_i}{2(1 + \eta_0)(1 + 2\eta_0)} \right\} \frac{2\delta_m\rho_0}{N_{ti}}, \quad (9.19b)$$

with $w_i = \rho_{0i}\phi_i$

$$\int c_{xi}\, d(\rho_{0i}\phi_i) = s_{xi}, \qquad \int s_{xi}\, d(\rho_{0i}\phi_i) = d_{xi}$$

$$\int d_{xi}\, d(\rho_{0i}\phi_i) = (w_i - s_{xi})/k_{xi}^2 \rho_{xi}^2.$$

Here, $A_{\delta i}$ equals $\int d_{xi}(d\rho_0\phi)$ and $A_{\delta i}N_{Ki}\delta_k$ or $A_{\delta i}N_{mi}\delta_m$ are the areas that would be obtained between particle trajectories characterized by $\delta_m = 0$, $\delta_K = \pm\delta_{K0}$ or $\delta_K = 0$, $\delta_m = \pm\delta_{m0}$.

Evaluating Eqs. (9.18c) and (9.18d) and (9.19a) and (9.19b) for all j stages of a complex optical system, we find with $(t_0|\delta_K) = (t_0|\delta_m) = 0$,

$$\sum_{i=1}^{j} \frac{A_{\delta i}}{2v_0\rho_{0i}} N_{Ki}N_{ti} = (t_j|\delta_K) + \frac{T_0}{2(1+\eta_0)(1+2\eta_0)}, \qquad (9.20a)$$

$$\sum_{i=1}^{j} \frac{A_{\delta i}}{2v_0\rho_{0i}} N_{mi}N_{ti} = (t_j|\delta_m) - \frac{T_0}{2(1+\eta_0)(1+2\eta_0)}, \qquad (9.20b)$$

where $T_0 = \sum w_i/v_0 = \sum \rho_{0i}\phi_i/v_0$ is the total flight time of a reference particle along the optic axis. This $\sum \rho_{0i}\phi_i/v_0$ also includes all field-free regions or quadrupoles of lengths $l_i = \rho_{0i}\phi_i$ so that $T_0 v_0 = l_0$ is the overall length of the optic axis. With $N_K + N_m = 1$ and $N_K - N_m = [h + \eta_0(1 + 2\eta_0)]/[(1 + \eta_0)(1 + 2\eta_0)]$ we find from Eqs. (9.20a) and (9.20b)

$$(t_j|\delta_m) + (t_j|\delta_K) = \sum_{i=1}^{j} \frac{A_{\delta i}}{2v_0\rho_{0i}} N_{ti}, \qquad (9.21a)$$

$$(t_j|\delta_m) - (t_j|\delta_K) = \frac{T_0}{(1+\eta_0)(1+2\eta_0)} + \sum_{i=1}^{j} \frac{A_{\delta i}}{2v_0\rho_{0i}} N_{ti}(N_{mi} - N_{Ki}). \qquad (9.21b)$$

Since $N_{mi} - N_{Ki}$ vanishes for $\eta_0 = h = 0$, we find for relativistically slow paticles ($\eta_0 = 0$) in purely magnetic systems ($h = 0$),

$$(t_i|\delta_m) - (t_i|\delta_K) = T_0. \qquad (9.21c)$$

For optical systems in which the particles are slow and do not move in electrostatic fields for a major time, Eq. (9.21c) is valid approximately.

9.3.1 Isochronous Optical Systems

Varying the parameters of an optical system, we can satisfy the condition

$$(t_j|\delta_K) = 0$$

in Eq. (9.20a). The more energetic and thus faster particles of the reference mass in this case are sent on a detour proportional to δ_K, so that at a final collector all particles arrive at the same time. For such so-called isochronous systems, Eq. (9.20a) becomes

$$T_0 v_0 \delta_K = \sum_{i=1}^{i} \frac{A_{\delta i} N_{Ki}}{\rho_{0i}} \delta_K [(1 + 2\eta_0)^2 + h_i] \frac{1 + \eta_0}{1 + 2\eta_0} \qquad (9.22a)$$

with $A_{\delta i} N_{Ki} \delta_K$ describing areas enclosed by the optic axis and by the trajectory characterized initially by $\mathbf{X} = (0, 0, 0, \delta_K, 0)$ in magnetic ($h_i = 0$) and electrostatic ($h_i = 1$) sector fields. If Eq. (9.22a) is fulfilled, i.e., if $(t_j | \delta_K)$ vanishes, Eq. (9.21a) reads

$$(t_j | \delta_m) = \sum_{i=0}^{j} \frac{A_{\delta i}}{2 v_0 \rho_{0i}} N_{ti} = \sum_{i=1}^{j} \frac{A_{\delta i} N_{Ki}}{v_0 \rho_{0i}} \cdot \frac{1 + \eta_0}{1 + 2\eta_0}. \qquad (9.22b)$$

Defining a time-of-flight mass resolving power

$$R_{tm} = (t_j | \delta_m)/2\delta_t,$$

Eq. (9.22b) is rewritten for $\eta_0 - 0$ as [Wollnik and Matsuo (1981)]

$$Q_t(\delta_m) = R_{tm}(2\delta_t, 2\delta_K) = \sum \frac{A_{\delta i} N_{Ki} \delta_K}{v_0 \rho_{0i}} \cdot \frac{1 + \eta_0}{1 + 2\eta_0}. \qquad (9.22c)$$

Thus, the product of the time-of-flight mass resolving power R_{tm} and $2\delta_t 2\delta_K$ is proportional to the area $A_{\delta i} N_{Ki} \delta_K$ shown in Fig. 9.10b and d divided by the product of ρ_{0i} and the particle velocity v_0. Note also that $\delta_t \delta_K$ is a constant as long as the energy of the reference particles stays unchanged.

9.3.2 Angle and Energy-Focusing Isochronous Optical Systems

For good isochronous time-of-flight spectrometers, we should postulate that particles of all energies, masses, and initial angles reach the final collector, i.e., that additionally to

$$(t_j | \delta_K) = 0, \qquad (9.23a)$$

the conditions

$$(x_j | a) = (x_j | \delta_K) = (x_j | \delta_m) = 0 \qquad (9.23b)$$

are satisfied. From Eq. (8.29b), we then find with Eq. (9.23b),

$$(t_j | a) = 0. \qquad (9.23c)$$

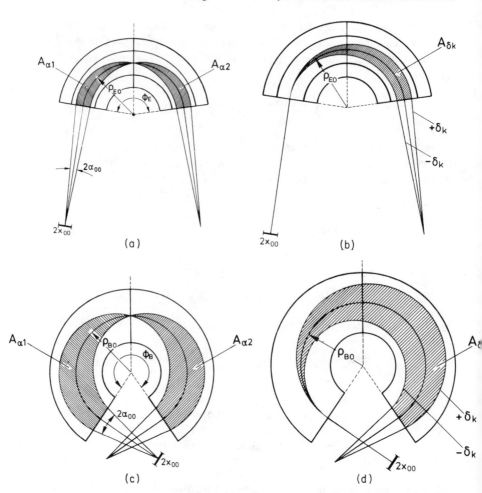

Fig. 9.10. Two laterally dispersion-free and stigmatic focusing $[(x|a) = (x|\delta_K) = (x|\delta_m) = (y|b) = 0]$ isochronous $[(t|\delta_K) = 0]$ time-of-flight spectrometers are shown. Note that the quantities $\Sigma A_{\alpha i}/\rho_{0i}$ vanish in (a) and (c) and that in (b) and (d) the quantities $\Sigma A_{\delta i}/\rho_{0i}$ determine the magnitude of $(t_j|\delta_m)$. Note also that such systems can be built with only inhomogeneous electrostatic sector fields [(a) and (b)] or with only inhomogeneous magnetic sector fields [(c) and (d)].

Postulating additionally $(a_j|\delta_K) = 0$, we find from Eq. (8.29a),

$$(t_j|x) = 0. \tag{9.23d}$$

Examples for which Eqs. (9.23a)–(9.23c) are satisfied are shown in Fig. 9.10 (see also Poschenrieder, 1970; Wollnik and Matsuo, 1981). In both cases, in addition to $(x_j|a) = (x_j|\delta_K) = (x_j|\delta_m) = 0$ the condition $(y_j|b) = 0$ is also satisfied. Thus, in both cases we have an isochronous time-of-flight mass spectrometer with a first-order stigmatic and achromatic image. Together with good higher-order properties, these conditions can also be achieved by more complex optical systems (Wouters *et al.*, 1984).

An interesting example is also that of Fig. 9.11 in which (Wollnik, 1986) Eq. (9.22a) is satisfied, with the side postulate that the overall angle of

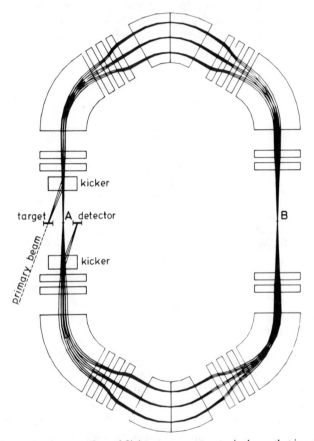

Fig. 9.11. An isochronous time-of-flight mass spectrometer is shown that is nothing more than an accelerator storage ring with the quadrupole excitations chosen such that after a 180° bend the matrix elements $(t|\delta_K)$, $(x|a)$, and $(y|b)$ vanish exactly and $(a|x)$ as well as $(b|y)$ vanish approximately. Particles with start simultaneously at the target thus arrive at the same time after a deflection of $n\pi$, with $n = 1, 2, 3, \ldots$, independent of x, y, a, b and δ_K. If introduced by the indicated first "kicker," they can be ejected from the ring by the second "kicker" after seveal turns. Introducing ac driven electrostatic deflectors at the points A and/or B, we can use such a ring also as a high-resolving mass separator, since charged particles can then pass only if the electrostatic fields go through zero. If this ac frequency is high, the achievable mass resolving powers can be very high.

deflection of the system should be 180° and that this system at least approximates a waist-to-waist transformation in both the x and the y plane, i.e.,

$$(x_j|a) = (a_j|x) = (y_j|b) = (b_j|y) = 0, \qquad (9.24a)$$

$$(x_j|x) = (y_j|y) = 1. \qquad (9.24b)$$

The advantage of this system is that many stages can be cascaded by simply passing the particles many times around the shown ring. In this case, all the $(t_j|\delta_m)$ values of Eq. (9.22b) add, so that finally high flight-time dispersions and, consequently, high mass-resolving powers can be achieved. However, to achieve infinitely many tuns is difficult, since in this case small fabrication inaccuracies would cause unstable particle motions (Section 6.4).

9.4 CORRECTION OF IMAGE ABERRATIONS

Having designed an optical system so as to obtain good first-order properties, we must investigate its image aberrations. To change the magnitudes of these aberrations, a correction element can be placed somewhere between object and image of the optical system (see Fig. 9.3). In this correction element the different trajectories of a bundle are deflected differently, with the goal that at the position of the final image these deflections cause offsets of the individual trajectories that counterbalance otherwise existing image aberrations.

9.4.1 Correction of Image Aberrations in Magnetic Systems

For the 180° sector magnet of Fig. 8.3, Section 8.2.2.2 stated that particles inclined under larger angles α_0 relative to the optic axis reach this axis not at A but at B, with the distance AB mainly proportional to a_0^2. To make these particles come to point A, so that the image aberration $(x|aa)a_0^2/2$ vanishes, the off-axis trajectories must be deflected a little less than the optic axis. Alternatively, the optic axis can be deflected a little more so that it reaches point B. The second case can be achieved (Balestrini and White, 1960) by a barrel-like section (Fig. 9.12) in which a slightly increased flux density exists as compaed to the main field. The central ray, i.e., the optic axis, then stays a little longer in this higher flux density than the off-axis trajectories so that the optic axis is additionally deflected as postulated above.

The procedure to correct image aberrations in all cases is analogous to the method outlined in Fig. 9.11. A correction element is always necessary

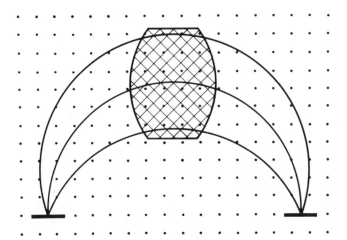

Fig. 9.12. The focusing of charged particles is shown for a 180° sector field. Different from Fig. 8.3, here a barrel-like region of increased field exists that deflects the optic axis a little more than the outer trajectories. Thus, all particle that started from a point are again focused to a point.

in which the different trajectories of a bundle are deflected differently, the magnitudes of these deflections being small. Such a correction element can have a constant magnetic flux density and an appropriately shaped boundary. Alternatively, it can have simple, straight boundaries and an appropriately varying magnetic flux density $B_y(x, y, z)$. A particle that enters this element at a point $P(x_1, y_1)$ in both cases is deflected by a small angle,

$$\Delta a(x_1, y_1) = \int_0^{l_c} \frac{(ze) B_y(x, y, s)}{mv} \, ds, \tag{9.25a}$$

$$\Delta b(x_1, y_1) = \int_0^{l_c} \frac{(ze) B_x(x, y, s)}{mv} \, ds, \tag{9.25b}$$

where $m/(ze)$ is the mass-to-charge ratio of the particle under consideration, and $v = |\mathbf{v}|$ is the particle velocity. At the position of an image, such a correction element displaces a particle trajectory under consideration slightly. In the case of an appropriately powered correction element, the displacement caused by some image aberration of the original optical system can be exactly counterbalanced.

To simplify the discussion, we first look only at geometric aberrations proportional to $\alpha_{00}, \alpha_{00}^2, \alpha_{00}^3, \ldots$, in the plane $y = 0$. According to Eq. (9.25a), the correction element can have a constant length Δz for all particle trajectories and a flux density that varies with x as

$$B_y(x, y = 0) = B_y(0, 0)[1 + xn_{B1} + \tfrac{1}{2}x^2 n_{B2} + \tfrac{1}{6}x^3 n_{B3} + \cdots]$$
$$= \sum (x^n/n!) K_{Bn}. \qquad (9.26a)$$

Alternatively, this correction element can have a constant $B_y(x, 0)$, where the length of the flight path in the correction element varies as

$$l = l_0[1 + xn'_{B1} + \tfrac{1}{2}x^2 n'_{B2} + \tfrac{1}{6}x^3 n'_{B3} + \cdots] = \sum (x^n/n!) K'_{Bn}. \quad (9.26b)$$

A correction element designed according to Eq. (9.26a) is a separate element (Fig. 9.13) that creates the required magnetic flux distribution by use of several specially designed magnetic poles. We call such an element a multipole correction element. A correction element designed according to Eq. (9.26b) is an appropriately curved boundary of a homogeneous or inhomogeneous sector field. In Fig. 9.14, the boundaries of such a magnetic sector field are shown. They are shaped such that in the plane of symmetry, the aperture aberration $(x|aa)a_0^2$ vanishes. Note, however, that in both cases, aberrations caused by particle trajectories that cross the correction element not in the plane of symmetry $(y = 0)$ are mainly increased (Section 9.4.2).

Both correction elements can achieve a correction of image aberrations. A curved sector field boundary is relatively simple to fabricate, at least if the field boundary is a simple circle. A big disadvantage is that once fabricated, it can not be altered. A multipole correction element, as shown in Fig. 9.13, on the other hand, is relatively difficult to build, normally requires an extra power supply, and last but not least, space. Its advantage is that it can be adjusted to the requirements of a realistic optical system.

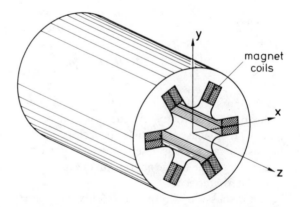

Fig. 9.13. A hexapole-correction element in which the magnetic flux density in the $y = 0$ plane is proportional to x^2.

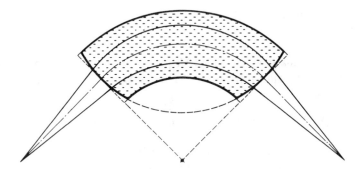

Fig. 9.14. A homogeneous sector field is shown, the field boundaries of which are formed such that all particles which originated from one point in the plane of deflection are focused back to one point. Normally, such field boundaries are approximated by circles that only in rare cases are modified by third-order functions.

Thus, it can compensate for fabrication inaccuracies and for poor adjustments as well.

Normally, we try to combine the advantages of both approaches. During the design state we combine different elements of the system such that for straight field boundaries, the system is sufficiently free of aberrations. Furthermore, we try to correct all important aberrations by curved or more complex shaped field boundaries. When the system is built, the remaining aberrations, possibly caused by improper machining or insufficient adjustments, can be corrected by one or several multipole correction elements. When systematic deviations from the design aberrations are observed one can also reshape the field boundaries and thus avoid the use of explicit multipole correction elements.

9.4.1.1 Procedure for the Correction of Image Aberrations by Magnetic Multipoles

In order to adjust an optical system optimally, we should try to provide several, possibly superimposed, multipole elements. Optimally, we should use at least

(1)   one element in which $B_y(x, 0)$ is an adjustable constant causing equal deflections for all particles of a beam in the x direction;

(2)   one element in which $B_y(x, 0)$ is proportional to x, with the proportionality constant adjustable so that the focusing power is varied, and the x image is shifted in the z direction;

(3)   one element in which $B_y(x, 0)$ is proportional to x^2, with the proportionality constant adjustable so that image aberrations of second order can be modified;

(4)   one element in which $B_y(x, 0)$ is proportional to x^3, with the proportionality constant adjustable so that image aberrations of third order can be modified, etc.

In this context it is important to note that a flux density distribution $B_y(x, 0)$ proportional to x^n influences only image aberrations of nth and higher order but no aberrations of smaller than nth order (Table 8.6). Thus, the adjustment procedure can be carried through as follows:

(1)   The dipole strength is adjusted, causing the particle beam to be deflected such that it can pass precisely through the exit aperture. Aberrations of order $n = 2, 3, 4, \ldots$, are modified slightly.

(2)   The quadrupole strength ($n = 1$) is adjusted, causing the x focus to move to the exact position of the exit aperture so that the overall coefficient $(x|a)$ vanishes exactly. This shift of the x focus also causes a shift of the y focus in the opposite direction. Aberrations of order $n = 2, 3, 4, \ldots$, are modified slightly. The aberration of zeroth order, i.e., the deflection of the particle beam, stays unchanged.

(3)   The hexapole strength ($n = 2$) is adjusted such that one aberration coefficient of second order vanishes exactly, for instance $(x|aa)$. By this operation, other aberrations of second order are also modified. (Normally, a decrease in one aberration of second order causes all others to increase.) Aberrations of third and higher order are modified slightly; aberrations of zeroth and first order, i.e., the deflection of the beam and the position of the image, stay unchanged.

(4)   The octopole strength ($n = 3$) is adjusted so that one aberration coefficient of third order vanishes exactly, for instance $(x|aaa)$. By this operation, other aberrations of third order are also modified. (Normally, a decrease in one aberration of third order causes all others to increase.) Aberrations of fourth and higher order are modified slightly; aberrations of zeroth, first, and second order stay unchanged.

Note here that a correction element that is supposed to modify the final aberration coefficient $(x| \cdots r_i^j r_k^e \cdots)$ must be placed at a position where both $(x|r_i)$ and $(x|r_k)$ are not zero, as was outlined in Section 8.4.3. An aperture aberration $(x|aa)$, $(x|aaa), \ldots$, thus can not be corrected by a multipole element at an intermediate x image, where $(x|a)$ vanishes. Advantageously, however, such a multipole element is placed at a y image where $(y|b)$ vanishes, since in this case, the aberrations $(x|bb)$, $(x|bbb)$, etc., stay unchanged.

9.4.1.2 Flux Distribution in a Multipole Correction Element

In Fig. 9.12 a magnetic multipole correction element is indicated, the z axis of which coincides with the optic axis of the particle beam. For a general description of the spatial flux density distribution $B_y(x, y)$ in such a correction element, it is advantageous to define a complex position variable

$$w = x + iy,$$

and describe the components of the magnetic flux density as $B_x(w) = B_x(x, y)$ and $B_y(2) = B_y(x, y)$. Thus, a complex flux density is described as

$$B(w) = B_x(w) + iB_y(w), \qquad (9.27a)$$

$$B^*(w) = B_x(w) - iB_y(w). \qquad (9.27b)$$

For a region without conductors, the Maxwell equations $\mathrm{div}(\mathbf{B}) = \mathrm{curl}(\mathbf{B}) = 0$ postulate

$$\partial B_y/\partial x = \partial B_x/\partial y \qquad \text{and} \qquad -\partial B_y/\partial y = \partial B_x/\partial x,$$

relations that are identical to the Cauchy–Riemann condition for $B^*(w)$, proving that $B^*(w)$ is a regular analytic function.

In Eq. (9.26a), a flux density distribution $B_y(x, 0) = K_n x^n$ was postulated in the correction element, with n the order of the image aberration to be corrected. The constants K_n in Eq. (9.26a) have the dimension T/m^n, which numerically is the flux density in teslas at $x = 1$ m from the optic axis. With Eq. (9.27) and $w = x + iy = re^{i\theta}$, Eq. (9.26a) implies

$$B^*(w) = iK_n w^n = iK_n(x + iy)^n, \qquad (9.28a)$$

$$B^*(w) = iK_n r^n e^{in\theta}. \qquad (9.28b)$$

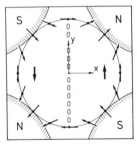 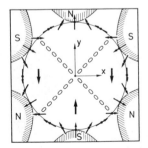 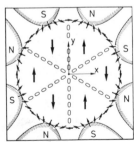

Fig. 9.15. Pole arrangements of magnetic quadrupoles, hexapoles, and octopoles are indicated. Also shown is a circle of radius r_0 along which the magnetic flux density is constant, and its direction varies as indicated. Finally, strings of zeros indicate lines along which B_y the y component of the magnetic flux density vanishes. These lines separate regions in which B_y is parallel or antiparallel to the y axis.

These relations state that for a given multipole, the magnitude of **B** is constant along a circle of radius r and only its direction varies (Fig. 9.15). From Eq. (9.28b) the components $B_x(r, \theta)$ and $B_y(r, \theta)$ of the magnetic flux density are determined for any r, θ as

$$B_x(r, \theta) = -K_n r^n \sin(n\theta), \tag{9.29a}$$

$$B_y(r, \theta) = K_n r^n \cos(n\theta). \tag{9.29b}$$

Since $\cos(n\theta)$ vanishes for $n\theta \pm \pi/2 = 0, \pi, 2\pi, \ldots$, the quantity $B_y(r, \theta)$ vanishes at certain values of θ for all r. In Fig. 9.1 corresponding lines are indicated by strings of zeros. These lines also separate regions in which the sign of $B_y(r, \theta)$ is reversed.

9.4.1.2.1 Multipole Correction Element Using Magnetic Poles

Instead of describing the x and y components of B, we can also describe the components B_r and B_t perpendicular and tangential to a circle of radius r. From simple trigonometry, we find from Eqs. (9.29a) and (9.29b),

$$B_r(r, \theta) = B_x(r, \theta) \cos \theta + B_y(r, \theta) \sin \theta = K_n r^n \sin[(n + 1)\theta], \tag{9.30a}$$

$$B_t(r, \theta) = B_y(r, \theta) \cos \theta - B_x(r, \theta) \sin \theta = K_n r^n \cos[(n + 1)\theta]. \tag{9.30b}$$

To form an nth-order multipole element, we can use either Eq. (9.30a) or (9.30b). In the first case, we should have the apex of a magnet pole at each point at which $B_r(r, \theta)$ is maximal (Fig. 9.16a). In the second case, we should rotate each of the permanent magnets of Fig. 9.16a by 90° and place them at those positions where $B_t(r, \theta)$ is maximal (Fig. 9.16b). In both

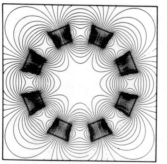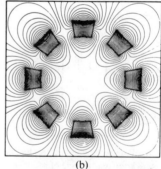

(a)   (b)

Fig. 9.16.   A eight-pole correction element built from eight permanent magnets. Note that there are two ways to arrange the eight permanent magnets in order to create the same multipole field distribution (Halbach, 1980).

cases, we find that a $2(n + 1)$-fold magnet pole symmetry creates an nth-order multipole. In other words, a $2(n + 1)$ pole produces a flux distribution for which the magnitude of $B_y(x, 0)$ is proportional to r^n, as postulated in Eq. (9.26a).

9.4.2.2 A Correction Element Using Extended Coils

The flux density distribution obtained in a system of n magnet poles can also be created by a special arrangement of current conductors (Wollnik, 1972). We assume that an iron tube surrounds a copper sleeve (Fig. 9.17) through which electric current flows in the z direction. In case the current density varies with the azimuthal angle θ, this copper sleeve creates a magnetic flux density $\mathbf{B} = \mu\mu_0\mathbf{H}$, with $\mu_0 = 4\pi 10^{-7}$ Vs/Am and μ a material constant that is 1 for vacuum. In detail, we find

$$\oint \frac{B\,ds}{\mu\mu_0} = I = jl, \tag{9.31}$$

where the integral over $\mathbf{B}$ is taken along the dotted lines in Fig. 9.17. Here, I is the total current encircled by the dotted lines and j is the current density in amperes per meter. For good iron with a large μ, we can neglect that contribution to the integral of Eq. (9.31) which is obtained along the part of these lines in the iron. Consequently, Eq. (9.31) reads for that part of the dotted line which is in vacuum, $jl\mu_0 = B_\parallel l$. Here $B_\parallel$ is the tangential component of $\mathbf{B}$ with respect to the iron surface. With $\mu_0 = 4\pi 10^{-7}$ Vs/Am, this relation may also be expressed as

$$j = (10^7/4\pi)B_\parallel, \tag{9.32}$$

where j is in amperes per meter (A/m) and $B_\parallel$ is in teslas ($T = $ Vs/m^2).

In order to obtain the magnetic flux density distribution described by Eqs. (9.30a) and (9.30b) in the element of Fig. 9.17a, the current density j in the copper sleeve must be proportional to $r^n \cos(n + 1)\theta$, according to Eq. (9.32). In order to obtain the magnetic flux density distribution described by Eqs. (9.30a) and (9.30b) in the element of Fig. 9.17b, the tangential component $B_\parallel$ of the flux density must vary with x and y according to Eq. (9.28a). Thus, the current density j in the rectangular copper sleeve must be

$$-j(\pm a, y) = I[iK_n(\pm a + iy)^n], \tag{9.33a}$$

$$j(x, \pm b) = R[iK_n(x \pm ib)^n], \tag{9.33b}$$

where R and I characterize the real and the imaginary parts of the expressions in square brackets.

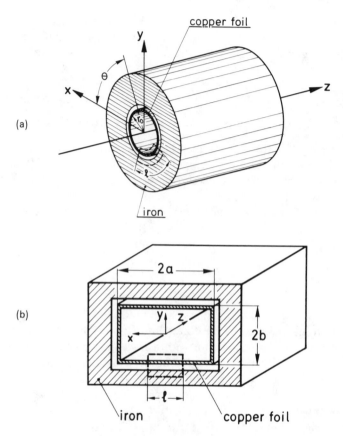

Fig. 9.17.   Two magnetic multipole elements formed by sheets of current bands. By varying the current density in the copper sleeves appropriately, the field distribution of any superposition of a 2, 4, 6, 8, . . . , pole arrangement can be achieved. Note that removing the round or square iron tubes only reduces the multipole field strengths by a factor of 2. Thus, these multipole arrangement can also very well be used in superconducting high field systems. They also can be used in the apertures of quadrupoles of other multipoles (see Figs. 9.13, 9.15, and 9.16), where the flexible current distribution can modify higher multipole field components.

For the multipoles of zeroth, first, second, and third order, Eqs. (9.32) and (9.33) postulate that the current densities in the copper sleeves of Figs 9.17 vary as indicated in Table 9.1. The current density $j(r_0, \theta)$ required for a multipole shown in Fig. 9.17a is listed with $2r_0$ being the diameter of the copper sleeve. The current densities $j(x, \pm b)$ and $j(\pm a, y)$ required for a multipole shown in Fig. 9.17b are also listed, where for $j(x, b)$ and $j(a, y)$ the upper signs are valid while for $j(x, -b)$ and $j(-a, y)$ the lower signs are applicable. For $a \gg b$, the vertical part of the copper sleeve of Fig. 9.17b

TABLE 9.1[a]

Variations of Current Densities in Multipoles

Multipole	n	$j(r_0, 0)$	$j(x, \pm b)$	$j(\pm a, y)$
Dipole	0	$K_0 \cos\theta$	0	$\pm K_0$
Quadrupole	1	$K_1 r_0 \cos 2\theta$	$-K_1 b$	$-K_1 a$
Hexapole	2	$K_2 r_0^2 \cos 3\theta$	$\pm 2K_2 bx$	$-K_2(a^2 - y^2)$
Octupole	3	$K_3 r_0^3 \cos 4\theta$	$\pm K_3 b(3x^2 - b^2)$	$\pm K_3 a(a^2 - 3y^2)$

[a] For multipoles of different orders the current density j is given as $j(r_0, \theta)$ for any angle θ in a cylinder of radius r_0, or as $j(x, \pm b)$ and $j(\pm a, y)$ in a box with sides of lengths $2a$ and $2b$.

produces only a very small effect and can be omitted as in Fig. 9.18 (Wollnik, 1970, 1972; Camplan and Meunier, 1981). An example of such a surface coil is shown in Fig. 9.19.

9.4.1.2.3 Inhomogeneity Correction Coils

In principle, all fabrication inaccuracies of a magnet can be compensated for by a superposition of multipole surface coils like the one in Fig. 9.19. However, there is a very effective alternative in which surface coils were constructed with very special coil patterns (Czok *et al.*, 1977; Herminghaus

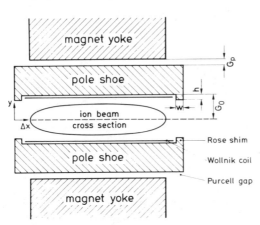

Fig. 9.18. A dipole magnet designed to obtain a wide homogeneous field distribution. The Purcell (1955) gap separates the pole shoes from the magnet yoke so that the magnetic forces leave the magnet poles undisturbed. The Rose (1938) shims compensate the fringe field fall-off, at least at its onset if $(h/G_0) \approx 0.16 \exp[-2.7(w/G_0) + 0.42(w/G_0)^4]$. A flat surface coil formed according to Fig. 9.21 or Fig. 9.19 compensates for a measured field inhomogeneity of adds multipole components to the dipole field.

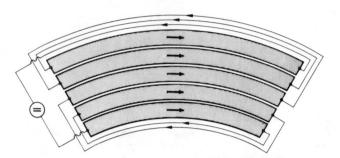

Fig. 9.19.   A quadrupole surface coil is shown built in the technique of etched electronic circuit boards. In the quadrupole coil shown, the current density is constant; ile., all leads have equal widths and minimal separations. If each copper strip is driven by a separate power supply, any current distribution can be achieved and thus also any superposition of multipoles of different order.

et al., 1981). We assume that we wish, to increase slightly the magnetic flux density in a fixed region of a given magnet. To do this, we could form wire loops with the shape of the said region, fasten these loops on both magnet poles (Fig. 9.20), and pass a certain electric current I through both loops. Determining $\oint H\,ds$ along the dotted line, we find $2I = 2G_0[B_\perp(x_1) - B_\perp(x_2)]/\mu_0 = 2G_0\Delta B_\perp/\mu_0$. With $B_\perp$ the perpendicular component of the

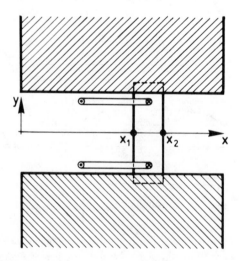

Fig. 9.20.   Two current loops are shown mounted in the air gap of a dipole magnet. There is a current I passing into the drawing plane between x_1 and x_2. The same current then comes out of the drawing plane to the left of x_1. The field increase at x_1 relative to x_2 equals $\Delta B_\perp = 4\pi \times 10^{-7}G_0I$, with G_0 in meters (m), I in amperes and $\Delta B_\perp$ in teslas.

magnetic flux density at x_1 and x_2 in Fig. 9.20, we find

$$I = (10^7/4\pi)G_0[B_\perp(x_1) - B_\perp(x_2)]. \qquad (9.34)$$

Here, the half-magnet air gap G_0 is in meters, I in amperes, and $\Delta B_\perp = B_\perp(x_1) - B_\perp(x_2)$ in teslas ($T = Vs/m^2$).

Using this principle to homogeneize the flux density of a given magnet, we must first measure the existing flux density distribution with the highest accuracy possible. As a second step one must transform these data points into a map of contour lines, i.e., lines of equal flux densities. Each contour line encompassing a certain region should then be replaced by a wire loop so that an appropriate electric current can reduce the high flux density in this region. In practice, we turn the map of contour lines into a copper pattern on a printed circuit board (Fig. 9.21), interconnect these lines at the least critical places and pass the same current through all "lines." If saturation effects of the magnet iron cause the contour map to vary with the flux density, we can place two such inhomogeneity coils on top of each other. In this case, each coil should be designed for one flux density; for

Fig. 9.21. An inhomogeneity surface coil is shown built by the technique of etched electronic circuit boards. To obtain this coil, the contour lines of a measured field distribution are all forced to close by postulating a constant flux density at some outer frame. These nonintersecting contour lines were then artificially cut open and interconnected at some uncritical position so that, finally, spiral-like lines were formed. The pattern thus obtained was transformed into copper. Consequently, a current can spiral from the outside to a center section, pass to the opposite side of the printed circuit board formed, and spiral outward again in a pattern for which copper-clad and copper-free regions are exchanged, compared to the pattern shown.

intermediate flux density values, each coil should be driven by a percentage of its design current.

9.4.2 Impossibility of Correcting Aberrations for Beams with Circular Cross Sections

In magnetic correction elements in which $B_y(x, 0)$ equals $K_n x^n$ Eq. (9.29b) causes $B_y(r, 0)$ to vanish along lines $n\theta = \pi(m + 1/2)$ for $m = 0, 1, 2, \ldots$. Particles that traverse a correction element at the position of one of these lines thus experience no deflecting force $v_z B_y$, and particles that traverse a correction element in neighboring sectors experience a deflection in opposite directions, since B_y has alternating signs. Both effects are independent of the magnitude of the flux density in the multipole element and originate from $B_y(x, 0)$ being proportional to x^n and from the validity of Maxwell's equations.

Consider an approximately parallel beam of almost circular cross section that passes through a multipole element perpendicularly to the drawing plane of Fig. 9.15. Some optical system may focus this beam to an x image, which is distorted by image aberrations in the x direction By choosing an appropriately varying $B_y(x, 0)$ in the multipole element, we can achieve that all particles passing through the multipole element in the plane $(y = 0)$ are deflected such that the initially existing image aberrations are eliminated. However, the deflections of particles passing through the multipole element off the midplane $(y = 0)$, cannot be corrected simultaneously, since the x deflection for these particles is zero (see the strings of zeros in Fig. 9.15) or has a value determined indirectly by the previously fixed $B_y(x, 0)$. Usually, thus the sum of all image aberrations is increased and not decreased if a round beam is passed through a multipole element. This argument holds not only for a multipole element but also for a curved sector field boundary, as shown by Brown (1967) and by Wollnik and Amadori (1967).

Regardless of the field strength used in a multipole element, the overall improvement in precise focusing is normally small so long as the beam cross section is more or less circular. In order to correct an aberration $(x|aa)a_{00}^2/2$ or $(x|aaa)a_{00}^3/6$, we must choose a beam that is as wide as possible in the x direction and as small as possible in the y direction at the position of the multipole element (Section 8.4.1.2). In this case, a reduction of the coefficients of image aberrations $(x|aa)a_{00}^2/2$ or $(x|aaa)a_{00}^3/6$ is feasible without increasing the aberrations $(x|bb)b_{00}^2/2$ or $(x|bbbb)b_{00}^4/24$ seriously. This situation can only be achieved in the case where some intermediate y image or y pupil is formed approximately at the position of the multipole element. For this reason, we often have split the poles (Fig. 9.3) of the magnet of a high-performance spectrometer, installed some correction element (Spencer and Enge, 1967; Ikegami, 1981) in this gap, and designed the system such that at this position, a y crossover was reached.

9.4.3 Correction of Image Aberrations in Electrostatic Systems

The possibilities for correcting aberrations in electrostatic lens systems are analogous to those in magnetic systems. The x and y deflections in a correction element here are for a particle traversing the multipole element at a point $P(x_1, y_1)$:

$$\Delta a(x_1, y_1) = \int_0^{l_c} \frac{zeE_x(x, y, s)}{mv^2} \, ds, \tag{9.35a}$$

$$\Delta b(x_1, y_1) = \int_0^{l_c} \frac{zeE_y(x, y, s)}{mv^2} \, ds. \tag{9.35b}$$

Here, m is the mass, (ze) the charge, and $v = |\mathbf{v}|$ the velocity of the particle under consideration. In order to correct the aberrations proportional to a_{00}, a_{00}^2, a_{00}^3, etc., a field strength distribution

$$E_x(x, y = 0) = E_x(0, 0) \left[1 + xn_{E1} + \frac{x^2}{2} n_{E2} + \frac{x^3}{6} n_{E3} + \cdots \right] = \sum \frac{x^n}{n!} K_{En}$$

$$\tag{9.36a}$$

is necessary. In the case of a correction element of constant field strength $E_x(x, 0)$ along the optic axis, the length of the flight path must vary as

$$l = l_0 \left[1 + xn'_{E1} + \frac{x^2}{2} n'_{E2} + \frac{x^3}{6} n'_{E3} + \cdots \right] = \sum \frac{x^n}{n!} K'_{En}. \tag{9.36b}$$

A multipole correction element built according to Eq. (9.36b) is a curved field boundary, as in the magnetic case. Here, however, the field boundary is curved not in the xz plane, as in Fig. 9.14, but in the yz plane, as shown in Fig. 9.22. At first sight, such a curved field boundary seems to cause different deflections only for particles that traverse the field boundary at different heights. However, because of div $E =$ curl $E = 0$ between the electrodes of the electrostatic field, a change of the field distribution with y also causes a change of the field distribution with x.

A multipole correction element built according to Eq. (9.36a) consists of many symmetrically arranged electrodes (Fig. 9.23), which cause a

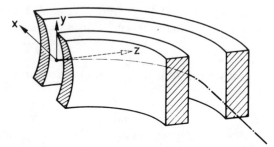

Fig. 9.22. An electrostatic sector field with cylindrical electrodes is shown. In order to influence image aberrations of second order, the entrance boundary of the electrostatic sector field is curved in the y direction. (This curvature should be compared to the magnet field boundary of Fig. 9.14, which is curved in the x direction.)

complex field strength distribution analogous to Eqs. (9.27a) and (9.27b):

$$E(w) = E_x(w) + iE_y(w), \tag{9.37a}$$

$$E^*(w) = E_x(w) - iE_y(w). \tag{9.37b}$$

In detail such a multipole correction element requires a field strength **E** the x component of which varies with x as $E_x(x, 0) = K_n x^n$ where K_n is in V/m^{n-1}. With $w = x + iy = re^{i\theta}$, this relation implies

$$E^*(w) = iK_n w^n = iK_n(x + iy)^n, \tag{9.38a}$$

$$E^*(w) = iK_n r^n e^{in\theta}, \tag{9.38b}$$

analogously to Eqs. (9.28a) and (9.28b). For a given multipole the magnitude of E is thus constant along a circle of radius r, and only its direction varies (Fig. 9.23).

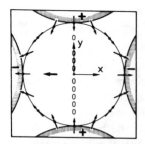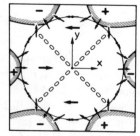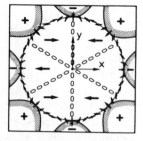

Fig. 9.23. Electrode arrangements of electrostatic quadrupoles, hexapoles, and octopoles are indicated. Also shown is a circle of radius r_0 along which the magnitude of the electrostatic field strength is constant, and its direction varies as indicated. Finally, strings of zeros indicate lines along which E_x, the x component of the electrostatic field strength, vanishes. These lines separate regions in which E_x is parallel or antiparallel to the x axis.

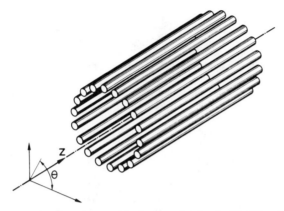

Fig. 9.24. Electrostatic multipole with 18 electrodes (optimally 32), which should take up potentials according to Eq. (9.39b). In case the electrode at the angle θ is at a potential $V_{01} \sin \theta$, the resuting field distribution is that of a vertically deflecting parallel-plate condenser. In case the electrode at the angle θ is at a potential $V_{10} \cos \theta$, the resulting field distribution is that of a horizontally deflecting parallel-plate condenser. In case this electrode is at a potential $V_{01} \sin \theta + V_{10} \cos \theta$, the resulting field distribution is that of a parallel-plate condenser inclined under an angle θ_0 with $\tan \theta_0 = V_{10}/V_{01}$. In general, the potential of the electrode at the angle θ can be at a potential $(V_{01} \sin \theta + V_{10} \cos \theta) + (V_{02} \sin 2\theta + V_{20} \cos 2\theta) + (V_{03} \sin 3\theta + V_{30} \cos 3\theta) + \cdots$, resembling a superposition of an effectively rotatable dipole, quadrupole, hexapole, etc.

To obtain the field distribution $E^*(w)$ of Eqs. (9.38a) or (9.38b), we need a potential distribution $V(w)$ that satisfies $E^* = -\mathrm{grad}\, V$. This potential distribution is the real (or the imaginary) part of

$$V(w) = \frac{iK_n}{n+1}\, w^{n+1} = \frac{iK_n}{n+1}\, (x + iy)^{n+1}, \tag{9.39a}$$

$$V(w) = \frac{iK_n}{n+1}\, r^{n+1} e^{i(n+1)\theta} = \frac{K_n}{n+1}\, r^{n+1} \sin[(n+1)\theta]. \tag{9.39b}$$

To generate this potential distribution one can either use $n + 1$ electrodes of alternating potentials as shown in Fig. 9.23 or arrange many electrodes symmetrically around a straight optic axis, as shown in Fig. 9.24, and supply them with potentials according to their angle-θ position in Eq. (9.39b). As in the magnetic case, the E_x or E_y components of the field strength vanish for certain values of θ. These θ values are marked by strings of zeros in Fig. 9.23. Note that these lines are identical to those of Fig. 9.15.

Because of the analogy of magnetic and electrostatic multipoles, it may be accepted that the considerations of Section 9.4.2 apply equally to both cases, requiring us also to place electrostatic correction elements only at positions at which an intermediate image, pupil, or waist exists, at least approximately.

REFERENCES

Aston, F. W. (1919). *Philos. Mag.* **38**, 709.

Balestrini, S., and White, F. (1960). Rev. Sci. Instrum. **31**, 633.

Brown, K. L. (1967). Private communication.

Brown, K. L., Carey, D. C., Iselin, C., and Rothacker, F. (1980). CERN-Report 80-04.

Camplan, J., and Meunier, R. (1981). *Nucl. Instrum. Methods* **186**, 445.

Chavet, I. (1966). *Nucl. Instrum. Methods* **45**, 340.

Czok, U., Moritz, G. and Wollnik, H. (1977). *Nucl. Instrum. Methods* **140**, 445.

Dempster, A. J. (1937). *Phys. Rev.* **51**, 67.

Drentje, A. G., Enge, H., and Kowalski, S. K. (1974). *Nucl. Instrum. Methods* **122**, 485.

Enge, H. (1958). *Rev. Sci. Instrum.* **29**, 885.

Enge, H. (1964). *Nucl. Instrum. Methods* **28**, 119.

Enge, H., and Kowalski, S. B. (1970). *Proc. Int. Conf. Magnet Technol.*, 3rd (*Hajmburg*), p. 366.

Ewald, H., and Hintenberger, H. (1953). "Methoden und Anwendungen der Massenspektroskopie." Verlag Chemie, Weinheim.

Halbach, J. (1980). Private communication.

Herminghaus, H., Kaiser, K. H., and Ludwig, U. (1981). *Nucl. Instrum. Methods* **187**, 103.

Ikegami, H. (1981). *Nucl. Instrum. Methods* **187**, 13.

Matsuo, T., Matsuda, H., Fujita, Y., and Wollnik, H. (1976) *Mass Spectroscopy* **24**, 19.

Mattauch, J., and Herzog, R. (1934). *Z. Phys.* **89**, 786.

Maurer, J. H., Brunnee, C., Kappus, G. Habfast, K., Schroeder, U., and Schulze, P. (1971). *Ann. Conf. ASMS*, 19th Paper K9. ASMS.

Nier, A. O., Roberts, T. R., and Franklin, F. J. (1949). *Phys. Rev.* **75**, 346.

Poschenrieder, W. (1971). *Int. J. Mass Spectr. Ion Phys.* **6**, 413.

Poschenrieder, W. (1972). *Int. J. Mass. Spectr. Ion Phys.* **9**, 357.

Purcell, N. E. (1955). U.S. Patent 5 436 876.

Rose, M. E. (1938). *Phys. Rev.* **53**, 715.

Shibata, T., Taya, S., and Yoshizawa, Y. (1968). *Nucl. Instrum. Methods* **64**, 29.

Spencer, J. E., and Enge, H. (1967). *Nucl. Instr. and Meth.* **49**, 811.

von Egidy, T. (1962). *Ann. Phys.* **9**, 221.

Wollnik, H. (1970). *Proc. Int. Conf. Electromagnetic Isotope Separators, Marburg*, H. Wagner und W. Walcher, (eds), p. 282. BMBW-Forschungsbericht K 70-28.

Wollnik, H. (1971). *Nucl. Instrum. Methods* **95**, 453.

Wollnik, H. (1972). *Nucl. Instrum. Methods* **103**, 479.

Wollnik, H. (1980). In "Applied Charged Particle Optics," A. Septier, (ed.), p. 133, Academic Press, New York.

Wollnik, H. (1981). *Nucl. Instrum. Methods* **186**, 413.

Wollnik, H. (1986). *Nucl. Instrum. Methods*, in press.

Wollnik, H., and Amadori, R. (1967). "Advances in Mass Spectrometry IV," p. 287. Adlard and Son Ltd., Dorking.

Wollnik, H., Brezina, J., Berz, M., and Wendel, W. (1984) *Proc. AMCO-7*, GSI-Report THD-26, 679.

Wollnik, H., and Matsuo, T. (1981). *Int. J. Mass Spectr. Ion Phys.* **37**, 209.

Wouters, J. M., Vieira, D. J., Wollnik, H., Enge, H., Kowalski, S., and Brown, K. L. (1984). *Nucl. Instrum. Methods* **A240**, 77.

Index